THE LARVAE OF THE GALL MIDGES
(DIPTERA, CECIDOMYIIDAE)

# The Larvae of the Gall Midges (Diptera, Cecidomyiidae)

*Comparative morphology, biology, keys*

B. M. MAMAEV & N. P. KRIVOSHEINA

*Translated by*
J. H. WIEFFERING

*Edited by*
J.C. ROSKAM
*University of Leiden, Department of Biology*

CRC Press
Taylor & Francis Group
Boca Raton London New York

CRC Press is an imprint of the
Taylor & Francis Group, an **informa** business
A BALKEMA BOOK

Published by:
CRCPress/Balkema
P.O. Box 447, 2300 AK Leiden, The Netherlands
e-mail: Pub.NL@taylorandfrancis.com
www.crcpress.com – www.taylorandfrancis.com

First issued in hardback 2017

*CRC Press/Balkema is an imprint of the Taylor & Francis Group, an informa business*



Original text:
B. M. Mamaev & N. P. Krivosheina: *Lichinki gallits (Diptera, Cecidomyiidae)*, Nauka, Moscow, 1965


Revised and updated edition in English:


ISBN 13: 978-1-138-42369-5 (hbk)
ISBN 13: 978-90-6191-787-8 (pbk)

Distributed in USA & Canada by:
A.A. Balkema Publishers, Old Post Road, Brookfield, VT 05036, USA

# Table of contents

# Editorial

The original edition of *The Larvae of the Gall Midges* appeared in 1965, ten years after Edwin Möhn's *Beiträge zur Systematik der Larven der Itonididae Mitteleuropas*. Why then, after more then twenty years, a translation of Mamaev and Krivosheina's book? There are several reasons. First, a bad one. Although several attempts were made, it was not earlier possible to find a capable translator, and, more important, an interested publisher. Second, together with Mamaev's book *Evolution of Gall Forming Insects – Gall Midges,* published in 1968 and translated into English by A. Crozy, in 1975, the present translation offers non-Russian readers an access to the major publications of a leading Russian cecidologist. Mamaev's highly original ideas regarding the polyphyletic evolution of gall inducing cecidomyiids are not only accepted, but also admired by many gall midge workers. Third, and main reason, Möhn's work and the Mamaev and Krivosheina book are highly complementary. Möhn mainly restricted himself to the gall inducers, which all belong to the advanced subfamily Cecidomyiinae. Mamaev and Krivosheina, however, concentrated on cecidomyiids living, at least during some phases of their life cycle, in the soil. Among these forms are mycetophagous species, mainly belonging to the more primitive subfamilies Lestremiinae and Porricondyliinae, and some gall inducing Cecidomyiinae.

Adult forms of gall inducing midges were earlier treated in Rübsaamen and Hedicke's standard work *Die Cecidomyiden und ihre Cecidien*. Edwards during the thirties, Pritchard during the late forties (Lestremiinae) and Panelius (Porricondylinae) in 1965, revised the mycetophagous groups. Together with Buhr's *Bestimmungstabellen der Gallen an Pflanzen Mittel- und Nordeuropas* and Skuhravá's recently published work: Cecidomyiidae, in Soós and Papp: *Catalogue of Palaearctic Diptera*, the present translation offers a firm basis for further taxonomic, geographical, evolutionary and phylogenetic studies of one of the most important and intriguing families of Diptera.

Since the original publication many changes occurred in gall midge nomenclature. We did not intend to revise the taxonomic ideas of Mamaev and Krivosheina, but added the current insights, adopting Skuhravá (1986) in the List of genera. Furthermore, the terminology of larval structures differs in some respects from that used by most non-Russian workers. These differences are listed in the caption of figure 1.

We are much indebted to the publisher, without whose efforts the publication of this translation would not have been possible, to Andy Gutow for his stimulating help with the English text, and to Paul Tolsma for his patient and kind technical assistance.

J.C. Roskam

# Note from the editor on nomenclature and taxonomy

The nomenclature and taxonomic views of Prof. Mamaev and Dr. Krivosheina as expressed in this volume date from 1965. As the study of gall midges has considerably progressed during the last twenty years, the nomenclature and taxonomic views have changed in a number of cases, as was already foreseen by the authors.

Incorporation of these changes into this translation would not be possible without making extensive rearrangements, and in fact rewriting, important parts and keys of the book. Such a procedure is, however, quite liable of introducing ideas and views alien to those of the original authors. Attempting to avoid this pitfall we have chosen to follow the original text as closely as possible in the translation proper and to add a list of changes in the nomenclature and taxonomy as used in Skuhravá (1986).

# Introduction

Traditionally, studies of insects, especially of the holometabolic groups, have always used the adult stage as their starting point. This situation is readily understood from the fact that it is easier to collect adults living openly than larvae with their hidden life. Making a collection of larvae is tedious labour. Not only must large series of larvae be collected, but, moreover, adults must be procured as well; for without adult specimens there will be no possibility to reliably identify the material.

Since, however, insects tend to be harmful especially during their larval life, the ability to handle and judge larval material is, from a practical point of view, more highly esteemed than knowing how to identify adults. Prof. Dr. Biol. Sci. M.S. Gilyarov's book *A determination key to the insect larvae living in soil* (1964), is of supreme importance for this kind of work; its value cannot be overestimated.

In this monumental work from the Laboratory of Soil-Zoology of the A.N. Severtsov Institute of Zoological Morphology of the Academy of Sciences of the U.S.S.R. the characteristics of all main groups of soil-bound insects are given. The book furnishes the very foundation upon which more detailed studies of paedo-zoological material are built. Of course its relevant keys may be refined and broadened as more precise and complete characterizations of separate groups of paedobiontic larvae are being prepared for publication. The present study of the gall midge larvae can be considered as one of the first attempts at such work.

There are many reasons why the study of gall midge larvae, which always form one of the largest components in the soil layer of a biocoenose, is necessary. First, many species of this family (400-500 in the fauna of the U.S.S.R., roughly estimated) complete their development in the soil or a similar substrate. In a thermo-eclector 150-200 gall midge larvae may be obtained per $m^2$ of soil, which clearly shows their essential role in soil communities. Zoological investigations on small soil-invertebrates, which, generally, take into account only mites and springtails, would remarkably benefit from the addition of some comment on the fauna of small larvae of beetles and diptera. Especially in damp biotopes the larvae of gall midges actually occupy, among the large group of paedobiontic arthropods, the third place after the mites and springtails. Hence, they should not be ignored in any reasonably thorough investigation on the complex of paedobiontic species.

Second, the species composition of the paedobiontic gall midges may serve as a very precise indicator of soil conditions. Since mycetophagous gall midges are often closely connected to a fixed set of fungus species, their species composition will, most importantly, reflect the composition of the microflora.

Finally, the larvae of many gall midge species pupate in the soil and often have a prolonged diapause there. Prognoses concerning the massive increase of gall midge pests are based on the direct estimation of the amount of larvae present in the soil. The accuracy of the prognosis depends, above all, on the ability to discriminate between the larvae of the various species. For, without this ability, there is no possibility of correctly calculating the number of noxious larvae, since noxious and neutral gall midge larvae co-occur in the soil.

The authors had this in mind when they decided to concentrate on those groups of gall midges whose larvae develop in the soil or a similar substrate, or which at least pupate in the soil. Therefore, they refrained from discussing the tribes whose representatives are mostly gall makers that complete their whole development inside galls. The delimitation of the subject accepted in this monography seems the more appropriate especially since the gall making cecidomyiid larvae from the Central European fauna have already been revised in a number of papers on individual species (Marchal 1897, Williams 1910, Vimmer 1925, Haseman 1931, Metcalfe 1933, Otter 1934, 1938, Anderson 1935a, b, Roberti 1946, 1953, 1954, Frediani 1955, Principi 1958, and others), as well as monographically (Möhn 1955b). A major shortcoming of the latter work is that the author does not at all use the highly important diagnostic characters found in the morphology of the head capsule of the larvae. His monography is, nevertheless, fully recommended as a manual for the identification of the gall making Cecidomyiidae. There exists no analogous manual for the study of the paedo- and xylobiontic gall midges in the entomological world literature. Kieffer's work (1913) only gives genus diagnoses and is obviously out of date.

The first descriptions of gall midge larvae, references to which can be found in Hennig's manual (Hennig 1948), are exceptionally fragmentary. As a rule they only give information on the shape and colour of the larvae, and most of them are useless for purposes of comparison. Bearing this in mind the authors have gone to great pains to provide exhaustive descriptions of the larvae in their collection. Recently, however, a number of descriptions meeting high standards have been published; the authors extensively consulted these, as the comments to the species diagnoses prove (see Chapter 4).

In the present monography special attention is given to items of practical importance, e.g., to predatory species that open perspectives for use in a system of biological control measures against aphids and mites, as well as to serious agricultural pests.

The authors based this monography on material collected by B.M. Mamaev from 1951-1962 in the Moskovskaya, Tul'skaya, Voronezhskaya, Luganskaya, Rostovskaya, and Volgogradskaya oblasts, as well as in the northern Caucasus and Transcaucasia; by N.P. Krivosheina from 1960-1962 in the Arkhangel'skaya and Vologodskaya oblasts; and by N.P. Krivosheina and B.M. Mamaev together in 1961 in the Yaroslavskaya oblast and in 1963 in Transcarpathia. Much material from the Moskovskaya oblast was granted to us for study by Kh.P. Mamaeva with her permission to make use of it in preparing this monography. M. Skuhravá and V. Skuhravý (Entomological Institute, Prague, Czechoslovakia), D. Wyatt (Institute for Plant Cultivation, Glasgow, Great Britain), W. Nijveldt (Institute for Phytopathological Research, Wageningen, Netherlands), and H. Stelter (Institute for Plant Cultivation, Grosz-Lüsewitz, G.D.R.) have, for reasons of comparison, put additional material at our disposal, for which the authors are deeply grateful.

We take advantage of the opportunity to express our gratitude to Prof. Dr. Biol. Sci. F.N. Pravdin, Prof. Dr. Biol. Sci. B.B. Rohdendorf, and Prof. Dr. Biol. Sci. E.S. Smirnov for many valuable comments during the reading of the manuscript.

The authors are also grateful to the collective of the Laboratory for Soil-Zoology, to the director of the Laboratory, Prof. Dr. Biol. Sci. M.S. Gilyarov, and to the managing board of the A.N. Severtsov Institute for Animal Morphology of the Academy of Sciences of the U.S.S.R. for their multifarious assistance.

# List of genera described in this monography

| FAMILY CECIDOMYIIDAE | Taxonomic changes (Skuhravá 1986) |
|---|---|
| SUBFAMILY LESTREMIINAE | |
| 1. *Acoenonia* Pritchard | |
| 2. *Aprionus* Kieffer | |
| 3. *Brittenia* Edwards | |
| 4. *Bryomyia* Kieffer | |
| 5. *Campylomyza* Meigen | |
| 6. *Catocha* Haliday | |
| 7. *Henria* Wyatt | |
| 8. *Heteropeza* Winnertz | |
| 9. *Heteropezina* Pritchard | |
| 10. *Leptosyna* Kieffer | |
| 11. *Lestremia* Macquart | |
| 12. *Miastor* Meinert | |
| 13. *Micropteromyia* Mamaev | |
| 14. *Monardia* Kieffer | |
| 15. *Mycopriona* Mamaev | |
| 16. *Peromyia* Kieffer | |
| 17. *Skuhraviana* Mamaev | |
| 18. *Trichopteromyia* Williston | |
| 19. *Xylopriona* Kieffer | |
| SUBFAMILY CECIDOMYIINAE | |
| 20. *Acaroletes* Kieffer | |
| 21. *Agevillea* Hubault | |
| 22. *Ametrodiplosis* Rübsaamen | |
| 23. *Anisostephus* Rübsaamen | |
| 24. *Aphidoletes* Kieffer | |
| 25. *Arthrocnodax* Rübsaamen | |
| 26. *Asynapta* H. Loew | |
| 27. *Atylodiplosis* Rübsaamen | 27. *Contarinia* Rondani |
| 28. *Buhriella* Stelter | |
| 29. *Camptomyia* Kieffer | |
| 30. *Clinodiplosis* Kieffer | |
| 31. *Colomyia* Kieffer | |
| 32. *Contarinia* Rondani | |
| 33. *Dicerura* Kieffer | |

| FAMILY CECIDOMYIIDAE | Taxonomic changes (Skuhravá 1986) |
|---|---|
| 34. *Dichaetia* Mamaev | |
| 35. *Dichrona* Rübsaamen | 35. *Planetella* Westwood |
| 36. *Dicroneurus* Kieffer | 36. *Porricondyla* Rondani |
| 37. *Diplolaboncus* Rübsaamen | 37. *Planetella* Westwood |
| 38. *Dishormomyia* Kieffer | 38. *Planetella* Westwood |
| 39. *Dyodiplosis* Rübsaamen | 39. *Planetella* Westwood |
| 40. *Echinella* Mamaev | |
| 41. *Endaphis* Kieffer | |
| 42. *Endopsylla* De Meijere | |
| 43. *Feltiella* Rübsaamen | |
| 44. *Gynapteromyia* Mamaev | |
| 45. *Haplodiplosis* Rübsaamen | |
| 46. *Harmandia* Kieffer | |
| 47. *Holoneurus* Kieffer | |
| 48. *Isodiplosis* Rübsaamen | |
| 49. *Itonida* Meigen | 49. *Cecidomyia* Meigen |
| 50. *Karschomyia* Felt | |
| 51. *Jaapiola* Rübsaamen | |
| 52. *Loewiola* Kieffer | |
| 53. *Lestodiplosis* Kieffer | |
| 54. *Macrodiplosis* Kieffer | |
| 55. *Massalongia* Kieffer | |
| 56. *Monarthropalpus* Rübsaamen | |
| 57. *Monobremia* Kieffer | |
| 58. *Mycodiplosis* Rübsaamen | |
| 59. *Parepidosis* Kieffer | |
| 60. *Paurosphondylus* Rübsaamen | 60. *Planetella* Westwood |
| 61. *Planetella* Westwood | |
| 62. *Plemeliella* Seitner | |
| 63. *Porricondyla* Rondani | |
| 64. *Profeltiella* Kieffer | 64. *Resseliella* Seitner |
| 65. *Proshormomyia* Kieffer | 65. *Planetella* Westwood |
| 66. *Pseudhormomyia* Kieffer | 66. *Planetella* Westwood |
| 67. *Putoniella* Kieffer | |
| 68. *Rhipidoxylomia* Mamaev | |
| 69. *Sitodiplosis* Kieffer | |
| 70. *Solntsevia* Mamaev | |
| 71. *Stelechodiplosis* Möhn | 71. *Cecidomyia* Meigen |
| 72. *Syndiplosis* Kieffer | 72. *Contarinia* Rondani |
| 73. *Synepidosis* Mamaev | |
| 74. *Taphodiplosis* Kieffer | 74. *Planetella* Westwood |
| 75. *Thecodiplosis* Kieffer | |
| 76. *Therodiplosis* Kieffer | |
| 77. *Tricholaba* Rübsaamen | |
| 78. *Thomasiniana* Strand | 78. *Resseliella* Seitner |
| 79. *Trichoxylomyia* Mamaev | 79. *Kronomyia* Felt |
| 80. *Triommatomyia* Mamaev | |
| 81. *Ulmomyia* Mamaev | 81. *Dicerura* Kieffer |
| 82. *Winnertzia* Rondani | |
| 83. *Xylodiplosis* Kieffer | |

CHAPTER 1

# Outline of the biology

The biology of the gall midges is in the first place the biology of their larvae. Compared with the life-span of the adults, which is measured in days only, the duration of the larval development is overwhelming, even in multivoltine species.

Adult gall midges do not feed. The eggs in the ovaria already mature before the adults emerge from the pupae. The primary function of the adult phase therefore consists of reproduction and dispersal. The basic function of the larvae is feeding, or feeding and reproduction if we consider the few paedogenically reproducing species.

It is especially by feeding that the larvae of many phytophagous gall midges cause damage to plants, and that the predatory species destroy the noxious ones. Thus, when we speak of damage or benefit in relation to gall midges we have in mind the harmful or beneficial qualities of their larvae. Therefore, in the study of the biology of gall midge larvae, the nature of their food holds a central position.

## THE TROPHIC GROUPS

The currently available knowledge of the biology of the gall midges leads us to consider soil as the primary food of their larvae. We use 'soil' here in a broad sense, meaning soil plus its contents of plant parts, as for instance litter, rotting wood, fruiting bodies of fungi etc., and living tissue of higher and lower plants. Though the larvae of some species are semi-aquatic they always occur in connection with certain plant species. There are no truly aquatic larvae of gall midges. A small group of species, though living on plants, do not feed directly on plant tissue. They develop at the expense of other insects, especially phytophages that occur in large colonies (aphids, scale-insects, mites).

The feeding specialization of those larvae of gall midges that feed on decaying plant parts, especially their development of extra-intestinal digestion, and the strong reduction of their jaws make it difficult to follow the authors (Möhn 1955b, Fröhlich 1960, and others) who include them among the saprophages or detritophages. The inclusive term 'saprophagy', and especially the narrower term 'detritophagy' are mostly used in connection with animals that feed on decaying plant parts by chewing and swallowing them. Experimental work on the breeding of paedogenic species and direct observation of larvae in nature have shown, however, that larvae of gall midges suck thin hyphae, or dissolve the film of mycelium by means of a digestive fluid. Therefore, the feeding specialization of these larve is better designated as 'mycetophagy' (Mamaev 1962).

In this text we use the term 'phytophages' in a restricted sense, excluding all but those that feed on living tissue of higher plants. The predatory and parasitic gall midges together make up the clearly delimited group of the zoophages.

We suggest a division of the larvae of gall midges in three trophic groups, namely mycetophages, phytophages, and zoophages. However, we should bear in mind that some larvae, at least those that live under bark in close contact with dying cambium (*Profeltiella* Kieffer, *Thomasiniana* Strand, and some others), clearly have a mixed mode of feeding. They feed on the mycelium growing there, as well as on the sap oozing from the cells. We think this feeding specialization is adequately expressed by the term 'phytomycetophagy'. We expect that this mode of feeding will also be found among species living in flower heads of Compositae and, possibly, in some other plant parts.

In the literature, it has often been asserted that some larvae, especially those of *Miastor metraloas* Meinert, are able to dissolve cellulose, and to feed on the products of hydrolysis. It is difficult indeed to imagine that the tremendous colonies of *M. metraloas*, found under bark and in rotting trunks, live entirely at the expense of mycelium. It is possible that these larvae have developed a mixed mode of feeding. However, the experiments of Nikolei (1961), who studied the influence of various feeding regimes on the developmental cycle of paedogenic gall midges, have shown that even then mycelium is the primary food. It is interesting to note that larvae of *M. metraloas* transplanted onto sterilized bark, from which mycelium is absent, perish (Harris 1923). Nevertheless, the suggestion that these larvae produce cellulase should be carefully verified.

The classification of gall midges and the mode of feeding of their larvae are clearly related. The subdivision of the family of the gall midges into subfamilies and tribes has not yet been definitively settled. Therefore, in table 1, we will use the system that, at the present stage of our knowledge, is the most commonly accepted one.

All larvae belonging to the subfamily Lestremiinae and to the tribe Porricondylini of the subfamily Cecidomyiinae are mycetophagous and can be found in soil, litter, rotting wood etc.

The larvae belonging to the tribes Lasiopterini and Asphondyliini and the bulk of the Oligotrophini are phytophagous. They develop at the expense of living plant tissue. In the vast tribe Cecidomyiinae, at last, we find phytophagous pests, harmless mycetophages, and useful predators. The latter are absent from all other groups of gall midges, a few genera in the Oligotrophini excepted.

Table 2 shows the interdependence between the larval mode of feeding and the classification.

Table 1

| Family Cecidomyiidae (= Itonididae)<br>Subfamily Lestremiinae<br>Tribes | | Subfamily Cecidomyiinae<br>Tribes |
|---|---|---|
| Acoenoniini | Lestremiini | Asphondyliini |
| Catochini | Miastorini | Cecidomyiini |
| Catotrichini | Micromyiini | Lasiopterini |
| Forbesomyiini | Moehniini | Oligotrophini |
| Heteropezini | Strobliellini | Porricondylini |
| Leptosynini | | |

Table 2

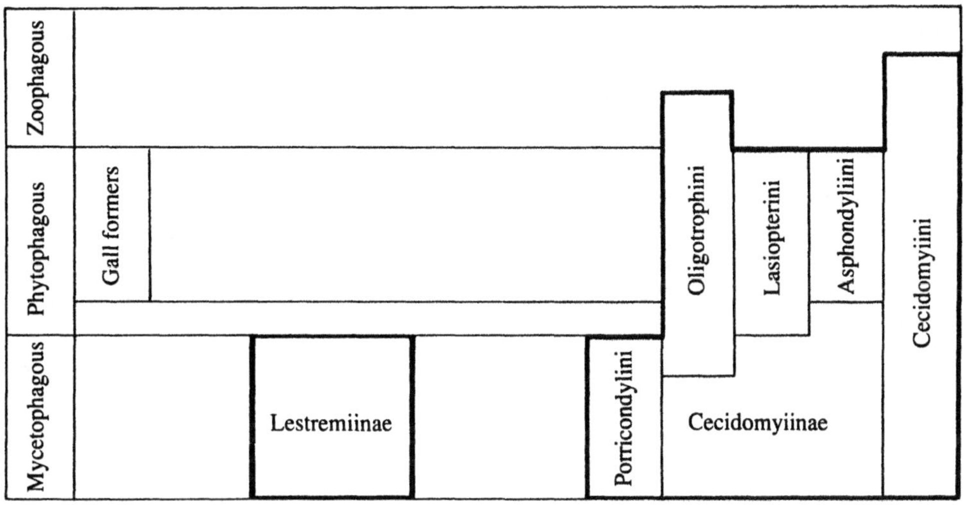


*The mycetophages*

The body of this group is formed by the subfamily Lestremiinae. species living on mycelium also occur in the tribes Porricondylini, Cecidomyiini, and Oligotrophini of subfamily Cecidomyiinae. In a more elaborate classification the mycetophagous species of the Oligotrophini may be set apart as Brachyneurini.

A considerable number of mycetophagous species develops in soil. Typical inhabitants of the soil under the canopy of the forest are, for instance, the large white larvae of *Xylopriona atra* (Meigen) that live at a depth of up to 50 cm. As many as 200 individuals per $m^2$ can be found in their colonies. A very high number of larvae of *Aprionus terrestris* Mamaev has been observed in the soil of oak forests near Tula. Another inhabitant of forest soil is *Aprionus flaviventris* (Winnertz). Under clover, in the upper layers of the soil of a Transcarpathian meadow we found the larvae of *Lestremia cinerea* Macquart. This species, however, has still many other possibilities for adaptation as its larvae have also been recorded from forest litter, decaying wood, mycelium of cultivated mushrooms etc.

The occurrence of gall midge larvae in the soil of the desert of Kyzylkum is ecologically extremely interesting. The larvae of *Winnertzia asiatica* Mamaev live there in the upper part of the thin humus layer near *Artemisa* shrubs. Larvae of this genus have not been observed in soil before (Mamaev 1963b).

The larvae of *Campylomyza ormerodi* (Kieffer) can be found en masse in the soil of clover and lucerne (alfalfa) fields. In contrast to all previously mentioned species they develop in close contact with the surficial roots of papilionaceous plants. In England this species is considered a serious pest of clover (McDugall 1913). We found larvae of *C. ormerodi* (Kieffer) in the soil of irrigated lucerne in the desert of Kyzylkum. Observations revealed that the larvae develop in decaying roots of lucerne, thereby stimulating a rapid expansion of the decay and causing the dying off of the roots. Obviously, the activity of these larvae is one of the factors slowing down the recovery of the subterreanean parts of lucerne plants that are periodically damaged by mowing. Larvae of *Monardia nigricans* Edwards have been found in the soil of an oats field (Edwards 1938).

Most noteworthy is the uneven distribution of these gall midges. The larvae are mostly encountered in niduses, where their number may be counted in hundreds of specimens per $m^2$. This is easily explained by the uneven distribution of the mycelium they need for their development. The scent of the mycelium attracts the ovipositing females. They therefore concentrate in sites rich in mycelium, where subsequently a high density of larvae will develop. The species composition of mycetophagous gall midges in accumulations of organic material – such as rotting wood and litter – is markedly more varied.

Litter of broad-leaved forests forms a rather special environment. The larvae live there in the narrow slit-like spaces between the more or less compressed leaves. Gall midges larvae from broad-leaved litter are strongly flattened, and are commonly provided with long bristles or hairs.

Typical litter species, characteristic of broad-leaved forests, are *Parepidosis venusta* (Winnertz) and *P. arcuata* Mamaev. Their yellow-orange larvae develop in the deeper layers of the litter. They are clearly marked off from related species by their covering of sparse, very long hairs. Morphologically akin to the larvae of *Parepidosis* Kieffer are those of *Dicroneurus montanus* Mamaev and *Porricondyla* sp. They also develop in litter, as do the larvae of *Bryocrypta indubitata* Mamaev.

We found the orange larvae of *Lestremia cinerea* Macquart and of *Campylomyza ormerodi* (Kieffer) in accumulations of rotting leaves with a profuse development of mould. A series of *C. dilatata* Felt has also been reared from larvae that developed in decaying plant parts (Pritchard 1947).

The larvae of *Rhizomyia* Kieffer, *Holoneurus* Kieffer, *Brachyneura* Rondani, and *Brachyneurina* Mamaev sometimes are very numerous in litter. They are not, however, attracted by rotting leaves, although they are also found on this substrate, but by rotting small twigs and root fragments. The larvae of *Rhizomyia perplexa* Kieffer and *R. circumspinosa* (Rübsaamen), interestingly, develop in roots and leaf sheaths of sedges.

The litter in coniferous forests has its own peculiarities: the layer is thicker than in broad-leaved forests, and the growth of mycelium is more intense. Therefore, though the litter fauna of coniferous forests is made up of the same genera as that of broad-leaved forests, the number of larvae is markedly higher in the former habitat. The occurrence of the larvae of *Aprionus dentifer* Mamaev in coniferous litter is interesting, as *Aprionus* Kieffer actually is characteristic of rotting wood.

Even such special litter as that under shrubs of tamarisk on the strongly insolated soils of the desert of Kyzylkum has its own fauna of gall midges. Most numerous are the larvae of *Winnertzia tamariciphila* Mamaev which develop in the lower layers of the litter. In the same desert gall midge larvae have also been found in the litter of *Haloxylon*.

The species composition of gall midges developing in moss has been relatively little studied. Among the most interesting species, whose larvae have been found in cushions of moss or in mosses on branches of trees, must be reckoned *Bryocrypta dubia* Kieffer, *Cryptoneurus muscicola* (Kieffer), *Bryomyia bergrothi* Kieffer, *Peromyia muscorum* (Kieffer), and *P. palustris* (Kieffer). Larvae of gall midges are among the main components of the fauna in primary soils on rocks under lichens and mosses (Stebaev 1962).

The group of gall midges that develop in wood is extremely rich in species. One of us has studied and published a series of papers on the species composition of xylobiontic gall midges (Mamaev 1960a, b, 1961b, c, f, 1963a, b, 1964b, d). We will therefore refrain from enumerating an exhaustive list of gall midges that have been found in wood, but we will pass in review some general problems. These elucidate the success of the species as measured from the extent of the damage to the wood, and from the main directions of specialization of their larvae.

The larvae of the large genera *Camptomyia* Kieffer and *Winnertzia* Rondani may be considered the most characteristic gall midge larvae connected with early stadia of degradation of wood. They have a thick cover and are provided with cuticular spines and supporting hooks at the end of the body that allow them to move about in bark that is still rather dense.

The species of *Camptomyia* Kieffer and *Winnertzia* Rondani that develop under bark in early stages of its degradation show some degree of food specialization. In most cases, however, this specialization is limited to the possibility of discrimination between gymnospermous and angiospermous bark. Thus, for instance, one of the most widely distributed species of xylobiontic gall midges, *Camptomyia corticalis* (H. Loew), develops under the bark of hornbeam, beech, oak, and birch, but its larvae are never found under the bark of spruce, pine, or fir. *Camptomyia maxima* Mamaev, on the other hand, is strictly bound to coniferous species – spruce and fir.

*Profeltiella meridionalis* Mamaev clearly also needs a special feeding environment. It develops in the mould on, and the juice oozing from, the chips of older stumps that remain after a tree has been felled. The larvae of the indigeneous species that settle under the bark of oak (*Profeltiella quercivora* Mamaev), in oculated grafts of fruit trees (*Thomasiniana oculiperda* Rübsaamen), under bark of twigs of shrubs (other species of *Thomasiniana* Strand and *Profeltiella* Kieffer, in tunnels of larvae of some phytophagous insects (*Profeltiella dizygomyza* Barnes), cause the death and decay of the bast and the wood. As we already remarked, they clearly use food of a mixed composition.

Tree stumps that have been splintered by the axe provide shelter for the development of some other groups of gall midges. Here live, for instance, the larvae of *Lasiopteryx* Stephens, *Brachyneurina* Mamaev, and *Colomyia* Kieffer. Observations on larvae of *Colomyia* Kieffer have shown that the young larvae, which develop in relatively fresh stumps, live in the depth of the cracks in the immediate vicinity of the still living bast. The adult larvae are more often found in places with profusely developed mould.

The larval biology of *Xylodiplosis praecox* (Winnertz) is remarkable. The larvae, which develop during spring in the pores in woody trunks left after felling in autumn combine a body length of 4.5 mm with a diameter of no more than 0.5 mm, and are well adapted to life in narrow spaces.

During the advancing destruction by fungi and insect larvae the structural qualities and chemical composition of bark and wood change. If the bark becomes crumbly, containing many holes and tunnels, its most characteristic gall midge inhabitants are larvae belonging to the species of *Peromyia* Kieffer. The larvae of *P. minutissima* Mamaev and *P. perpusilla* (Winnertz), for instance, develop in porous bark. The larvae of *Skuhraviana triangulifera* Mamaev live under comparable conditions.

When the bark of mouldering tree stumps and logs becomes more or less detached a film of mycelium of various fungi generally develops beneath it. The larvae of *Aprionus similis* Mamaev, *Winnertzia nigripennis* Kieffer, *Ulmomyia fungicola* Mamaev, and *Camptomyia maxima* Mamaev live in or below such films on trunks and logs of conifers. *Micropteromyia ghilarovi* Mamaev, *Karschomyia aceris* Mamaev, *Winnertzia nigra* Mamaev, and other species are found in comparable habitats on broad-leaved trees.

Under the bark of logs and trunks of broad-leaved trees, especially of birch, aspen, lime, oak, and beech, vast colonies of the paedogenic larvae of *Miastor metraloas* Meinert, *Heteropeza pygmaea* Winnertz, and *Nikandria brevitarsus* Mamaev may be found. Though we have doubtlessly seen larvae of *Heteropeza pygmaea* Winnertz under the bark of spruce and yew, we have never found such extensive colonies of paedogenic species under the bark of remains of felled conifers. The larvae of other paedogenic species that develop under bark *Brittenia fraxinicola* Edwards, *Heteropezina cathistes* (Pritchard) do not form extensive colonies.

Gall midge larvae are rarely found in strongly mouldered bark. Decaying wood proper is only invaded by gall midge larvae, the above mentioned genus *Xylodiplosis* Kieffer excepted, if its firmness has been strongly diminished by the action of wood destroying fungi.

Corroded wood shows numerous cracks and holes. Wood that has been destroyed by fungi, on the other hand, remains compact, though it has lost much of its firmness. Depending on the species of fungi either the lignin or the cellulose is destroyed. Wood turns brown if its cellulose is affected, but retains its light colour if its lignin is affected.

Since corroded wood is less common than destroyed wood, the composition of the species in the tunnels and cracks of corroded wood has been studied less extensively. The larvae of *Lestremia cinerea* Macquart occur, for instance, in porous oak wood, whereas those of *Peromyia monilis* Mamaev and *Aprionus flavidus* (Winnertz) are found in porous fir wood.

Among the gall midges that occur in connection with the destruction of wood, the group connected with phosphorescent destruction is especially diversified and rich in species. One of the most interesting discoveries is that of the previously unknown larvae of the genus *Trichopteromyia* Williston in phosphorescent hornbeam wood. The larvae of *Mycopriona caucasica* (Mamaev), *Aprionus smirnovi* Mamaev, *Diallactes ocellaris* Mamaev, and *Triommatomyia barnesi* Mamaev have been found in phosphorescent beech wood, those of *Monardia longicauda* Mamaev and *Winnertzia* sp. in willow wood. In the phosphorescent, rotting wood of broad-leaved trees the larvae of a whole group of closely related genera occur: *Winnertzia* Rondani, *Camptomyia* Kieffer, *Rhipidoxylomyia* Mamaev, *Trichoxylomyia* Mamaev, *Synepidosis* Mamaev etc. The larvae of *Solntsevia nigripes* Mamaev develop in trunks of spruce. Since the larvae of *Dicerura* Kieffer previously were known only from the leaf sheaths of some plants, the discovery of the larvae of *Dicerura* sp. in strongly weakenend phosphorescent beech wood is of special interest for the knowledge of the ecology of the gall midges.

Larvae of gall midges are very rarely found in brown, destructively decayed wood. The larvae of *Aprionus corniculatus* Mamaev and *Winnertzia rubricola* Mamaev have been recorded from brown oak wood; those of the curious *Aprionus dispar* Mamaev have been found in the brown wood of a willow trunk. In elm wood, where the process of rotting proceeds somewhat atypically, the larvae of *Ulmomyia rossica* Mamaev have been found.

Though several species of *Porricondyla* Rondani and *Clinorhytis* Kieffer have been described from material originating from wood, we have not succeeded in finding representatives of these species under bark or in rotting wood. The larvae of *Lasiopteryx flavotibialis* (Felt) and of some other Nearctic species also develop in wood.

The larvae developing on the surface of leavings of wood are remarkable. As a rule they are found in the luxuriantly growing mycelium on the lower side of logs. Thus, for instance, *Peromyia monilis* Mamaev and *Dichaetia pusilla* Mamaev have been reared from larvae collected in the mould on the lower side of fallen lime trunks. The white larvae of *Acoenonia europaea* Mamaev have been seen on top of beech logs, as well as in the mould on their lower side.

The flattened larvae of *Gynapteromyia carpatica* Mamaev, which were seen on the surface of beech logs in Transcarpathia, are very remarkable. The larvae of *Holoneurus fulvus* Kieffer and *H. pini* Mamaev develop on the surface of oak and pine trunks and, probably, logs. Adults of *Brachyneura fungicola* Mamaev have been reared from the mycelium growing on alder branchlets lying on the ground. The mycetophagous species, whose larvae live in the tubular hymenophores of polyporaceous species, make an ecological group of their own. Their larvae apparently feed on the growing spores. The attain large numbers in the young fruiting bodies of *Polyporus sulphureus*, *Fomes marginatus* etc.

Gall midge larvae also invade decaying fruiting bodies of fungi. The larvae of *Winnertzia lugibris* (Winnertz), for instance, have been seen in old fruiting bodies of *Fomes marginatus*. We have reared a series of adults of *Brachyneura fungicola* Mamaev from dead fruiting bodies of *Polyporus sulphureus*.

The larvae of *Camptodiplosis boleti* (Kieffer) live in a semifluid environment in the fruiting bodies of *Polyporus confluens*. The development of another species of this genus, *C. auriculariae* Buxton et Barnes, is bound to the fruiting bodies of *Auricularia mesentherica* (Buxton and Barnes 1953).

The gall midges that cause damage to the spawn and fruiting bodies of cultivated mushrooms have been studied in more detail. Cultivated mushrooms are damaged, not only by the larvae of *Lestremia cinerea* Macquart, which are able to live on the mycelium of many species of fungi, but also by the larvae of *Mycophila fungicola* Felt, *M. speyeri* (Barnes), *M. barnesi* Edwards, *Monardia stirpium* Kieffer, and *Henria psalliotae* Wyatt. The larvae of *Campylomyza pumila* Winnertz have also been collected from the fruiting bodies of toadstools (Rübsaamen 1899).

The larvae of *Mycodiplosis* Rübsaamen feed on the spores of rusts, and occur, as a rule, on the surface of the affected plant tissue. These larvae are found in rust on the leaves of elder, willow, lady's mantle, nettle, and on the needles of pine. They obviously belong to one of those species that occur en masse. Some species of *Clinodiplosis* Kieffer also feed on rusts. *C. pucciniae* Pritchard, for instance, develops at the expense of *Puccinia evadens* (Pritchard 1948).

Larvae of gall midges can be found in the cones of spruce (*Asynapta strobi* (Kieffer), *Winnertzia conorum* Kieffer, *Clinodiplosis piceae* Kieffer), in rotting olives (*Asynapta furcifer* Barnes), in old galls of other gall midges (*Cecidophila artemisiae* Rübsaamen, *Winnertzia deserticola* Mamaev), and similar substrates.

As far as known, some species belong to strictly mycetophagous groups, but are nevertheless bound to certain species of higher plants. Naturally such species attract our attention. Thus, for instance, the larvae of *Porricondyla cerealis* (Santer) develop in the leaf sheaths of rye, oats, and wheat, and are supposed to cause the death of the affected stems (see Barnes 1937). The larvae of *Asynapta phragmitis* Giraud live in subepidermal cavities in reed stems (Giraud 1863). *Colomyia caricis* Kieffer and *C. hordei* Barnes, which are representatives of another genus of the tribe Porricondylini, also develop in leaf sheaths. The harmless larvae of *C. hordei* Barnes, for instance, live in the sheaths of the lower leaves of barley at the surface of the soil, where they also pupate (Barnes 1927). The larvae of *Asynapta thuraui* Rübsaamen, which have been described from the leaf sheaths of *Calamagrostis*, do not cause damage either.

The semiaquatic larvae of *Dicerura scirpicola* Kieffer and *D. kaltenbachi* (Rübsaamen) are also bound to leaf sheaths; the larvae of the latter species live particularly on irises. According to Rübsaamen (1899), the larvae of *D. kaltenbachi* (Rübsaamen) leave a small strip of dead tissue when moving. In order to decide whether the species of the tribe Porricondylini, enumerated above, are already phytophages or still mycetophages, needing the mycelium that develops on dead parts of their host plants, a close examination of the nature of their food is necessary.

Not only porricondyline larvae live in leaf sheaths, but also some representatives of the morphologically and ecologically primitive subfamily Lestremiinae use this habitat. In this context the larvae of *Wasmaniella aptera* Kieffer, which have been found in the leaf axils of *Scirpus sylvaticus* may be mentioned (Kieffer 1899, Pritchard 1951). Leaf axils, especially those of dying and decaying leaves, are clearly quite suitable for the successful development of mycetophagous larvae of gall midges. The larvae of a number of species that develop in leaf sheaths, and are considered as phytophages, do not cause a substantial deformation of the plant tissue. This will be illustrated by some examples from the group of the phytophages.

### *The phytophages*

As we remarked when we discussed the division of the gall midges in trophic groups, we consider as phytophages only those species the larvae of which feed on living tissue of higher plants. The gall midges owe their fame especially to this group. First because the quaint galls, which their larvae form on plants, have always attracted, and still attract, the curiosity of investigators from various disciplines; and secondly because some phytophagous species are serious pests, the fight against which uses up much of the energy and means of practical agriculturists.

Not all phytophagous larvae act upon the plant tissue in the same way. On the lowest level of ecological specialization stand those species whose larvae are not able to induce the formation of galls on the plants they attack. These larvae do not stimulate its growth when they act upon the tissue of a plant, but on the contrary bring about its death. However, most phytophagous gall midges develop in pathological outgrowths of living plant tissue, i.e. in galls. It is to these phytophagous species that the entire family of gall midges owes its name.

Gall making phytophages are found in four tribes of the subfamily Cecidomyiinae: Lasiopterini, Oligotrophini, Asphondylini, and Cecidomyiini) two of these (Lasopterini and Asphondyliini) comprise, with a few exceptions, only gall makers.

Galls are induced by vegetable and animal organisms alike. Depending on the causing agent, galls are called phyto- or zoocecidia. Not only do representatives of many groups of insects, as for instance gall wasps, bugs, plant lice, leaf fleas, beetles, moths, and flies cause zoocecidia; they are also caused by some other invertebrates, notably nematodes and mites.

The multiformity of the agents and of the galls themselves does not make it easy to frame a precise definition of the concept 'gall'. According to Küster (1911) a gall is a pathological formation caused by an alien organism, resulting from the growth reaction of the plant to the stimulus exerted on it by that alien organism. The alien organism maintains a physiological (nutrimental) bond with the plant. Ross and Hedicke (1927) consider as gall a pathological formation that is actively induced by any alien organism, living parasitically or symbiontically on a plant. Slepyan (1962) divides the pathological neoplasms on plants in galls and teratoses. According to this author, galls are neoplasms, arising on vegetative and generative organs of plants as a result of a stimulus exerted by a gall maker; teratoses are neoplasms arising from the transformation of whole vegetative or generative organs as a result of the destruction of their normal growth pattern by a teratose maker.

These general definitions, of course, do not reflect the peculiarities of the gall that are caused by insects, and in particular by gall midges. Therefore, they should be moulded into a more concrete formula. The ability of insects to induce the growth of galls is an adaptation that not only solves problems of food supply, but one that serves as a reliable insurance against desiccation as well. Thus, the gall of arthropods provides the gall maker with a more or less closed chamber the atmosphere of which is saturated with water. Therefore, we will not designate as gall every deformation of the tissue, but only such which lead to the growth of a closed chamber in which the gall maker develops.

Galls caused by insects are pathological neoplasms on a plant. They arise as a reaction to a stimulus, exerted by the gall maker. The latter actively acts upon the morphogenetic processes of the plant, thus creating around itself a more or less closed chamber, consisting of plant tissue. According to this description those species that develop underneath living bast, which do not cause an outgrowth but do destroy the cambium, are excluded from the group of the gall making phytophages. Those species are also excluded whose larvae live in flowers or fruits, openly on the surface of leaves or in their axils, within galls of other gall midges etc., without causing a significant deformation. All these groups will be discussed after the gall makers, which constitute the bulk of the phytophages.

The ability to induce galls represents a very high level in the evolution of the gall makers. The environment inside galls under circumstances that are favourable for their development, as a rule, goes hand in hand with the flourishing of the whole group. This is clearly exemplified by the gall midges, which nowadays represent one of the most prospering families of Diptera.

According to the peculiarities of their structure and development, galls are divided into organoid and histoid galls (Küster 1911, Ross 1932). Organoid galls arise, as is indicated by their name, as a result of the overall deformation of a whole plant organ, with which the galls retain a resemblance. As a matter of fact, the origin is also reflected by the structure and components of the gall. Flowers, for instance, are frequently transformed into organoid galls. In this case the ovary, stamens, and pistil swell, and the corolla does not unfold. All these parts can, however, be easily recognized in the developing gall, even though no neoplasm whatsoever is formed. Histoid galls arise as a result of a local deformation of the tissue of some plant organ. In this case the organ is only partially deformed.

Only after a gall maker has exerted an initiating stimulus which activates the regenerative mechanisms of the plant does a gall develop. This stimulus is exerted either when the plant tissue is being damaged by the deposition of the egg or, if the egg is laid on the surface of some plant organ, when the hatching larva exerts its influence on the tissue.

The development of the structurally most complex galls has been divided into two stages (Slepyan 1961a). The first stage starts at the moment when the cells of the injured tissue begin to divide quickly, and leads, irrespective of the characteristics of the plant and the gall maker, to the development of a parenchymatous regenerative blastema. This stage comes to its end when a zone of sclereids differentiates. During the second stage the ultimate formation of three specialized layers takes place: a layer of trophic tissue, internally clothing the chamber that is being formed by the agent; a mechanically protecting sclereid layer; and a peripheral zone of primary gall parenchyma, which, as a rule, constitutes the most strongly developed zone. During the growth of less specialized galls the cell division in the injured tissue proceeds chaotically. Therefore, it is not possible to discern structurally different stages in the development of galls of this type.

The formation of a gall is the result of a close cooperation between the plant and the stimulating agent. During this cooperation the plant procures the building material, i.e. the embryonic tissue with its quickly dividing cells, while the gall maker transforms this tissue into a typical gall. Up to the present time the mechanisms of the influence exerted by the gall maker on the plant tissue have not been adequately studied. The possibility that these mechanisms are not identical in the various groups should not be excluded.

We must appreciate as a special suitability for gall making the fact that gall midge larvae digest their food externally and have only weakly developed mouthparts that are clearly not appropriate for the energetic destruction of cell walls.

Observations on the development of *Contarinia medicaginis* Kieffer (Fröhlich 1960) have shown that the first changes in the flower are seen only after the hatching of the larvae. This is evidence for the absence of a stimulating influence on the growth of the plant tissue by the eggs of gall midges during the embryonic development, and by the secretion of the egg laying female. In those cases, however, in which the female gall midges have a piercing ovipositor and wound the plant tissue during oviposition, the initiation of the gall formation must be placed at the moment of oviposition and at the site of the wound, where the regenerative blastema is also formed. Later on the young larva will act upon the plant tissue at this site.

Investigations of Jensen (1948, 1952) with *Mikiola fagi* (Hartig) have shown that growth substances from the group of the auxines, stimulating the cell division in plant tissues and secreted by the larva, have a decisive influence on the formation of galls.

Nolte (1957) has adduced interesting facts, testifying to the assertion that the effect of the substances secreted by some larvae shows strong resemblance to the effect of β-indole acetic acid. Nolte does not infer, however, that the substances secreted by the larvae are identical with growth substances of plants. He thinks that the larval substances act as specific catalysts for the plant growth hormones. In the same work we find support for the view that gall midge larvae can only induce deformations of young plant organs. Thus, for instance, larvae of *Contarinia nasturtii* (Kieffer) that are transplanted onto old leaves of Cruciferae, do not induce their deformation.

If we take into account the mode of digestion of gall midge larvae and the structural peculiarities of their salivary glands, which are substantially more complicated in the gall making phytophages than in the mycetophages (White 1948, Slepyan 1960), we may assume that the substances, secreted by the larvae and stimulating the growth of plant tissues, are synthesized in particular by the salivary glands.

We must note, however, that the development and the structural composition of the gall anticipate the development of the larvae considerably; a fact which, up to this time, has apparently escaped the attention of investigators. Rather frequently, on opening a fully developed gall, the investigator finds inside a still very small and premature larva, the development of which is still very far from its completion. Such observations bear witness to the fact that the life of gall midge larvae is roughly divided into two periods. During the first one the still very young larvae have an extremely strong stimulating effect on the plant tissue. As a result a gall quickly appears and develops to nearly normal proportions. The second period is the feeding stage of the larvae. Whereas the gall does not grow substantially during this period, its walls lignify, and the larvae in multichambered galls encapsulate etc. We must assume that during this period the production of growth substances decreases.

Though the biochemical side of the gall forming process still requires persistent investigation, the mechanical side is easily understood. Gall midge larvae, settling in the inner parts of a twig or stem, cause its growth to become undetermined and tumorlike. Only when a single larva settles in a stem can the shape of the gall be specific for the species (for example, the solitary gall midge larvae that damage *Haloxylon* and other desert plants). Since larvae, settling on a leaf blade, feed on the tissue, they slow down the cell growth in their immediate vicinity. Therefore, the side of the leaf opposite to the larvae shows a higher growth rate, especially if we keep in mind the local, stimulating action of the larva on the tissue. This means, generally speaking, that the leaf rolls up to the side where the larvae are. Thus, if the larvae happen to be in the zone of the midrib, a pod-like gall will be formed. If the larvae are localized near the leaf margin, this will roll up. If the larvae are distributed more or less evenly over the surface, the leaf blade will shrivel. Solitary larvae, finally, cause a local swelling of the leaf blade.

Larvae settling in flowers or on the tip of offshoots act analogously upon these organs. By inhibiting the growth at the inner side of the base of petals and sepals, gall midge larvae prevent the opening of the flower. By a similar action at the leaf bases, at the tip of an offshoot a spindle of leaves is formed. More complex types of galls are comparatively rarely caused by gall midges.

At the end of the survey of gall making we will go into the problem of the development of fungus mycelium inside the flower and fruit galls of the genera *Asphondylia* H. Loew and *Schizomyia* Kieffer. This mycelium, which changes from white to grey, covers the walls of the gall, but attacks neither the larva nor the pupa. The mycelium does not occur in all galls formed by one and the same species. A part of the galls does not contain mycelium, but the larvae inside them, nevertheless, develop normally. The mycelium starts growing energetically and soon fills up the whole gall only after the pupa has completed its development and the imago has flown away.

Ross (1932) holds the view that the co-occurrence of the larva and the mycelium does not have the character of a symbiosis. The spores of the fungus fall randomly in the gall. The larva does not use the developing mycelium. The larva, however, does inhibit the growth of the mycelium. Our knowledge of the assortment of fungus species, occurring inside galls, is still very scanty. Inside the galls of various species of Asphondyliini occur the mycelia of: *Cladosporidium* sp., *Macrophoma coronillae*, *Coniothyrium leguminum*, *Diplodia* sp., *Aureobasidium* (*Pullaria*) *pullulans* etc. (Batra and Lichtward 1963).

A survey of the damage caused by gall making phytophagous gall midges may be conveniently given if we successively consider all organs and the damage which is inflicted on them.

A considerable number of gall midges develops in leaf galls. In the most primitive cases, the larvae, living in colonies, cause the shriveling of the leaf and can be found in the resulting folds. Thus the larvae of, for instance, *Dasyneura crataegi* (Winnertz) on *Crataegus*, *D. plicatrix* (H. Loew) on *Ribes*, *D. thomasiniana* (Kieffer) on *Tilia*, *D. acercrispans* (Kieffer) and *Contarinia acerplicans* (Kieffer) on *Acer*, *C. heraclei* (Rübsaamen) on *Heracleum sphondylium* etc. do not cause true deformations.

If the larvae are distributed in the central part of the leaf, near the midrib, more regular deformations result. In this case galls of two types may arise. The larvae of *Wachtliella rosarum* (Hardy) on *Rosa*, *Dasyneura acrophila* (Winnertz) on *Fraxinus*, and *D. trifolii* (F. Loew) on clover, cause the folding along its midrib of a leaflet of a compound leaf. Often the result is a thickening of the blade of the leaflet and the formation of a pod-like gall. Sometimes, however, not the whole leaflet is deformed, but a narrow, thick-walled, shallow fold, running along the midrib, is formed. This type of damage is characteristic for *Dasyneura fraxini* (Kieffer) on *Fraxinus* and *Putoniella marsupialis* (F. Loew) on plum and blackthorn etc.

The development of very many species is restricted to leaf margins. Thick-walled, fleshy galls, resulting from the rolling up of the leaf margin, are formed by colonies of larvae of *Wachtliella persicariae* (Linné) on *Fagopyrum* sp., *Dasyneura mali* (Kieffer) on *Malus*, *D. pyri* (Bouché) on *Pyrus*, *Rhabdophaga marginemtorquens* (Winnertz) on *Salix*, the solitary larvae of *Dasyneura tiliamvolens* (Rübsaamen) on *Tilia* etc. The larvae of *Macrodiplosis dryobia* (F. Loew) on *Quercus*, *Dasyneura populeti* (Rübsaamen) on *Populus tremula* etc. cause a bending of the leaf margin without substantially thickening it.

Within the thick-walled invaginations on the leaves of *Urtica* develop a few larvae of *Dasyneura urticae* (Perris), whereas in those on the leaves of *Filipendula* some larvae of *D. engstfeldi* (Rübsaamen) co-occur in every gall. The solitary larvae on the leaves of *Populus* form more regular galls: *Harmandia cavernosa* (Rübsaamen) makes rounded, fleshy galls protruding from both sides of the leaf and opening at the upper side by means of a slit; *Lasioptera populnea* Wachtl makes similar galls with a circular opening; *Harmandia globuli* (Rübsaamen), *H. populi* (Rübsaamen) and *H. loewi* (Rübsaamen) live in rounded galls at one side of the leaf. *Janetiella lemei* (Kieffer) and *Craneiobia corni* (Giraud) form respectively on Ulmaceae and on *Cornus*, fleshy, pitcher-shaped galls, that, though protruding from both sides, open at the lower side of the leaf. The larvae of *Dasyneura ulmariae* (Bremi) form similar, but smaller galls on the leaves of *Filipendula*.

The larvae of *Didymomyia reaumuriana* (H. Loew) develop in more complex galls on the leaves of *Tilia*. Within a fleshy outgrowth, a capsule containing the larvae is formed. In autumn this capsule falls onto the ground and the larva hibernates within its capsule in the litter. Complex, ornamented, or hairy leaf galls are formed by: *Dryomia circinnans*

(Giraud) on *Quercus, Hartigiola annulipes* (Hartig) on *Fagus, Rondaniola bursaria* (Bremi) on *Glechoma hederacea, Diarthronomyia chrysanthemi* Ahlberg on *Chrysanthemum*, etc.

On the leaves of various plants real parenchyma galls are caused by gall midges. In outward appearance these galls remind us of mines. They appear as pockets in the parenchyma of a leaf or between the parenchyma and the epidermis. The formation of a parenchyma gall is accompanied by a considerable deformation of the surrounding tissue. In most cases a parenchyma gall is formed within the parenchymal layer and presents itself as a large, thick-walled, blister-like swelling, as, for instance, the galls of *Anisostephus betulinum* (Kieffer) on *Betula*, of *Loewiola centauriae* (F. Loew) on *Centaurea*. In other cases a compact parenchymatous formation with a very small chamber is formed, as is done by *Polystepha quercus* Kieffer on *Quercus*. On leaves of *Betula* sometimes galls, closely resembling parenchyma galls, develop along nerves, for example the galls of *Massalongia rubra* (Kieffer).

Sometimes parenchyma galls of another type are found. For these the galls formed by the larvae of *Cystiphora sonchi* (F. Loew) on the leaves of *Sonchus* may serve as an example. The larvae of this species develop in rounded galls between the parenchyma and the lower epidermis. Other species of this genus cause considerable damage: *C. hieracii* (F. Loew) on *Hieracium*, *C. taraxaci* (Kieffer) on *Taraxacum*. The larvae of a pest of *Buxus*, *Monarthropalpus buxi* (Laboulbène), develop within the parenchymal layer, and on maturing loosen a part of the upper epidermis from the parenchyma, thus causing so-called 'windows'.

Apart from the species mentioned above, the larvae of the following species develop in galls of the leaf parenchyma: *Dasyneura vitrina* (Kieffer) and *Drisina glutinosa* Giard on *Acer*, *Laubertia schmidtii* on *Chondrilla*, *Phlyctidobia solmsi* (Kieffer) on *Viburnum*, *Physemocecis ulmi* Rübsaamen on *Ulmus* etc. Within the parenchyma of coniferous needles also gall midge species develop, put prominent among them *Agevillea abietis* Hubault, a well-known pest of abies.

Some gall midges damage the leaf stalk. Very characteristic are, for instance, the galls of *Syndiplosis petioli* (Kieffer) on the petioles of *Populus tremula*. Similar galls are caused by *Contarinia tiliarum* (Kieffer) on the petioles of *Tilia*, and on those of *Acer* by *Atrichosema aceris* Kieffer.

A considerable number of gall midges develops in galls on shoots of trees and shrubs, and in deformed stems of herbs. Deformation of shoot-tips is one of the most commonly encountered disorders inflicted by gall midges. The most primitive type of galls at the tip of shoots consists of a loose accumulation of malformed leaves. Some of these loose galls have no definite shape, as, for instance, the galls induced by the larvae of *Dasyneura violae* (F. Loew) on *Viola*, *D. crataegi* (Winnertz) on *Crataegus*, and *Contarinia quercina* Rübsaamen on *Quercus* etc. Most of these galls, however, do have a characteristic shape. Thus, long, fusiform, rather loose galls are formed by the larvae of *Rhabdophaga terminalis* (H. Loew) on shoots of *Salix*, and by *Dasyneura tortrix* (F. Loew) on *Prunus spinosa*. Similar galls can be found on *Euphorbia* (Dombrovskaya 1940) and on some other plants.

Shorter, but also rather loose, tip-galls are induced by the larvae of *Dasyneura hyperici* (Bremi) on *Hypericum*, *D. vincae* Rübsaamen on *Vinca*, *Jaapiella veronicae* (Vallot) on *Veronica* etc. Compact, bulb-shaped galls, consisting of many leaves placed together at the tip of shoots, are characteristic of *Bayera capitigena* (Bremi) on *Euphorbia*, *Bouchéella artemisiae* (Bouché) on *Artemisia*. *Taxomyia taxi* (Inchbald) on *Taxus*, *Oligotrophus juniperinus* (Linné) on *Juniperus* etc. form similar, though more elongated galls. *Rhabdophaga (Rabdophaga) rosaria* (H. Loew) and *R. heterobia* (H. Loew) induce at the terminal shoots of willow peculiar galls, nicknamed 'rosette galls'. Externally, these galls show some resemblance to rose flowers indeed. The spring geneneration of these gall midges develops in willow catkins. Gall midges that damage the stem of shoots of trees, shrubs, and some herbaceous plants, cause structurally uniform, fusiform or more or less rounded stem tumors. *Rhabdophaga salicis* (Schrank) develops in such galls on the shoots of *Salix*, *Lasioptera rubi* (Heeger) on raspberries and brambles, *Stefaniella** *atriplicis* Kieffer on *Atriplex*, *Lasioptera artemisiae* Dombrovski on *Artemisia*, *Contarinia jacobaeae* (H. Loew) on *Senecio*, and *C. melanocera* Kieffer on *Genista*. In recent years a considerable number of gall midges of stem galls on *Haloxylon* and *Tamarix* has been described (Marikovskiy 1953a, b, 1955a, etc., Mityaev 1961).

The gall midges that develop in the stem bases of grasses cause their deformation and death. To this group belong serious agricultural pests, such as: *Mayetiola destructor* (Say), *M. secalis* Bollow, *M. avenae* (Marchal) and other species of the genus *Mayetiola* Kieffer. The stem galls of a number of species are structurally more elaborate than simple outgrowths of plant tissue. The larvae of *Giraudiella inclusa* (Frauenfeld), for instance, develop in grain-shaped galls protruding from the stem wall of reed. The larvae of *Hybolasioptera cerealis* (Lindeman) live in shallow excavations covered by a shiny black membrane on the stems of rye and wheat. The larvae of *Rhopalomyia baccarum* Kieffer form round, fleshy, one-chambered galls on the stems of *Artemisia*.

Comparatively few gall midges transform buds, which have not yet sprouted, into galls. The larvae of *Asphondylia prunorum* (Wachtl), for instance, develop in swollen buds of plums; *A. sarothamni* H. Loew inflicts analogous damage to *Sarothamnus*; *Misospatha tubifex* (Bouché) transforms the buds of *Artemisia* into tubular galls; *Dasyneura salicariae* (Kieffer) transforms the axillary buds of *Lythrum salicaria* into galls etc.

Considerably more often flowers are transformed into galls. The gall midge larvae cause a swelling of the flower, they prevent the opening of the corolla, and develop within the thus formed chamber. One flower mostly lodges several larvae, in some cases only one. Anthophilous species may be represented by the following examples: *Contarinia craccae* Kieffer in distorted flowers of *Vicia*, *C. anthobia* (F. Loew) on *Crataegus*, *C. medicaginis* Kieffer on lucerne, *Gephyraulus raphanistri* (Kieffer) on Cruciferae, *Schizomyia galiorum* Kieffer on *Galium*, *Ametrodiplosis viciae* Mamaev has been described from *Vicia*, *Dasyneura salviae* Kieffer in deformed flowers of *Salvia*, *Asphondylia verbasci* (Vallot) in non-opening flowers of *Verbascum* etc.

The fruit and seed damaging gall midges also form a well defined ecological group. The larvae of *Kiefferia pimpinellae* (F. Loew) transform the seeds of Umbelliferae, in which they live, into globular galls. The larvae of *Clinorhyncha tanaceti* Kieffer live in

*Note from the editor: Mamaev uses the spelling '*Stephaniella*'; in the original description, however, Kieffer correctly used '*Stefaniella*' as he named the genus after the Italian entomologist Stefani (Stefani in: Kieffer 1898: 55).

enlarged fruits in the inflorescences of *Tanacetum*; the larvae of *C. millefolii* Rübsaamen do the same in the fruits of *Achillea*. The larvae of *Contarinia marchali* Kieffer have been recorded from the fruits of *Fraxinus*, whereas the larvae of *Dasyneura geranii* (Kieffer) develop in swollen fruits of *Geranium*. The larvae of *Asphondylia melanopus* Kieffer damage the pods of *Lotus*; those of *Lathyrus* are enlarged by *A. lathyri* Rübsaamen. *Semudobia betulae* (Winnertz), *Plemeliella abietina* Seitner, and *Resseliella piceae* Seitner develop in the seeds of respectively *Betula*, *Abies*, and *Picea*. The larvae of these species take the place of the contents of the seed, and cause only a slight deformation. Gall midge larvae are found in fleshy fruits as well. *Contarinia pyrivora* (Riley), for example, is a serious pest of young pears.

Very few gall midges develop on the subterranean parts of plants, inducing galls on rhizomes. *Ametrodiplosis auripes* (F. Loew), which damages *Galium*, may serve as an examle for these. Phytophagous gall midges are very rarely found on roots (Wyatt 1963b).

Let us now consider those species the larvae of which do not induce the formation of galls. Thereby we should note that the division of the phytophages into gall makers and non-gall makers reflects only an ecological concept. In most cases it has nothing to do with evolutionary primitivity of the non-gall making phytophages. Frequently closely related species, belonging to one and the same genus, strongly differ in their ability to influence the tissue of their host plant. They thereby show the idiosyncrasies of the species formation within the family of the gall midges (Mamaev 1964c). As opposed to the typical gall makers, we will call the phytophages whose larvae do not induce the formation of galls 'free-living gall midges'. The term 'frei lebende Gallmücken', used in German literature, thus unites all gall midges that are unable to induce the formation of galls, and that in most cases develop outside galls. The group of the free-living gall midges contrasts with the group of the gall makers, and comprises mycetophages, non-gall making phytophages, and zoophages alike.

Among the free-living gall midges, species developing within flowers (anthobionts), especially of Compositae and Leguminosae, occur most frequently. Only one of the species that live in the flowers of clover, *Dasyneura flosculorum* (Kieffer) can be considered as a phytophagous gall maker. Besides this species seven phytophagous species, not causing galls, have been recorded from the flowers of clover: *Dasyneura leguminicola* (Lintner), *D. gentneri* Pritchard, *Tricholaba trifolii* Rübsaamen, *Giardomyia britannica* Milne, *Clinodiplosis leguminicola* Milne, *Ametrodiplosis viciae* Mamaev, and *Contarinia* sp. Two predatory and two mycetophagous species apparently also develop in the already withering flower heads of clover (Milne 1960). Of the enumerated species, *Dasyneura leguminicola* (Lintner) only develops successfully on red clover. The females of this species lay their eggs in the still green flower heads, and the larvae strongly deform the flowers. Barnes (1946-1956), therefore, thinks it possible that *D. flosculorum* (Kieffer) and *D. leguminicola* (Lintner) are identical. *D. gentneri* Pritchard shows no such food specialization and accepts any species of clover. The females of this species lay their eggs in the already opening flowers, and the larvae, feeding on the completely developed ovary, do not cause a substantial deformation. The flower heads that are inhabited by the larvae become, however, rougher to the touch. The possibility that other phytphagous gall midges also occur in the flower heads of clover should not be excluded. How widely the gall midges are distributed in flowers may be judged from the species list compiled by Kh.P. Mamaeva (1964b). In this list the following plant species are represented: *Achillea millefolium, Arctium tomentosum, Artemisia vulgaris, Brunella vulgaris, Carduus crispus, Centaurea cyanus, C. jacea, Chrysanthemum leucanthemum, Fagopyrum sagittatum, Knautia arvensis, Leontodon hispidus, Rumex crispus, Solidago virgaurea, Stachys palustris, Trifolium pratense, T. arvense*, etc.

The gall midges whose larvae live in flowers without causing the formation of galls generally belong to genera comprising representatives of various life forms, and counting among their numbers typical gall makers as well as phytophagous non-gall makers and inquilines. Within the genus *Contarinia* Rondani the following species live in flowers but do not induce gall formation: *C. artemisiae* Rübsaamen on *Artemisia*, *C. avenae* Kieffer on oats, *C. dipsacearum* (Rübsaamen) on *Hypochaeris* sp. and *Crepis biennis*, *C. tanaceti* Rübsaamen on *Tanacetum*, *C. tritici* (Kirby) on rye, wheat etc. In the genus *Clinodiplosis* Kieffer a similar life is led by the larvae of *C. cilicrus* (Kieffer) in the flower heads of several composites, of *C. scorzonerae* Rübsaamen in the inflorescences of *Scorzonera*, and of *C. schlechtendali* Rübsaamen in the wilting flowers of *Convolvulus* etc. Species from the genera *Dasyneura* Rondani and *Macrolabis* Kieffer are also widely distributed in flowers. From the other species that feed on the ovary of flowers we must mention the pest of wheat *Sitodiplosis mosellana* (Géhin), the pest of millet *Stenodiplosis panici* Plotnikov, the pest of *Alopecurus Stenodiplosis geniculati* Reuter; and *Lamprodiplosis rhopalothrix* Kieffer and *Jaapiella rubicunda* (Rübsaamen) which develop in the flowers of *Rumex*, etc. The study of the biology of the gall midges that develop in flowers is still very far from its completion. First of all it will be necessary to get a clear idea of the character of the food specialization of the species belonging to this group. For, if it turns out that the anthobiontic species are not restricted in their choice of food to the flowers of a single plant species, many of the hitherto described species will be found to be identical with others, and many names will be reduced to synonyms.

The species that live openly on leaf blades form an ecologically interesting group of free-living gall midges. The larvae of *Dasyneura pustulans* (Rübsaamen), which develop on the lower side of the leaves of *Filipendula*, protect themselves from drying out in an original way. The larva exerts an influence on the tissue of the leaf, causing a continuous exosmosis of plant sap. This accumulates in a small hollow resulting from the deformation of the tissue. The droplet of sap provides the larva with an environment that not only secures its food supply, but that also serves as a reliable protection against drying out. The affected tissue becomes yellow and dies only after the full-grown larvae have migrated to the soil. The larvae of other species live in a similar way: *Mikomyia coryli* (Kieffer) on the leaves of *Corylus*, *Massalongia aceris* (Rübsaamen) on the leaves of *Acer*, and some other species.

The species composition of free-living gall midges occurring in the leaf sheaths of sedges and grasses is extremely diverse. In the review of the mycetophages we have already discussed some species belonging to this ecological group. In a list of species whose development is bound to the stems of sedges and grasses, Skuhravá and Skuhravý (1960) mention 38 names. About half of the enumerated species do not form galls. The most characteristic species on *Phragmites* are: *Thomasiella flexuosa* (Winnertz), *Asynapta phragmitis* (Giraud), and *Antichiridium striatum* Rübsaamen. These species may also be found on some other plants. The larvae of *Thomasiella calamagrostidis* (Rübsaamen) develop on *Calamagrostis*. *Mayetiola moliniae* (Rübsaamen) develops on *Molinia*, causing only a slight deformation of the stem. On sedges occur: *Brachydiplosis caricum* Rübsaamen, *Thurauia aquatica* Rübsaamen, *Trichodiplosis caricis* (Kieffer), and *Sterrhaulus corneolus* (Rübsaamen). The larvae of *Herbomyia scirpicola* (Kh. Mamaeva) develop at the stem base of *Scirpus sylvaticus*, where they occur en masse under the epidermis of the leaves.

The larvae developing in the resin flowing from coniferous trees have a remarkable ecology. Until recently only one species – *Itonida* (*Cecidomyia*) *pini* (De Geer) – was known whose larvae are able to develop in the droplets of resin at the base of the needles of *Pinus* as well as in the resin flowing from the trunks. Recently, two more species have been described whose larvae have been found in liquid resin: *Stelechodiplosis magna* Möhn in the resin of *Picea*, and *Wichmanniella crassa* Möhn in the resin of *Picea* and *Abies*.

The inquiline gall midges form a group that is biologically and systematically close to the denizens of flowers. This kinship of inquilines and anthobionts is evident from the fact that the species developing within the galls of other gall midges generally belong to the very same genera that are most frequently encountered in flowers. Moreover, sometimes one and the same species has been found in flowers as well as (as an inquiline) in the galls of other gall midges.

*Tricholaba trifolii* Rübsaamen may serve as an example. The larvae of this species develop in the flowers of clover and in the galls of *Dasyneura trifolii* (F. Loew), which galls present themselves as longitudinally folded and thickened leaflets of a compound leaf. A gall former on *Salix*, *Rhabdophaga heterobia* (H. Loew), sometimes develops in the galls of a related species, *Rh. rosaria* H. Loew. The genera *Dasyneura* Rondani and *Contarinia* Rondani contain relatively few inquiline species. The larvae of *Contarinia inquilina* Rübsaamen which develop within the galls of *Kiefferia pimpinellae* (F. Loew) and of *Diodaulus traili* (Kieffer) on Umbelliferae are inquilines.

Several species comprise gall makers as well as inquilines. One third of the European species of *Clinodiplosis* Kieffer is inquiline.* Among them are: *Clinodiplosis urticae* Kieffer in the galls of *Dasyneura urticae* (Perris) on *Urtica*; *Clinodiplosis sorbicola* Rübsaamen in the galls of *Contarinia sorbi* Kieffer on *Sorbus*; *Clinodiplosis sarothamni* Kieffer coexists in the galls of *Contarinia pulchripes* (Kieffer) on *Sarothamnus* and *Genista*; *Clinodiplosis cerricola* Rübsaamen in the galls of *Syndiplosis quercicola* Rübsaamen on oak; *Clinodiplosis botularia* (Winnertz) in the galls of *Dasyneura fraxini* (Kieffer) on *Fraxinus*; etc. In the genus *Macrolabis* Kieffer we also find gall makers along with inquilines. Thus we find, for instance, that: *M. luceti* Kieffer develops in the galls of *Wachtliella rosarum* (Hardy) on roses; *M. jaapi* Rübsaamen inhabits the galls of *Dasyneura aparines* (Kieffer) on *Galium*; *M. incolens* Rübsaamen, one of the commonest species, is found in the galls of *Jaapiella veronicae* (Vallot) on *Veronica*; *M. saliceti* (H. Loew) is known from the galls of *Rhabdophaga heterobia* (H. Loew); etc. Since the larvae of *Ametrodiplosis medialis* Mamaev develop in the galls of *Dasyneura crataegi* (Winnertz) on *Crataegus*, inquilines also occur among the species of the genus *Ametrodiplosis* Rübsaamen. The larvae of *Monodiplosis liebeli* (Kieffer) live in the galls of *Macrodiplosis dryobia* (F. Loew) on oak. *Amerapha gracilis* Rübsaamen is an inquiline

*Note from the editor: Skuhravá (1974) reduced most species to *Clinodiplosis cilicrus* Kieffer.

found in the galls of *Kiefferia pimpinellae* (F. Loew). The larvae of *Lasioptera tiliarum* Kh. Mamaeva have been observed in the galls of *Contarinia tiliarum* (Kieffer). The larvae of *Camptoneuromyia petioli* Kh. Mamaeva have been found in the galls of *Syndiplosis petioli* (Kieffer).

Some genera as, for example, *Trotteria* Kieffer, comprise only inquilines. *T. umbelliferarum* (Kieffer) and *T. inquilina* Rübsaamen are known from the galls of *Kiefferia pimpinellae* (F. Loew) on Umbelliferae; *Trotteria sarothamni* (Kieffer) from the galls of *Contarinia pulchripes* (Kieffer) on *Sarothamnus*; *Trotteria galii* Rübsaamen from the galls of *Schizomyia galiorum* Kieffer on *Galium*; *Trotteria ligustri* (Rübsaamen) from the galls of *Placochela ligustri* (Rübsaamen) on *Ligustrum*; etc. Until some time ago all species of the genus *Tricholaba* Rübsaamen were considered inquilines. Quite recently, however, Borg (1960) published an article stating that the larvae of *T. trifolii* Rübsaamen are able to independently induce gall formation on the leaves of clover. According to our own observations, *T. similis* Rübsaamen, which is known in the literature as an inquiline of *Dasyneura viciae* (Kieffer), can also develop in its own galls.

Though most inquiline gall midges live in the galls of phytophagous gall midges, some species have ensconced themselves in the galls of other insects. Thus, for instance, in the galls of *Andricus foecundatrix* (Hartig) (Cynipidae) on oak the larvae of *Clinodiplosis galiicola* Rübsaamen and *Arnoldia gemmae* (Rübsaamen) develop between the scales of the gall; the larvae of *Parallelodiplosis galliperda* (F. Loew) are bound to the fleshy galls of various species of *Diplolepis* Geoffr. (Cynipidae) on oak species; the larvae of *Clinodiplosis rhynchitou* Rübsaamen live in the leaf-rolls of the weevil *Rhynchites betuleti*; and the species *Profeltiella dizygomyzae* Barnes has been described from the leaf mines of the fly *Dizygomyza barnesi* Hendel.

Inquilinism, leading to the co-existence in one gall of larvae belonging to different species of gall midges, is rather frequently considered a prerequisite for the transition to zoophagy.

### *The zoophages*

This trophic group of gall midges is considerably less rich in species than are the mycetophages and the phytophages. Predatory gall midges do not nearly occur in all subfamilies and tribes. The predatory way of life is predominantly characteristic for the highly organized representatives of the tribe Cecidomyiini and for three genera of the tribe Oligotrophini, both tribes belonging to the subfamily Cecidomyiinae. From the subfamily Lestremiinae no zoophages have been recorded.

Gall midge larvae, being weakly mobile, could only adapt themselves to food consisting of sedentary victims that, moreover, form large colonies. Therefore, zoophagous gall midges are fundamentally bound to plant lice, pine and spruce aphids, scale insects, mites living en masse, and larvae of other gall midges. In the following paragraphs these main groups of zoophagous gall midges will be reviewed.

Gall midge larvae living in colonies of plant lice were first observed by Rondani who described the species *Cecidomyia aphidimyza* Rondani (1842). Subsequently, biologists began to vigorously collect information on this interesting group of gall midges, which were described, together with mycetophages, in the genus *Bremia* Rondani. In 1904 Kieffer (1904) erected the independent genus *Aphidoletes* Kieffer, in which he included

16 species, for the aphidophagous gall midges. By 1913 *Monobremia* Kieffer, *Isobremia* Kieffer, *Cryptobremia* Kieffer, *Phaenobremia* Kieffer, the latter numbering no less than 27 species, had been split off as independent genera.

All data pertaining to the species observed in large colonies of aphids by various authors have been summarized, without critical examination, by Barnes (1929). In his list Barnes mentioned 14 genera, including the predators of pine and spruce aphids. The most substantial contribution to the system of the aphidophagous gall midges has been made by Nijveldt (1952, 1954, 1955, 1957). He not only studied all data from the literature and the pertaining collections, but he also performed biological experiments that proved the absence of a narrow food specialization in several species belonging to this group. Until the publications of Nijveldt (1954, 1955), investigators, when describing new species, assumed a narrow food specialization. Thus, every species found in colonies of aphids for which no previous reports of predatory gall midges were known, was considered new, without careful comparison with other species. The denial of a narrow food specialization, and the absence of substantial morphological differences between the majority of the previously described species led to doubt concerning the existence of many of them. In a list of aphidophagous genera Nijveldt (1957) reduced seven genera to synonyms.

A new revision of these genera, taking into account the material collected in the European part of the U.S.S.R. (Mamaeva 1964a), has shown that the genera *Isobremia* Kieffer and *Phaenobremia* Kieffer had been excluded from the genus *Aphidoletes* Kieffer on insufficient grounds. In Europe, in fact, only three widely distributed and unspecialized species of aphidophagous gall midges exist: *Monobremia subterranea* (Kieffer), *Aphidoletes aphidimyza* (Rondani) and *A. urticariae* (Kieffer). These three species are morphologically clearly distinct. Their biology, however, is so similar that they are often encountered in one and the same colony of plant lice. Their life cycles are also extremely alike. The independence of *Aphidoletes meridionalis* Felt, described from North America, should be sufficiently supported by a comparison with *A. aphidimyza* (Rondani). The area of distribution of *A. aphidimyza* (Rondani) might encompass both continents.

The females lay their eggs, from which the larvae hatch after 3-4 days, either on leaves carrying colonies of aphids, or on the bodies of the aphids themselves. The first instar larvae pierce their victims mostly from below, between the legs, through the chitin between the segments. After having been successfully pierced, the aphid quickly becomes paralyzed, and the larva leisurely sucks its contents. The larvae crawl through the colony of lice, quickly moving about with the pointed front end of the body in search of a fresh victim. Observations on the larvae of *Aphidoletes meridionalis* Felt have shown that the full grown larvae sometimes paralyze the lice by piercing one of the joints of the leg, mostly that between the femur and the tibia. An aphid that is being attacked in this way generally does not react to the predator; whereas, if the abdomen is being pierced the victim has a few seconds to try to ward off the gall midge larva with its legs.

The larvae kill considerably more aphids than they need for food, since they often throw their victim away before it has been sucked dry. The colour of the larvae depends on the species of lice on which they feed. The larvae are mostly orange, but sometimes colonies of yellow larvae are found. In the Moscow Region co-existing orange and yellow larvae belonging to one and the same species, *Aphidoletes urticariae* (Kieffer), have been observed among aphids belonging to the species *Dentatus sorbi* Kalt. The duration of the development of the larvae varies with the circumstances (temperature, humidity, food species of aphids) from 10 to 19 days. In Rübsaamen (1899) we find, for the first time, experimentally supported indications that some species of aphidophagous gall midges are not narrowly specialized to certain aphid species as a source of food. Rübsaamen transplanted aphidophagous gall midges from a colony of *Chaetophorus versicolor* Koch on *Populus tremula* to a colony of *Vacuma dryophila* Schrk., and observed their successful development in the colony of this aphid species. Nijveldt (1952) repeated the experiment on a large scale. He was able to show that the larvae, hatched from eggs laid by the females of *Aphidoletes urticariae* (Kieffer) in colonies of various species of aphids, nearly always develop successfully. The females of this species have, however, their preferences, since they lay their eggs rather in a colony of *Aphis urticata* Gmel. than in a colony of *A. fabae* Scop. or of *Chaitophorus populeti* Panz. The development proceeds, however, always normally. The full grown larvae pupate in the soil, more rarely on the leaves, in a silky cocoon.

To conclude we present a list, enlarged with our own observations, of the aphid species from whose colonies aphidophagous gall midge larvae have been recorded (Nijveldt 1955, 1957; Mamaeva 1964). *Aphidoletes aphidimyza* (Rondani) occurs in colonies of: *Aphis fabae* Scop., *A. farinosa* Gmel., *A. frangulae* Kalt., *A. nerii* Fons., *A. pomi* de Geer, *A. rumicis* L., *A. urticata* Gmel., *Brevicoryne brassicae* L., *Brachycaudus cardui* L., *Cavariella pastinacae* L., *Chaitophorus beuthani* Börn., *Ch. populeti* Panz., *Cryptomyzus ribis* L., *Hyalopterus pruni* Geoffr., *Myzus ornatus* Laing., *M. sanborni* Gyll., *M. persicae* Sulz., *Pentatrichopus fragaefolii* Cock., and *Periphyllus acericola* Wlk. *Aphidoletes urticariae* (Kieffer) occurs in colonies of: *Acyrthosiphon pisum* Harr., *Aphis fabae* Scop., *A. farinosa* Gmel., *A. pomi* de Geer, *A. sambuci* L., *A. urticata* Gmel., *Chaitophorus populeti* Panz., *Dentatus sorbi* Kalt., and *Myzus persicae* Sulz. *Monobremia subterranea* (Kieffer) occurs in colonies of: *Aphis farinosa* Gmel., *A. urticata* Gmel., *Sappaphis plantaginea* Pass. (*Anuraphis roseus* Baker), *Symydobius oblongus* Heyd., and *Toxopterina vandergooti* Börn.

In the galls of pine and spruce aphids develop only two species of gall midges, belonging to two different genera. One of the species, *Adelgimyza strobilobii* Del Guercio, has been described from the galls of *Adelges laricis* Vall. (= *A. strobilobius* Kalt.) on *Larix*. The second species is the rather common, omnipresent *Aphidoletes abietis* (Kieffer). We have found this species in the galls of *Adelges laricis* Vall. and *A. (Sacchiphantes) abietinis* L. on *Picea abies*. Möhn (1954) has described still another species, *Aphidoletes thompsoni* Möhn, that feeds on Adelgidae. The independent status of this species, however, should be verified.

During the first half of the summer, the females of *Aphidoletes abietis* (Kieffer) lay their eggs on the galls of pine and spruce aphids. The hatched larvae penetrate into the galls and, while sucking the lice dry, grow rapidly. According to counts performed in the Moscow Region, on an average 3-5 (up to 33) gall midge larvae live in one gall. The full grown larvae are large and have an orange colour. They leave the gall to pupate in the soil in a white cocoon, in which they hibernate. Though there seems to be only one generation yearly, the hatching of the imagines is spread over such a large interval of time that larvae can always be found in the galls of pine and spruce aphids. The species is an efficient adelgid hunter. The biology of the other species of this group is so far unknown.

Among the predatory gall midges the group that develops at the expense of scale insects is richest in species. The representatives of more than a score of genera, some of them comprising four to seven species, develop in the colonies of scale insects (Barnes 1929, 1930). Nearly all coccidophagous species have been described from the tropics. Often several species develop at the expense of the same coccid species. Thus, for instance, the larvae of the following species occur as predators in the colonies of *Pseudococcus citri*: *Coccodiplosis citri* Barnes (South Africa), *C. pseudococci* De Meijere (Java), *Acaroletes pseudococci* Felt (Sicily), *Dicrodiplosis coccidarum* Felt (Puerto Rico), *Lobodiplosis coccidarum* Felt (West Indies). *Tricontarinia ciliatipennis* Kieffer (Japan), *Silvestrina silvestri* Kieffer (South Africa), and *Arthrocnodax diaspidis* Kieffer (Italy) are bound in their development to *Aulacaspis pentagona*, whereas *Lobodiplosis pseudococci* Felt (Hawaii), *Diadiplosis pseudococci* Felt (Africa), and *Schizobremia formosana* Felt (Taiwan) have been described from colonies of *Pseudococcus brevipes*. The largest coccidophagous genus, *Diadiplosis* Felt, comprises seven species. Representatives of monotypic genera are: *Microperrisia pulvinariae* Felt, whose larvae develop at the expense of *Pulvinaria* sp. on citrus plants (Taiwan); *Streptodiplosis indica* Felt, described from colonies of *Mytilaspis piperis* (India); *Karschomyia cocci* Felt, a predator of *Trionymus sacchari* (Puerto Rico); *Androdiplosis coccidivora* Felt, occurring in colonies of *Aspidiotus* sp.; *Xyphodiplosis fulva* Felt, recorded as feeding on *Saisettia nigra* (Ceylon); *Adelgimyza dactylopii* Del Guercio on *Dactylopius* sp.; *Cecidomyiella aulacaspis* Del Guercio on *Aulacaspis rosea*; and *Feltodiplosis hirta* (Felt) and *Triommata coccotroctes* Barnes, whose hosts have not been precisely identified. Apparently certain representatives of some genera, as *Mycodiplosis* Rübsaamen, *Clinodiplosis* Kieffer, and a few others, have been accidentally included in the list of predatory gall midges. The species belonging to these genera are mycetophages or inquiline phytophages.

Interestingly, it is especially in the group of the coccidophages that we find the predatory representatives of the tribe Oligotrophini (subfamily Cecidomyiinae), in particular species of the genera *Coccidomyia* Felt, *Microperrisia* Kieffer, and *Coccomyza* Del Guercio. The publications by Pritchard (1952), and Barnes and Nijveldt (1954) give some information concerning the biology of these species (of *Coccomyza*). We do not, however, know of a single species in the whole extensive group of coccidophages whose biology has been adequately studied. Nevertheless, these species play an important role in limiting the unbounded reproduction of coccids.

The number of gall midge species developing in colonies of mites is only surpassed by that of the coccidophages. The representatives of some eight genera develop to the detriment of mites. Each of these genera, the large genus *Arthrocnodax* Rübsaamen excepted, comprises, as a rule, only one to three species (Barnes 1933, 1951). The acariphagous species may, according to their ecology, be divided into three groups. The first group consists of species whose larvae live openly on leaves, where they feed on spider mites, mostly of the genus *Tetranychus*. To the second group belong the species whose larvae develop within galls of mites; and to the third the species whose larvae feed on mites occurring in stored cereals and flour. When presenting a summary of this group, we must point out at the onset that the study of gall midges, known to develop in colonies of mites, is still far from its completion. It seems highly probable that many of the hitherto described species will be reduced to synonyms, the more so if we bear in mind our previous experiences with revisions of gall midge classifications.

The only two species whose biology has been studied in sufficient detail, and that are connected with spider mites, are *Therodiplosis persicae* (Kieffer) (Roberti 1954) and *Arthrocnodax wissmanni* kieffer (Wissmann 1926). The larvae of *Therodiplosis persicae* (Kieffer) live in the colonies of *Tetranychus* sp. on the leaves of several plants. The larvae kill mites in all stages of development, from eggs to adults. Eggs and small larvae are sucked dry in 2-3 minutes, adult mites in 30-60 minutes. At favourable temperatures the larval lifespan is 3-4 days, during which period up to 30 mites are sucked dry daily. The number of larvae in colonies of mites is very high, up to 9 individuals a leaf. The activity of these predatory larvae is doubtlessly an important factor in cutting down the proliferation of spider mites. The larvae pupate on the leaves in white cocoons. Several generations are bred yearly.

The larvae of *Arthrocnodax wismanni* Kieffer are found on the leaves of fruit trees infested with the spider mite *Tetranychus bimaculatus*. The number of larvae per leaf, however, is not high (1-2). The larvae are very mobile and move rapidly about in search of victims. If the food abounds, a larva may suck 13 mites dry in 5 minutes. Under natural conditions the larvae can be found in mite colonies during the entire summer season and in autumn up to October. Their development takes, dependent on the weather, 14 to 21 days. They pupate in the soil. There the adults appear after 15-28 days. Another species of this genus that should be mentioned is *Arthrocnodax mali* Kieffer. Nearly all summer its larvae can be found on apple trees that are infested with mites. These larvae apparently pupate in the soil as well.

The genus *Feltiella* Rübsaamen also comprises species developing at the expense of spider mites. The European species *F. tetranychi* Rübsaamen has, for instance, been reared from larvae found in colonies of *Tetranychus* sp. on leaves of *Humulus* (Rübsaamen 1911). The larvae feed on the mites and complete their development in two weeks. *Feltiella venotoria* Felt has been described from a colony of *Tetranychus telarius* which occurs on Ulmaceae in the U.S.A.

The sole representative of the genus *Acaroletes* Kieffer, *A. tetranychorum* (Kieffer), is, according to the collections of I. Vasil'ev, known to feed on mites occurring on cotton in the southern part of Central Asia. Four of the species that are connected with spider mites belong to the genus *Mycodiplosis* Rübsaamen. If we, however, take into consideration the fact that the larvae of this genus are mycetophages developing in rust fungi, records of predatory habits should be carefully verified.

The species composition of the gall midges developing in the galls of mites is more uniform, and consists almost exclusively of representatives of the genus *Arthrocnodax* Rübsaamen. As a rule each species of this genus has been recorded from the galls of only one mite species. This may be interpreted as the result of a very highly developed food specialization. We should, however, also consider another possibility: if the species of *Arthrocnodax* Rübsaamen are able to feed on various mite species a considerable number of the presently recognized species will turn out to be synonymous. The answer to this question still awaits conclusive empirical proof. According to the literature only one species, *A. peregrina* (Winnertz), can develop on two different mite species; namely in mite galls on the leaves of *Salix* and on those of *Prunus spinosa* (the mite species have not been precisely identified). Other species known to occur in Europe are e.g.: *A. coryligallarum* (Targioni-Tozzetti) in the bud galls of *Eriophyes avellanae* on *Corylus*, *A. gemmarum* Kieffer in galls of *Eriophyes stenaspis* on *Fagus*, *A. fraxinella* (Maede) in the flower galls of *Eriophyes fraxinivorus* on *Fraxinus*, *A. incana* (Rübsaamen) in the galls of *Dasyneura populeti* (Rübsaamen) on *Populus tremula* at the expense of the mites living therein, *A. vitis* Rübsaamen in the galls of *Eriophyes vitis* on vine, and *Lestodiplosis tarsonemi* Rübsaamen in the galls of *Tarsonemus* sp. on *Phragmites*.

The biology of the third group of acariphagous gall midges, the species that are encountered in stored seed and flour, has been well studied.

*Silvestrina tyrophagi* Dombrovski, which develops at the expense of the mite *Tyrophagus castellanii* (= *T. noxius*), is found together with the mite in stored hempseed (Elago 1940). The hatched larvae fiercely kill the mites, sucking up their contents. A pierced mite gradually stops moving, while the larva sucks it dry until only a shriveled skin remains. The larvae are able to move about energetically. Sometimes they hold themselves at the substrate, lifting only the front part of their body and making rotating movements in search of a prey. After 14 days a larva is full grown and spins a white cocoon in which it pupates. The larvae of another species of this genus, *Silvestrina farinicola* Barnes, have been found in stored flour infested with mites, in Naples. The tiny (–1 mm) orange larvae of *Trisopsis tyroglyphi* Barnes prey on the flour mite *Tyroglyphus farinae* (Sorokin 1938, Barnes 1951). As a rule the larvae stay put, lifting only the head end of the body in search of a victim. They attack the mites at the underside of the body, near the legs. When seized, the mites stop moving and are completely sucked dry. The larvae of *Lestodiplosis raphani* Barnes, representing the other genera of gall midges, inhabit stored seed of cabbage, radish, and turnip, and feed on the mites living there.

Concluding our outline of the biology of the predatory gall midges, we will dwell, for a moment, on a few special cases of food specialization. In this connection the large genus *Lestodiplosis* Kieffer stands out. The larvae of this genus are specialized in preying on the larvae of other genera of gall midges (Barnes 1928). Thus, for example, the larvae of *L. rosarum* develop in the galls of *Wachtliella rosarum* (Hardy) on *Rosa*, those of *L. pyri* Barnes in the galls of *Dasyneura pyri* (Bouché) on apple trees, those of *L. plicatrix* Barnes in the galls of *Dasyneura plicatrix* (H. Loew) on *Ribes*, those of *L. urticae* Nijveldt in the galls of *Dasyneura urticae* (Perris) on *Urtica*, those of *L. gracilis* Nijveldt in the galls of *Giraudiella incurvans* Nijveldt on *Phragmites*, etc. The larvae of *L. pallidicornis* Kieffer develop to the detriment of the larvae of *Dasyneura genteri* Pritchard in the flower heads of clover. Larvae of *Lestodiplosis* sp. are also frequently found in colonies of free-living gall midge larvae, in particular those of *Miastor metraloas* Meinert. Only a few species of this genus show a different type of food specialization. The larvae of *Lestodiplosis polypori* (H. Loew), for instance, apparently prey on mites in the fruiting bodies of *Polyporus confluens*. The larvae of *L. raphani* Barnes, which develop in stored seed of cabbage, radish and turnip, also feed on mites. *L. tarsonemi* Rübsaamen is known to occur in mite galls; *L. pini* Barnes has been observed in colonies of plant lice; and some species of *Lestodiplosis* Kieffer are apparently bound to scale insects. The closely related *Chiliodiplosis vasta* Möhn is extremely specialized. Its larvae develop in the intersegmental faults of the myriapod *Polyxenus lagurus*, causing the death of their host (Möhn 1955d).

The transition of tiny predatory larvae to a life within the body of their host has led to the genesis of a small group of endoparasitic gall midges. The endoparasitic gall midge larvae of *Endaphis perfidus* Kieffer have been observed for the first time in colonies of the aphid *Drepanosiphum platanoides* Schr. in Lorraine (Kieffer 1896b, 1900). Close inspection of aphid colonies may lead to the discovery of the tiny reddish eggs, singly attached to the thorax or the abdomen of some odd wingless aphids. On hatching the larvae penetrate into the aphids, but do, at first, not cause any externally visible sign of their presence. The larvae complete their whole life cycle within a single aphid. When full-grown, the larvae attain a bright orange colour and become visible through the skin of their host. Then the larvae leave their host and migrate to the soil to pupate. They posses a well developed spatula which enables them to jump. There are several generations a year.

In the European part of the U.S.S.R. gall midge larvae, endoparasitic in aphids, have been observed in colonies of *Aphis farinosa* Gmel. on *Salix* (Yaroslav and Moscow regions), and in those of *Dentatus sorbi* Kalt. on *Sorbus* Moscow region). We consider them as belonging to *Endaphis perfidus* Kieffer. The endoparasitic larvae that have been found in Europe in colonies of the aphids *Macrosiphus jaceae* L. on *Carduus* and *Chaitophorus* (*Pseudomicrella*) sp. on *Salix*, apparently belong to this species too. The data on these findings have been compiled by Barnes (1954). In the same publication Barnes described a new species and a new genus, also with endoparasitic larvae, found in the island of Trinidad. The larvae of this *Pseudendaphis maculans* Barnes develop in the body of wingless individuals of *Toxoptera aurantii* (Boy). Their biology has been described by Kirkpatrick (1954). The larvae preferably attack wingless individuals of *Toxoptera aurantii* (Boy), but are sometimes found in considerable quantities in colonies of *Pentalonia nigrinervosa* Coq., *Aphis gossypii* Glov., and *A. craccivora* Koch as well. In a single aphid only one parasitic larva develops. The development takes about 6-7 days. When the larvae mature the affected aphids become patchy, and are easily distinguished from the sound ones. The full-grown larvae measure up to 0.25 mm and, by the time they leave their host, become orange. The larvae pupate in the soil out of which the adults appear after ten days. Small aphid colonies may be affected in their entirety; in large colonies the parasite may kill up to 50% of the individuals. Interestingly, the affected aphids are not eaten by the larvae of coccinellids and syrphids.

Endoparasitic larvae of gall midges have also been found in Psyllidae. De Meijere (1907) was the first to publish on these larvae and to describe the corresponding new species, *Endopsylla agilis* De Meijere, which he based on material from the Netherlands. The larvae of this species develop within the larvae of *Psylla foersteri* Flor, and pupate in the soil. Sometimes a few parasites develop in the same host specimen. In the same year (1907) Kieffer described another endoparasitic species belonging to the same genus, but parasitizing *Stephanitis pyri* F. in Portugal (Kieffer 1907). Contrary to all above mentioned species the larvae of this *Endopsylla endogena* (Kieffer) pupate within the body of their host. Larvae probably belonging to *E. agilis* De Meijere have been found in Scotland, where they develop at the expense of *Psylla mali* Schm., *P. pyricola* Foerst., and *P. melanoneura* Foerst., killing up to 20-40% of the individuals of these pests (Lal 1934). According to Lal the larvae hatch from orange eggs that, as a rule, are singly attached to individual psyllids. Interestingly, the larvae feed the first 3-4 days as ectoparasites. Then they penetrate through the intersegmental membranes into the body of the host. After 6-12 days the full-grown larvae leave their host through an opening at the base of the abdomen. They remain on the leaves for 1-2 days before migrating to the soil to pupate. There are two generations a year. This species is expected to be of wide occurrence in Europe and the U.S.S.R.

## THE FOOD SPECIALIZATION

The character and the degree of food specialization differ greatly among the various trophic groups, as was shown by the topics discussed in the previous section.

Generally, phytophagous gall makers show a narrow food specialization. In the majority of cases the larvae of each species develop in specific galls on one particular plant species or on a few closely related species, thereby complying with the rule that the more primitive the type of gall making, the wider the choice of host plants. The larvae of the four most important gall makers on grasses – *Mayetiola destructor* (Say) (the hessian fly), *Contarinia tritici* (Kirby), *Sitodiplosis mosellana* (Géhin), *Haplodiplosis equestris* Wagner) – not only infest 3-4 species of cultivated cereals (see table), but are also found on a number of wild grasses (e.g. hessian fly on *Agropyron*, *Roegneria* etc). Typically, these gall midges developing on grasses are strikingly primitive in their way of gall making: the larvae of *Mayetiola destructor* (Say) cause deformations of the stem, the larvae of *Haplodiplosis equestris* (Wagner) develop in the hollows covered by the leaf sheaths, and the larvae of *Contarinia tritici* (Kirby) and of *Sitodiplosis mosellana* (Géhin) live in the flowers. *Dasyneura brassicae* (Winnertz) which also has a wide choice of hosts damages turnips, rape, wild and garden radish, wild mustard, and other wild Cruciferae.

As opposed to these ecologically primitive gall midges, the species that develop in structurally highly developed galls are strictly bound to one distinct plant species: *Mikiola fagi* (Hartig) and *Hartigiola annulipes* (Hartig) are only found on *Fagus*, *Didymomyia reaumuriana* (F. Loew) occurs only on *Tilia*, *Oligotrophus juniperinus* L. only on *Juniperus*, *Taxomyia taxi* (Inchbald) only on *Taxus*, etc.

An interesting mode of food specialization is shown by the representatives of some groups of gall midges that, as a rule, affect one and the same organ of several plants. The species of the group consisting of *Thomasiniana* Strand and *Profeltiella* Kieffer, for example, develop in wounds in the cambial zone of several plants, the species of *Lasioptera* Meigen make predominantly stem galls, those of *Asphondylia* H. Loew are found nearly exclusively in flowers and fruits etc.

The opposite phenomenon is also well-known: some species may cause galls on various organs of the same plant. Thus, though the larvae of *Rhabdophaga heterobia* (H. Loew) develop in the catkins of *Salix* in spring, they damage the tips of the twigs in summer; in Cruciferae the flowers as well as the tips of the branches may be deformed by the larvae of *Contarinia nasturtii* (Kieffer); and in *Ranunculus*, *Dasyneura ranunculi* (Bremi) causes leaf galls as well as flower galls; etc.

Table 3. Damage to cultivated cereals caused by gall midges

| Culture | *Mayetiola destructor* | *Contarinia tritici* | *Sitodiplosis mosellana* | *Haplodiplosis equestris* |
|---|---|---|---|---|
| Wheat | + | + | + | + |
| Rye | + | + | + | + |
| Barley | + | + | + | + |
| Oats | – | ? | + | + |

In a number of cases a parallel evolution of a group of gall midges and their food plants prevails, as is expressed by the ties between particular gall midge genera and certain plant genera or clusters of closely related genera. Thus, the species of *Planetella* Westwood and a few related genera are, for instance, bound to sedges; the species of *Rhabdophaga* Westwood develop predominantly on *Salix*; those of *Rhopalomyia* Rübsaamen show a preference for the composite genera *Artemisia*, *Tanacetum*, and *Achillea*. The gall midges developing in deserts and semideserts on *Salsola* and *Haloxylon* form an exceptionally well defined complex.

The mycetophages and the predators show a less clearly expressed food specialization. For example, paedogenic gall midges are able to succesfully develop on the mycelium of several species of fungi but far from all fungi are suitable for their development. The predatory gall midges of the genera *Aphidoletes* Kieffer and *Monobremia* Kieffer can feed on aphids not only belonging to several species, but even to different genera. However, if there is a possibility of choice, the females show a preference for a particular species of aphids and lay their eggs especially in colonies of this species (Nijveldt 1955, 1957). Interestingly, the representatives of these two genera feed only on plant lice and pine and spruce aphids; those of *Arthrocnodax* Rübsaamen on mites; and those of *Lestodiplosis* Kieffer mainly on the larvae of other gall midges. Thus, we find a clear-cut food specialization, though distinctly less narrow than in the phytophages, in the group of the predatory gall midges as well.

Food specialization has also been clearly observed in some free living mycetophages. The larvae of *Micropteromyia ghilarovi* Mamaev are found, often directly on the mycelial film, in wood affected by *Fomes fomentarius*. The group of xylobiontic gall midges is, as we saw before, bound to a specific type of fungal decay. Presently, we are badly in need of thorough investigations on the mutual connections between mycetophagous gall midges and particular fungal species. We will be in a position sufficiently founded to evaluate the level and the character of their food specialization only after the gap in our knowledge of this aspect of the biology of the gall midges has been filled.

## THE LARVAL DEVELOPMENT

The study of the life cycles of the various species demands the clarification of some of their developmental idiosyncrasies. First of all we must distinguish between two groups:

1. The univoltine species, which have a clearly rhytmical development and which produce one generation per year;
2. The multivoltine species, in which the development of the successive generations often overlaps in time, making it difficult to recognize the individual generations.

*Mikiola fagi* (Hartig), which occurs in the distributional area of *Fagus* is a good example of an univoltine species. The females lay their eggs from the end of March to the beginning of April in the buds that are then just sprouting. The hatched larvae exert an influence on the tissue of the leaf causing a swelling, which later transforms into a gall. According to observations made in Transcarpathia during the summer of 1963, the galls of *Mikiola fagi* (Hartig) are already formed in June and July, but the larvae inside them grow very slowly and only reach maturity in late autumn, in October. They subsequently hibernate in the fallen galls in the litter to pupate in spring. The protracted active life of the larvae, lasting all summer, in connection with a diapause that more or less coincides with the winter period, is a peculiarity of the life cycle of this species.

The larvae of some other univoltine species have a remarkably rapid development as compared to the former, and enter into diapause long before winter conditions begin. *Syndiplosis petioli* (Kieffer), a common species on *Populus tremula*, may serve as an example. Its larvae hatch from the eggs in spring and already complete their development in the second half of June. They then leave their galls, migrate to the soil, and, ensconced in silken cocoons, enter into their diapause which lasts until the following spring.

Interestingly, in some desert species, living under special conditions, a peculiar life cycle has evolved (Marikovskiy 1955b). The larvae rapidly reach maturity in spring, when also the adults fly and the eggs are laid. The young larvae enter into an extended (8-10 months) diapause when their galls have only hardly been formed. The diapause coincides with the summer and winter resting periods of the desert plants, and is broken twice: First, in autumn when the desert plants start vegetative life anew and when the rudimentary galls that were formed in spring slightly gain in size; second, in spring when the plants themselves as well as the galls on them grow vigorously. The larvae of some species, especially of many gall midges affecting *Haloxylon*, manage to complete their development in spring and have a diapause in their galls till the following spring. The gall midges on *Anabasis*, on the other hand, develop mostly in autumn, as some peculiarities of the damage afflicted to the plants show.

Among the multivoltine species, we must first distinguish those species that have no diapause. The xylobiontic larvae of *Camptomyia* Kieffer, *Winnertzia* Rondani, etc. enter into a state of cold rigor at the falling of temperature in autumn. Every rise in temperature will induce them to recover from their rigor. Larvae of these species, when collected late in autumn and kept indoors, will complete their development and pupate, skipping diapause.

The phytophagous gall midges, however, have a diapause fixed in their life cycle. According to Barnes (1935), *Dasyneura pyri* (Bouché) may have up to four generations a summer. Already in the second generation 2-7% of the larvae may enter into diapause until the following spring. In the third generation the percentage of larvae entering into diapause rises to 16-78%. The larvae of the fourth generation, with a few stray exceptions, all enter into diapause.

The absence of diapause in many mycetophages as well as its presence in the phytophages can easily be explained in terms of their ecology. The mycetophages can dispose of their food, i.e. fungal mycelium, throughout the year. But the phytophages must attack a plant during particular periods favourable for the development of the larvae. Therefore, a diapause is intercalated between the end of the larval development and the period optimally suited to the flight of the adults, thus regulating the whole life cycle.

What kind of developmental rhythm is the most characteristic one for the phytophagous gall midges? Independently of the number of generations in summer, in the majority of cases the full grown larva, which enters into diapause under the influence of the fall of temperature in autumn, is the hibernating stage of this group. In arid regions the larvae respond with diapause not only to the falling temperature in autumn, but also to the rising of temperature if accompanied by a shortage of moisture.

The hessian fly, *Mayetiola destructor* (Say), which in our country (in the temperate zone) produces two generations a season, has an interesting developmental rhythm. The larvae of the first (spring) generation, completing their development under moderate temperatures, generally skip the summer diapause. The larvae that hatch somewhat later, experience in the decisive part of their development the influence of the high summer temperatures, and respond by entering into a long diapause. This ends in autumn when the adults fly, thereby especially endangering the winter corn (Pavlov 1958). Moisture is of no small importance in the complex of conditions leading to the beginning of diapause in the larvae of the hessian fly. In arid regions, for instance in Povolzh'e, all larvae of the spring generation enter into diapause and only awake after the autumn rains.

The larvae of *Stenodiplosis panici* Plotnikov also have a long diapause. This ends in spring under the influence of temperatures between 18 and 25°C and a simultaneous increase of the humidity of the soil in which the larvae hibernate.

The same complex of conditions that regulates the development of their food plants, controls the beginning and end of the diapause of the larvae. Thus, the high degree of adaptation of the various phytophagous gall midges to a life on strictly defined plant species, the phenomenon that brought about the flourishing of the gall midges, becomes understandable.

In order to fully appreciate the adapative significance of the larval diapause, we must direct our attention especially to the phenomenon of the prolonged diapause which has already been observed in many species. When rearing larvae in the laboratory under circumstances closely resembling natural conditions, it turns out that in spring, after hibernating, only a limited number of larvae awakes from diapause to pupate. Generally, a part of the larvae enters into a state of prolonged diapause that may last for several years. Thus, for example, a part of the larvae of *Contarinia tritici* (Kirby) has a diapause in the soil lasting two to three years; in *C. pisi* (Winnertz) cases of prolongation of the diapause up to two years have been recorded. *Sitopdiplosis* mosellana (Géhin) holds a remarkable record for duration of the diapause: the larvae of this species may rest for up to twelve years (Barnes 1943, 1952, Barnes and Arnold 1960). Therefore, the soil of fields where for six to seven years no cereals have been sown, may, nevertheless, contain larvae living on gramineous plants in diapause. These larvae may serve as an extant reservoir augmenting the number of their species (Colightly 1952). The prolonged diapause is a biological adaptation ensuring the survival of the species for the duration of several seasons, even if adverse conditions during one of these seasons cause the wholesale annihilation of the gall midges. In planning measures to control gall midges we must necessarily keep in mind the fact that a generation of larvae, present in the soil in diapause, may cause a sudden explosion of a species, even if its abundance during the preceding season was low.

Gall midge larvae moult several times during their development. Establishing the number of moults in phytophagous larvae living inside galls is, however, so difficult that up to this time no trustworthy data pertaining to this aspect are available. Free living gall midge larvae, as a rule, moult twice and have three instars. The larvae of *Heteropeza pygmaea* Winnertz form an interesting exception to this rule. They do not moult at all and only gain in size at the cost of stretching their larval skin. The large larvae of *Miastor metraloas* Meinert that reproduce paedogenically, also do not moult after hatching from the egg. They moult a single time when the larvae develop into sexual individuals. More details on the number of moults in the paedogenic gall midges will be given in the section on their life cycles.

## THE PAEDOGENIC REPRODUCTION OF GALL MIDGE LARVAE

In gall midges a phenomenon of rare occurrence among insects may be observed: larval reproduction, or paedogenesis. Paedogenic reproduction, as a mode of reproduction without fertilization, may be considered as a kind of parthenogenesis. In paedogenesis, however, the unfertilized eggs develop much faster than the individual as a whole and mature already during the larval stage, or during a special period of larval development, called pseudopupa, which replaces the pupa in the normal sequence of stages. From the evolutionary point of view the acquisition of the possibility to reproduce during the larval stage, which greatly enhances the reproductive vigour of a species, is extremely important. Paedogenesis has, however, not arisen in all groups of gall midges, but only in those that develop in an environment containing large supplies of food (Gilyarov 1945). Such environments are provided by decaying organic remains on which mycelium, that may serve as a feeding substrate, thrives luxuriantly. Consequently, paedogenesis does not occur in larvae that develop in galls, or in flowers or fruits. This mode of reproduction has only developed in some representatives of the subfamily Lestremiinae that are found in soil, litter, and – most of all – in damp, decaying wood.

A survey of the paedogenically reproducing European species has been given in a special paper (Mamaev 1964b). Paedogenesis has been recorded in species of the following genera: *Miastor* Meinert, *Heteropeza* Winnertz, *Heteropezina* Pritchard, *Aprionus* Kieffer (?), *Tekomyia* Möhn, *Mycophila* Felt, *Moehnia* Pritchard, *Leptosyna* Kieffer, *Frirenia* Kieffer, *Henria* Wyatt, *Britteni* Edwards, and *Nikandria* Mamaev.

The history of the study of paedogenesis in the gall midges is very interesting. This original phenomenon has attracted the attention of many entomologists and cytologists. Paedogenesis was first discovered by the Russian scientist N. Wagner, and described by him in a number of publications in Russian (1862) and in German (1863, 1865). Afterwards the most important contributions were made by Gabritschewsky (1928, 1930), Ulrich (1934, 1936, 1938, 1940, 1943), and more recently by Hauschtek (1959, 1961), Nikolei (1958, 1961), and Camenzind (1962). Our description of the biology of the paedogenic gall midges is based on the data of the latter two authors, augmented by our own observations. In one of these papers (Nikolei 1961) a very complete survey of the literature on paedogenesis is given.

Up to the present time the life cycles of five paedogenic species have been studied quite well.

*Heteropeza pygmaea* Winnertz

Comparatively large (up to 1 mm) larvae hatch from the eggs that have been laid by the females. We will call these 'primary larvae'. Only two eggs, which take up nearly the whole abdomen, develop in the ovaries of the female. The primary larvae grow rapidly and, without a singel moult, transform into reproducing larvae.*

*Note from the editor: In all investigated paedogenic species the undifferentiated reproducing larvae are blind and have no spatula.

The l-reproducing larvae* occupy a central position in the life cycle of the paedogenic gall midges. Though they develop from the primary larvae, they have no metamorphosis within a pupa, but they function completely as reproducing individuals. Within their rapidly growing ovaries some dozens of descendants develop, each into an independent larva. We call these 'daughter larvae'; in this species their number may amount to 36 per l-reproducing larva.

A few days before the ultimate maturation of the daughter larvae the reproducing larva loses its mobility. At last the daughter larvae rupture the skin of the reproducing larva to begin their independent existence. Their further fate depends largely on the conditions of life. As a rule, the daughter larvae grow and each of them is transformed into a l-reproducing larva. So the cycle is closed and the species rapidly steps up its numbers.

Under certain conditions, however, other developmental potentialities are realized. In such a case the daughter larvae develop into 'imaginal' larvae which pupate to become adult gall midges. In one developmental line the daughter larvae develop directly into ♀-imaginal larvae. They are distinguished from reproducing daughter larvae by their well-developed separate eye spots and the presence of a one-toothed spatula. These larvae may pupate and develop into females. Pupation, however, does not always set in. If the conditions change, the ♀-imaginal larvae may again be transformed into l-reproducing larvae, which subsequently keep on reproducing paedogenically.

The production of male adults from undifferentiated daughter larvae is somewhat more complicated. In this line of development the daughter larvae become reproducing larvae with slightly deviating characteristics. A part or all of the embryo's developing within such reproducing larvae tend to grow into daughter larvae that are somewhat larger than they normally are. These larger daughter larvae subsequently develop into ♂-imaginal larvae which pupate and become male adults. Already before they are completely mature and leave the maternal organism, the ♂-imaginal larvae develop well-differentiated and widely separated eyes and a one-toothed spatula. We propose to call reproducing larvae, that only produce larger, future ♂-imaginal larvae, '♂-reproducing', as distinct from '♂♀-reproducing' ones. Within the latter, ♂-imaginal larvae develop as well as undifferentiated daughter larvae which seem to be of the same type as the larvae that are, for example, produced by l-reproducing larvae. Under natural conditions, in colonies of paedogenic larvae, ♂- and ♀-imaginal larvae occur synchronously; they pupate practically at the same time, and the adults hatch from the pupae very well-mixed and in large numbers.

The theoretical possibility that was suggested about ten years ago (Mamaev 1955) of a genuine parthenogenic reproductivity of adult gall midges, especially of the paedogenic species, lacked experimental support until very recently. In 1962 Camenzind not only succeeded in obtaining, but also in rearing, parthenogenic larvae from the very species under discussion, *Heteropeza pygmaea* Winnertz. Unfertilized eggs as well as the larvae that hatched from these eggs, proved to be far less viable than their respective fertilized versions. Thus, if for instance 63% of the primary larvae that hatch from fertilized eggs

*Note from the editor: According to Mamaev (in litteris) the terminology of paedogenic larvae corresponds in the following way to that of Camenzind (1966): l-reproducing larva – WM (Weibchenmutter), undifferentiated daughter larva – TL (Tochterlarve), ♂-reproducing larva – MM (Männchenmutter), ♂♀-reproducing larva – MWM (Männchenweibchenmutter).

develop into reproducing larvae, only 5% of the parthenogenic larvae will survive and develop. The lower viability of the parthenogenic larvae is still apparent in the fourth generation. Nevertheless, a part of the larvae do develop normally, proving the existence, at least in some species, of genuine parthenogenesis.

Concluding this section we may mention still another interesting phenomenon: sometimes a few reproducing larvae (of *H. pygmaea* Winnertz) slow down the rate of their development, entering into a state reminiscent of the diapause of other insect larvae. The presence of such larvae in a colony ensures the survival of the species when unfavourable conditions, causing the death of most individuals, suddenly occur. The paedogenic cycle of *H. pygmaea* Winnertz is the most specialized one. Another species, *Miastor metraloas* Meinert. shows basically the same cycle. The developmental cycle of this species, however, has its own idiosyncrasies, as we will see below.

*Miastor metraloas* Meinert

As a rule, two eggs mature within the ovaries of the females of this species. The primary larvae originate, however, only from fertilized eggs. According to the findings of Nikolei (1961) unfertilized eggs do not develop. The primary larvae develop into l-reproducing larvae which produce daughter larvae. The maximum number of daughter larvae developing in a single reproducing larvae is about 47.

The daughter larvae have different potentialities. They may develop into l-reproducing larvae, ♂-reproducing, and ♂♀-reproducing larvae, and also into ♀-imaginal larvae, as in the life cycle of *Heteropeza pygmaea* Winnertz. The duration of the development of the daughter larvae within a reproducing larva takes only six days. The paedogenic cycle of this species is complicated by the fact that the daughter larvae may develop into particular orange larvae. These differ from the normal white larvae not only in their colour, but also in the hypertrophic development of their eyes. The orange larvae either arise directly from the white daughter larvae or they are formed within intermediate larvae. The fate of an orange larva may be one out of various possibilities. The larva may change back into a reproducing larvae or it may enter into a prolonged resting stage.

In this species the imaginal larvae have highly developed eyes and a one-toothed spatula. Though the change of the daughter larvae into reproducing larvae occurs without moult, a moult is inserted during the development of the ♀-imaginal larvae from daughter larvae, showing another difference between the life cycles of *Heteropeza pygmaea* Winnertz and *Miastor metraloas* Meinert. Although the larvae moult during the development of ♀-imaginal larvae from daughter larvae, they are, under certain circumstances, not prevented from changing back into reproducing larvae.

Adult males develop from two other types of reproducing larvae, i.e. from ♂-reproducing and from ♂♀-reproducing larvae. Within these larvae larger daughter larvae develop which leave the maternal body without any additional moult to become ♂-imaginal larvae. The latter pupate to hatch as male adults. In any case the reproducing larvae lose their mobility about two days before the ultimate maturation of the young, at which time their heart tube stops contracting. The daughter larvae leave through a tear in the body wall of their 'mother'.

The life cycles of three other, morphologically and biologically less specialized species, have been more or less thoroughly investigated.

*Mycophila nikoleii* Möhn
The white or orange, 0.8 mm long, primary larvae hatch from fertilized eggs. Unfertilized eggs do not develop.

During the first day after hatching the young larva lies immobile, then it moults and starts feeding, sucking young fungal hyphae. It moults a second time ten days after the first moult. During the third instar the ovaria can be easily discerned. Within these larvae, which reach a length of about 3 mm, up to 54 daughter larvae develop in six days. A few hours before the daughter larvae are to leave the maternal body, the reproducing larvae stop moving. By that time all their internal organs become dissolved. The daughter larvae leave by breaking through the body wall of the reproducing larvae, after which they themselves become reproducing larvae, thus the cycle is closed and the number of larvae rises rapidly. Under certain circumstances, however, the daughter larvae change into (in most cases) ♀-imaginal larvae.

Males are only very rarely found in populations of this species, and their ways of development have not been traced. The imaginal larvae have a three-toothed spatula and well-developed eyes. As is the case with the previously discussed species, the ♀-imaginal larvae may fall back into paedogenic reproduction.

*Tekomyia populi* Möhn
Upon hatching, the tiny larvae of this species measure 0.7-0.8 mm. Under experimental conditions they develop from fertilized and unfertilized eggs. Only larvae hatched from fertilized eggs are sufficiently viable to complete their development.

The first moult is after 1.5 days, the second 1.5 days after that. In full-grown larvae the ovaria are well-discernable through the skin. The larvae have well-developed eyes and a three-toothed spatula. After six days daughter larvae are formed within these larvae which are equivalent to l-reproducing larvae. The reproducing larvae, however, moult still another time to enter into a special stage: the pseudopupa. This stage shows a mosaic of features. Some features that are characteristic for the pupa (strong bristles, breathing tubes, etc.) represent, in a somewhat distorted way, larval characters, whereas some other larval characters (sclerotized mouthparts, antennae, etc.) tend to be reduced.

The paedogenic reproduction takes place during the peudopupal stage. The daughter larvae appear very soon, since they are already formed during the third larval instar. They do not leave the maternal body through a rent in the wall, but through a special opening at the ventral side of the terminal segment of the pseudopupa. When a larva crawls out, the opening is widened, but afterwards it closes again. The pseudopupa retains the ability to move, even after the 'birth' of the first daughter larvae. Under favourable conditions the daughter larvae reach the pseudopupal stage after seven to nine days, thereby closing the paedogenic cycle.

Dependent on the conditions, however, not all daughter larvae enter the reproducing phase, but a part of them develops into ♀-imaginal and, much more rarely, into ♂-imaginal larvae which pupate and start a sexual generation. As in all discussed species, the ♀-imaginal larvae may again turn into reproducing larvae, in which case they moult one time and transform into pseudopupae. Up to the present time it has not been possible to establish the way of development of the ♂-imaginal larvae with sufficient reliability.

*Henria psalliotae* Wyatt (Wyatt 1961)
The larvae of this species, upon hatching, have a length of 0.8 mm. They already have well-discernable eye spots, but no spatula. The larvae feed on the contents of young hyphae in the mycelial film. On the third day they moult and reach a length of 2.3 mm. After another four days they moult a second time, after which the eyes are large and well-separated and a clearly three-toothed spatula has been formed.

At 12-15 days, when the larvae are 3.0 mm long, the developing daughter larvae, though still inside the eggs, are clearly visible. The full-grown larvae of the third instar build a small cradle that they entwine as a cocoon in which the larvae transform into pseudopupae. Shortly before they moult, the embryos developing within a larvae break their eggshells and enter into the body cavity.

The duration of the development of the daughter larvae within the pseudopupa varies strongly. Sometimes the daughter larvae succeed in completing their transformation while still within the third instar larva and then immediately leave the pseudopupa; sometimes the whole embryonic development takes place within the pseudopupa. The larvae leave the pseudopupa through a rent in the body wall. On the average 30-60, maximally 96, daughter larvae develop within a single pseudopupa.

## SOME REMARKS ON THE EVOLUTION OF PAEDOGENESIS

Though up to the present no more than half of the life cycles of the known paedogenic species have been worked out, a comparison of the already known life cycles allows us to understand the lines along which paedogenic reproduction has evolved to its completion. Doubtlessly, parthenogenesis, i.e. the successful development of progeny from eggs laid by an unfertilized female, was the starting point of the evolution of paedogenesis. In the simplest case the displacement of the completion of the embryonic development in the egg from the imaginal stage to the pupal stage caused the arising of pupal paedogenesis. This, in its turn, led to changes in the pupa itself, from which the pseudopupa originated.

In *Henria psalliotae* Wyatt the development of the daughter larvae is completed during the pseudopupal stage. In *Tekomyia populi* Möhn the development of the larvae is displaced still further back; the daughter larvae are wholly formed within the reproducing larvae of the third instar and they enter into the 'outside' environment immediately after a moult leading to the pseudopupal stage.

The slight displacement of the development of the eggs that is observed in the larvae of the above mentioned species, does not fundamentally influence the morphology of the larval stages, i.e. substantial differences between imaginal and reproducing larvae have not yet been found. Both, for instance, have well-developed eyes and a three-toothed spatula, and both moult twice before metamorphosis into a pupa or pseudopupa. It is also of interest that in these least specialized species the number of daughter larvae that develop within a pseudopupa is comparatively high (up to 86-90 individuals).

*Mycophila nikoleii* Möhn represents a typical intermediate stage. The daughter larvae are already wholly formed within the third instar. The pupal and pseudopupal stages have been abandoned, but the two larval moults are retained; the differences between the reproducing and imaginal larvae, though present, are not very pronounced; the maximum number of daughter larvae per reproducing larva is lowered to 54.

The greatest perfection of the paedogenic cycle has led to the further speeding up of the development of the eggs. The daughter larvae mature fully within the reproducing larva, which does not moult at all. The reproducing larvae have no spatula and the development of their eyes is less elaborate than in the imaginal larvae. The maximum number of daughter larvae, maturing within a single reproducing larva, falls to 47 (*Miastor metraloas* Meinert) or even to 36 (*Heteropeza pygmaea* Winnertz).

The perfection of the paedogenic reproduction in the gall midges is thus accompanied by: first, displacement of the completion of the formation of the daughter larvae to ever younger stages of the post-embryonic development, as is expressed in the lowering of the number of moults. Second, the lowering of the number of daughter larvae developing per reproducing larva. Third, the arising of differences between reproducing and imaginal larvae.

Which external factors regulate the paedogenic reproduction and lead to the development of sexual individuals? Nikolei (1961) went rather deeply into this matter. When the temperature rises from 18 to 25°C the growth of the larvae of the paedogenic species is sped up. When the temperature rises still further the growth rate falls, and at a critical temperature, which for the various species lies somewhere between 29 and 33°C, the larvae begin to die. In colonies of *Tekomyia* Möhn and *Mycophila* Felt the highest number of adults is reached at 25°C.

An increase of the content of nutritious substances in the environment stimulates the development of the larvae, since the fungal mycelium grows more intensively on more nutritious substrates. Moreover, the reproducing larvae become larger and the number of their daughter larvae is raised. Not all species of fungi, however, are suitable as substrates for paedogenic gall midge larvae. Thus, for instance, the larvae of *Miastor metraloas* Meinert, *Heteropeza pygmaea* Winnertz, *Mycophila nikoleii* Möhn, and *Tekomyia populi* Möhn grow very well on the mycelium of *Peniophora albula*. Good results are also reached with the fungi *Schizophyllum commune*, *Stereum purpureum*, *Coniophora cerebella*, and some strains of *Trichoderma viride* (Nikolei 1961). The larvae of paedogenic gall midges develop badly, and partly even die, when feeding on the mycelium of *Fomes annosus*, *F. betulinus*, *Trametes radiciperda*, *T. serialis*, *Polyporus betulinum*, *P. sulphureus*, and *Endothia parasitica*.

Interestingly, the species of fungus on whose mycelium the larvae feed exerts its influence on the whole paedogenic cycle. Thus, for instance, three days old mycelium of *Peniophora albula*, kept on agar at 25°C, is recommended for rearing reproducing larvae of *Heteropeza pygmaea* Winnertz. To obtain males of this species one should put mature ♀-reproducing larvae into a petri dish with seven days old mycelium of *Peniophora albula* at 25°C. In such a case most of the daughter larvae will turn into ♂-reproducing larva. For rearing of females mycelium of *Trichoderma viride*, *Polystictus zonatus* etc. should be used. Direct observations in nature show the strong influence of the conditions of development during the larval phase on the sex ratio. Often colonies of *Miastor metraloas* Meinert or *Heteropeza pygmaea* Winnertz with many hundreds of pupae, all of the same sex, are encountered. Colonies with representatives of both sexes at the same time and in roughly equal quantities, are found more rarely. The strong developmental synchronism within the vast colonies of paedogenic gall midges, comprising many thousands of specimens, is fascinating. With a high rate of probability we may suggest the existence of a close humoral bond between all members of the colony, allowing them to directly influence the development. Under natural conditions the larvae in the colonies pupate from the end of May till the beginning of July; spring being the season for intensive paedogenic reproduction.

Before the massive pupation the appearance of the colony changes. The more lively imaginal larvae with their black spatulae crawl together, and form separate aggregations. All larvae of the same aggregation pupate nearly simultaneously; and the period of flight of the adults is also highly synchronized. They all fly off in the course of one or two days at a fixed time, mostly between 9 and 11 o'clock a.m. If we take into account that the duration of adult life is measured in hours, their simultaneous appearance in large numbers is an important asset to assure a high quality progeny.

After the adults have flown away, only dying or dead larvae and pupae are left, and the colony has ceased to exist. These observations show the existence in the colonies, apart from external factors that exert their influences on the development of paedogenic larvae, of only partially understood internal rules of development. These, under natural circumstances, ensure a certain developmental rhythm and an alternation of periods of exclusively paedogenic reproduction with periods of complete metamorphosis and dispersal.

To conclude the section on the biology of the gall midges let us examine some special problems related to the life of their larvae.

## OTHER BIOLOGICAL PROBLEMS

The mobile abilities of the gall midge larvae are different and are determined by their ways of life. The larvae of species whose entire development is confined to a gall, from which also the adults emerge, cannot move along. When removed from their gall these larvae may wriggle as a result of uncoordinated muscular contractions, but they have lost the ability to perform directional movements. The larvae of species that, though developing inside, pupate outside their gall, on the other hand, can move along, not only by way of crawling from place to place, but in some cases also by jumping distances of several dozens of centimeters. To perform a jump, a larva bends as a ring, it hooks the end of the body behind the spatula, and all of a sudden stretches powerfully. As a result of the shock it jumps. Some authors, however, deny the participation of the spatula (Milne 1961), as this mode of locomotion is not characteristic for all species whose larvae posses a spatula.

Observations on the locomotion of larvae in wood suggest an active role of the spatula in the digging of tunnels and the widening of naturally occurring holes. Larvae digging in the soil use the spatula to push soil particles aside.

In a number of cases it is precisely the larvae of a species that may be passively dispersed over distances of thousands of kilometers. Gall midge larvae are dispersed by rafting and transportation of wood, transport of seeds, food, etc. Even on isolated, sown fields of lucerne (alfalfa) with artesian irrigation in the central Kyzylkum the whole complex of gall midge pests develops in the first two or three years of cultivation. Hessian fly (*Mayetiola destructor* (Say)) has been imported into Europe from America with straw; other species living on grasses and legumes are dispersed with seed material; garden pests are often introduced with scions etc. Therefore, as a rule, all groups of gall midges, plant pests, mycetophages, and predators alike characteristically have vast areas of distribution.

We should consider pupation in the soil and comparable substrates, litter and wood, as the most primitive mode of pupation. It is characteristic for the larvae of all members of the subfamily Lestremiinae, and – in the subfamily Cecidomyiinae – for the vast majority of species of the tribes Porricondylini and Cecidomyiini, as well as for many Oligotrophini; but in the tribes Asphondyliini and Lasiopterini it is found in only a few species.

As a rule, the larvae of gall midges spin a cocoon before pupating. Some xylobiontic larvae from the genera *Camptomyia* Kieffer, *Winnertzia* Rondani, *Diallactes* Kieffer, *Rhipidoxylomyia* Mamaev, *Solntsevia* Mamaev, etc. that pupate in holes in the wood do not make a cocoon. The paedogenic larvae and the larvae of the ecologically more specialized species that pupate within their galls also have no cocoon. Though many of the latter do not make a cocoon, some of them do spin a very flimsy cocoon: as do, for example, the larvae of *Stefaniella atriplicis* Kieffer which pupate in stem galls on *Atriplex*. The larvae of zoophagous gall midges pupate in a cocoon in the soil; only some endoparasites are known to pupate within the body of their host (see above).

A puparium occurs in only a few species. The puparium of the phytophagous gall midges can, however, hardly be considered as equivalent to the puparium of the Cyclorrapha. In a true puparium the skin of the last larval instar is, by a complicated process, incrusted with protein. It then becomes barrel-shaped and is loosened from the body. Inside the puparium an exarate pupa is formed. The Stratiomyiidae and the Scatopsidae also have puparia formed from the larval skin of the last instar. The pupa inside is, however, obtect and its wing sheaths are fused with the extremities to form a single complex.

In *Mayetiola destructor* (Say) and the other species of this genus as well as in *Contarinia medicaginis* Kieffer the skin of the penultimate larval instar is transformed into a puparium. In this case the larva of the last instar first has a diapause inside the puparium; subsequently it pupates into an obtect pupa, this being the usual type of pupa in the gall midges.

In *Gynapteromyia carpatica* Mamaev it seems that the only weakly modified larval skin of the last instar turns into a puparium, which contains the obtect pupa. Thus, only in the latter species does the formation of the puparium proceed in conformity with the scheme of the Brachycera.

Concluding the outline of the biology of the gall midges we especially want to emphasize that, though many problems have not yet been definitively solved, the accumulated knowledge already bears evidence to the fact that it will be hardly possible to find another insect family that is biologically as many-sided as the gall midges are. The gall midges are a promising object of investigations designed to solve a whole series of, in the first place evolutionary, questions of general biological interest.

CHAPTER 2

# Outline of the morphology

## GENERAL CHARACTERISTICS OF THE LARVAE OF THE GALL MIDGES

The larval body (fig. 1) comprises a head, a neck, three thoracic and nine abdominal segments. The majority of larvae has on the ventral side of the first thoracic segment a special sclerotized structure, called the 'spatula'. The anal opening is found on the last abdominal segment; the spiracles are located on the first thoracic and the first eight abdominal segments. The head is connected to the trunk by a clearly recognizable elongated segment, called the 'neck' or 'supernumerary' segment, which bears the eye spots. Then three thoracic segments follow: the pro-, meso-, and metathorax; and the nine abdominal segments. The body of the gall midge larvae thus consists of a head and 13 visible segments.

The head (fig. 2,1) is formed by a completely developed head capsule and a mouth apparatus, which consists of a labrum and a labium (the 'upper' and 'lower lip'), mandibles and maxillae. As a rule there is no clear suture between the dorsal and the ventral side of the head capsule. The tentorial arms, which function as a reinforcement, are situated within the head capsule. The two anterior tentorial arms arise from the clearly visible epistomal suture and run through the whole head capsule to fuse with the posterior arms. In some cases well-developed transverse arms are present in the middle of the head capsule.

The antennae are attached to the sides of the head on the anterior part of the head capsule. Generally they consist of two segments, a very short first and a longer second segment. Around the base of the antennae the head capsule is often very weakly sclerotized, which results in an annular membranous zone.

The spatula (fig. 2,2) is morphologically divided into two parts: the head formed by the widened anterior part, and the elongated shaft. The front margin of the head is mostly one-, two-, or three-toothed.

Outgrowths specific to the larvae of the gall midges, the so-called 'papillae', are found on the dorsal and ventral surfaces of the body (fig. 2,3-4). A completely developed papilla resembles a bristle, reinforced with variously shaped (knobs, warts, platelets, etc.) cuticular formations. Sometimes the bristles are modified or reduced (fig. 2,3). Dependent on their distribution over the body, several groups of papillae are discerned. All body segments, including the thoracic segments, carry 'dorsal' papillae. They form a single row on the middle or the posterior part of each segment. On the sides of the body we find

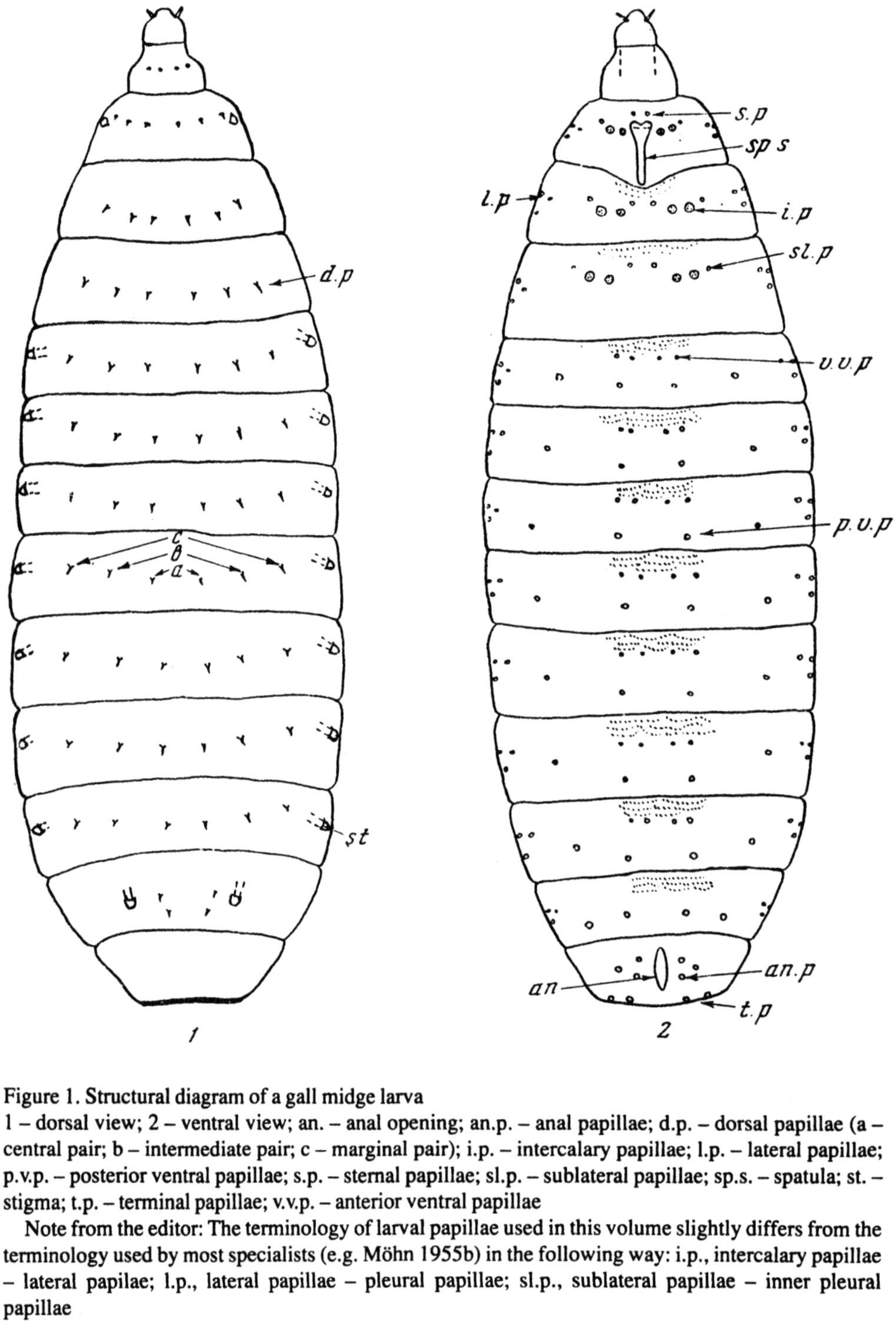


Figure 1. Structural diagram of a gall midge larva
1 – dorsal view; 2 – ventral view; an. – anal opening; an.p. – anal papillae; d.p. – dorsal papillae (a – central pair; b – intermediate pair; c – marginal pair); i.p. – intercalary papillae; l.p. – lateral papillae; p.v.p. – posterior ventral papillae; s.p. – sternal papillae; sl.p. – sublateral papillae; sp.s. – spatula; st. – stigma; t.p. – terminal papillae; v.v.p. – anterior ventral papillae

Note from the editor: The terminology of larval papillae used in this volume slightly differs from the terminology used by most specialists (e.g. Möhn 1955b) in the following way: i.p., intercalary papillae – lateral papilae; l.p., lateral papillae – pleural papillae; sl.p., sublateral papillae – inner pleural papillae

the 'lateral' papillae, which occur in groups of two to three papillae on both sides of the segments.

On the ventral surface of the thoracic segments some groups of papillae occur symmetrically. Along the middle line of the segment the 'sternal' papillae are well-recognizable. The 'intercalary' papillae are located somewhat closer to the sides of the segment. The 'sublateral' papillae are situated between the intercalary and the lateral papillae.

The ventral surface of the first seven abdominal segments bears two rows of papillae: the 'anterior' and the 'posterior ventral' papillae. On the eighth abdominal segment a single row of 'ventral' papillae is found.

The distribution of the papillae on the terminal segment differs somewhat from that on the other segments. The papillae that are found on the dorsal surface of the central part of the segment correspond with the dorsal papillae. The papillae surrounding the anal opening on the ventral surface are called the 'anal' papillae; and those situated at the very end of the terminal segment or on its various appendages are called the 'terminal' or 'end' papillae.

The dorsal and lateral sides of the segments may, apart from the papillae, be covered with cuticular formations as protuberances, platelets, warts, etc. On the anterior third part of the dorsal and ventral surfaces of the meso- and metathoracic and abdominal segments fields consisting of some parallel rows of tiny spines are situated. Moreover, the larvae of several species can have sclerotized hook-shaped processes of the cuticle ('uncinuli') on the ventral surface of the segments.

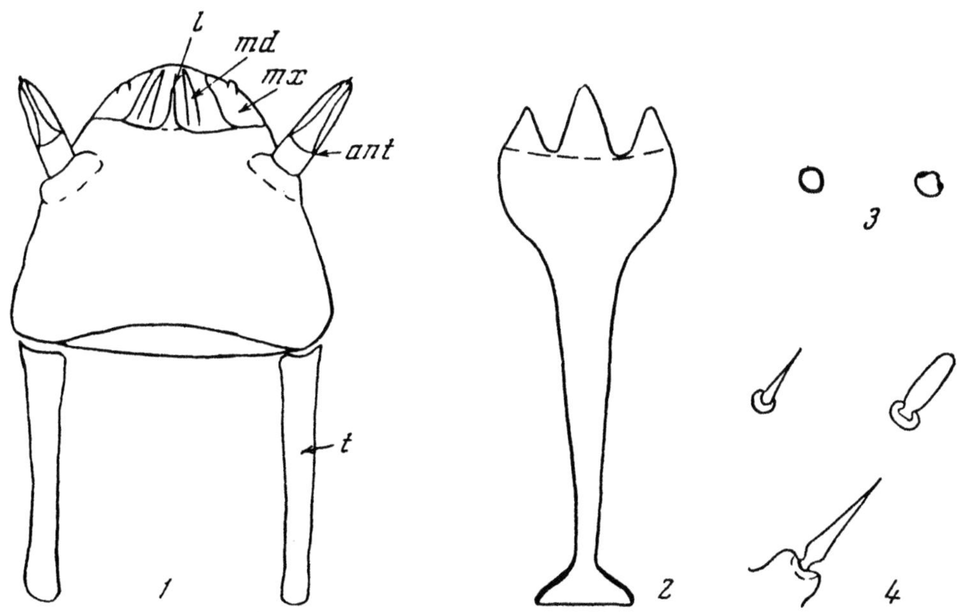


Figure 2. Structural details of a larvae
1 – head; 2 – spatula; 3 – simple papillae; 4 – complex papillae; md. – mandibulae; mx. – maxillae; l. – labrum; ant. – antennae; t. – tentorial arm

The terminal segment of the larvae may be rounded or provided with variously shaped, soft, unsclerotized, mostly large appendages, which seem to be part of the segment. The strongly flattened short appendages we will call 'plates'. The sclerotized formations on the terminal segment are, according to their shape, denoted as processes, teeth, or hooks.

The anal opening is either a narrow slit on the ventral side of the terminal segment or it is rounded and situated at the end of the body. In the latter case the anal opening is found either directly at the end of the body or on a special tubular formation that is slightly retractable into the terminal segment.

The stigmata are mostly located on top of small sclerotized protuberances.

## THE COMPARATIVE MORPHOLOGY OF THE LARVAE

The shape and the size of the body, the structure of the cuticle, the distribution of the stigmata, the type of anal opening and many other morphological peculiarities are extremely variable. Nevertheless and in spite of the large amplitude of variability in these characters, the gall midge larvae share a distinctive set of common traits. They all have a well-developed head capsule and mouth apparatus, and most species have a spatula at the ventral side of the prothorax. The body is covered with variously shaped papillae, whose distribution is, however, rather uniform. The stigmata are well-developed, while the first and the last pair are largest.

The body shape of the larvae is very different and highly dependent on the substrate in which they live. The majority of the larvae belonging to the tribes Cecidomyiini and Oligotrophini, being phytophages developing within galls, have an oval, comparatively short, slightly dorso-ventrally flattened body. In most cases they are 2.5-3 times as long as wide. Larvae that pupate within a gall are soft bodied and barrel shaped. As a result of the pluriformity of their habitats the morphological variation is far greater among the free living species than among the gall makers. The larvae that develop within decaying wood have a cylindrical body without any obvious appendages or processes; whereas their cross section is rounded, as, for example, is the case in *Aprionus* Kieffer, *Winnertzia* Rondani, *Camptomyia* Kieffer, *Rhipidoxylomyia* Mamaev, *Solntsevia* Mamaev, *Leptosyna* Kieffer etc. (fig. 3,1,3). The length of such larvae surpasses their width 4-5 times. The body of the larvae of *Trichopteromyia* Williston is exceptionally elongated (fig. 3,2).

Larvae living in various lightly packed substrates with slit-like spaces (litter, bark, chips of stumps etc.) have a peculiar shape. Thus, the larvae of *Porricondyla* Rondani, *Holoneurus* Kieffer, and *Rhizomyia* Kieffer are strongly dorso-ventrally flattened and remarkably short and broad, and, as a rule, provided with various processes and plates at the sides. They are less than two times as long as broad (fig. 4,1).

Various appendages, processes, and papillae may give the larvae a very peculiar appearance. Thus, the larvae of *Gynapteromyia carpatica* Mamaev have on all body segments large, wide, blade-shaped appendages. The last segments of the larvae of *Parepidosis* Kieffer are strongly elongated, and especially the anal segment has large blade like appendages at the end (fig. 4,2). In some cases the papillae (especially the dorsal ones) attain very large dimensions, and may assume weird shapes. The dorsal papillae of *Parepidosis* Kieffer are setaceous, but the setae are very long, considerably surpassing the width of the body segments, and are found on high conical protuberances (fig. 12,2). The dorsal papillae of *Peromyia* Kieffer (fig. 5) look like broad conical processes that sometimes are as long as the segment is wide. The whole surface of the body of these larvae and all its appendages are thickly covered with various scale-shaped processes (fig. 12,7). The larvae of *Holoneurus* Kieffer, *Dicroneurus* Kieffer, and *Dichaetia* Mamaev have very peculiarly shaped bodies, carrying comparatively large and weird papillae.

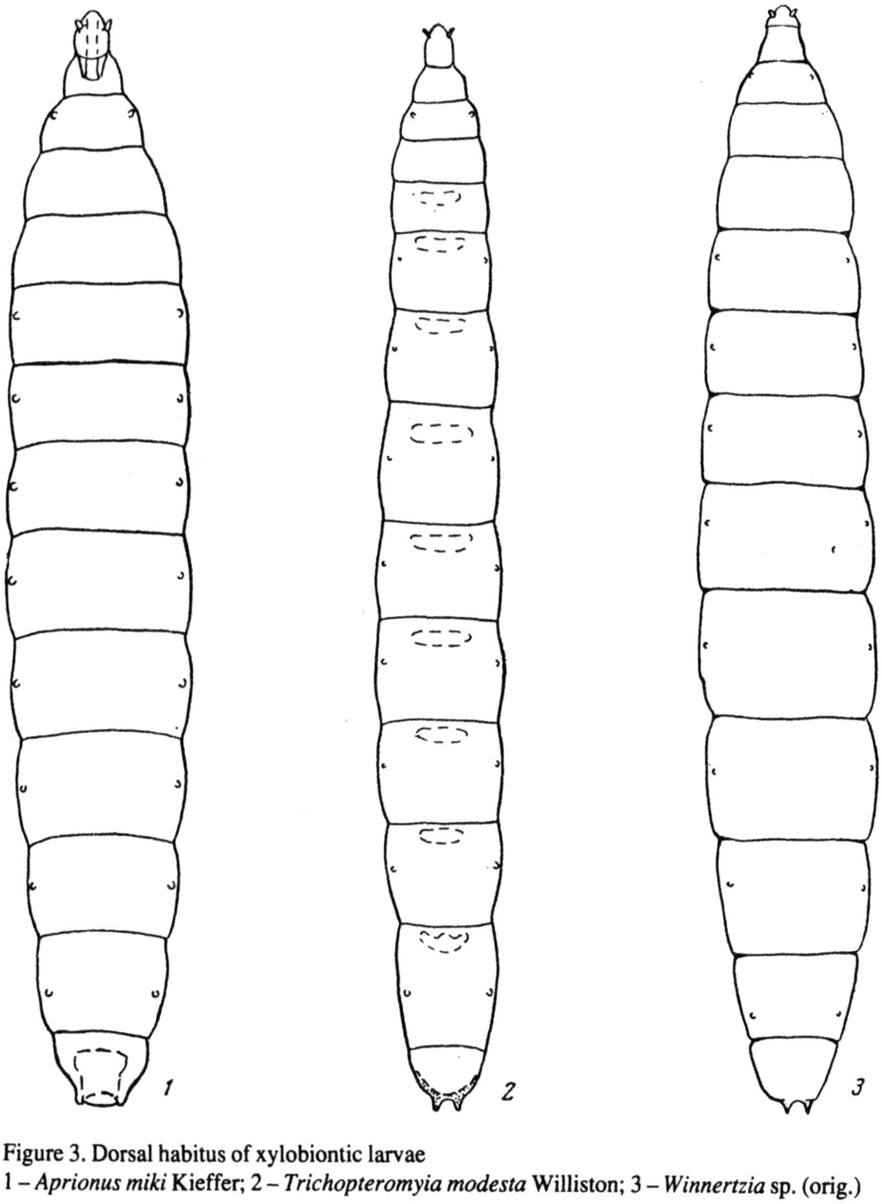


Figure 3. Dorsal habitus of xylobiontic larvae
1 – *Aprionus miki* Kieffer; 2 – *Trichopteromyia modesta* Williston; 3 – *Winnertzia* sp. (orig.)

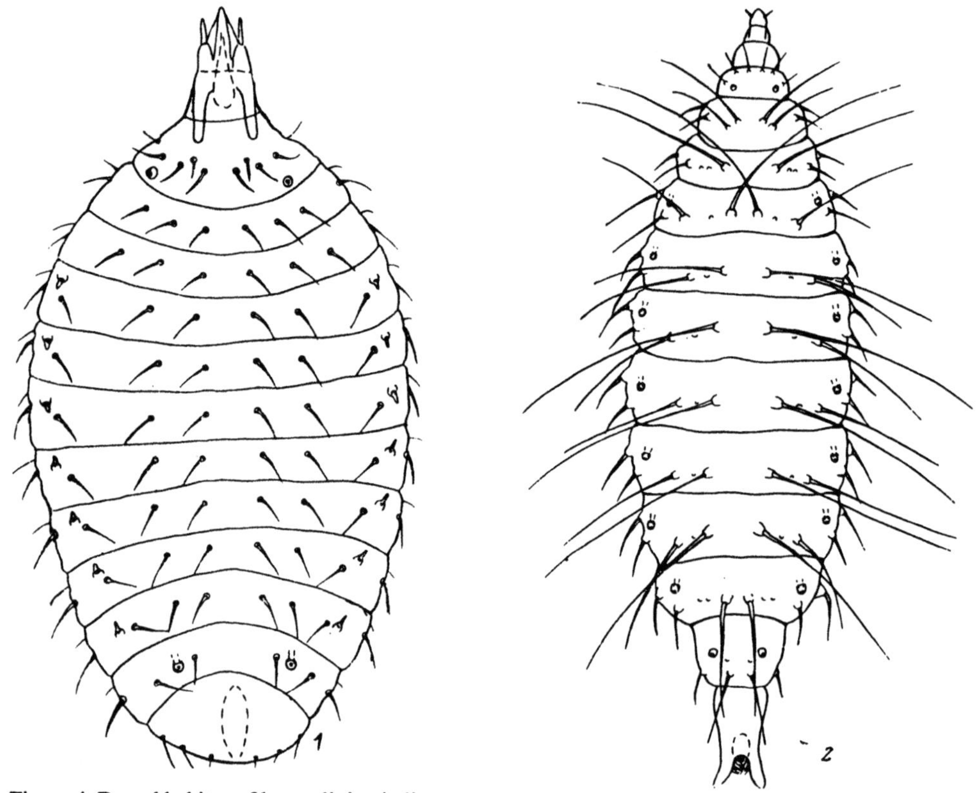


Figure 4. Dorsal habitus of larvae living in litter
1 – *Rhizomyia* sp.; 2 – *Parepidosis ulmicorticis* Mamaev (orig.)

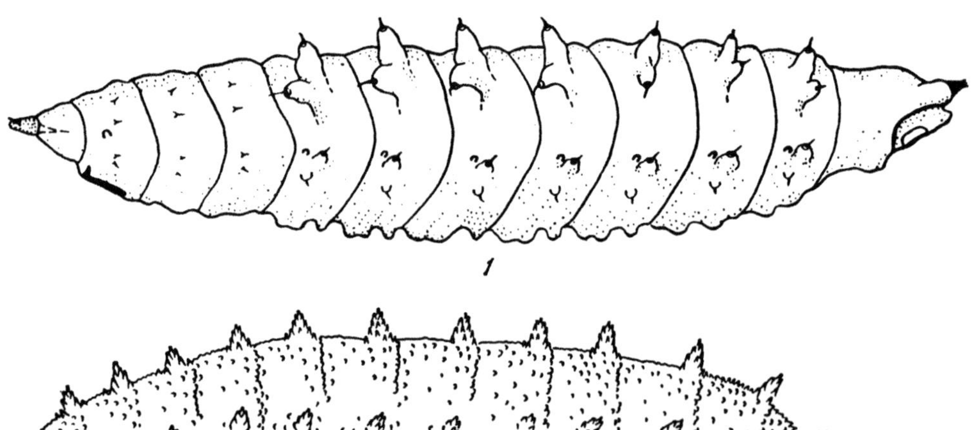


Figure 5. Lateral habitus of larvae
1 – *Itonida pini* (De Geer); 2 – *Peromyia perpusilla* (Winnertz) (orig.)

The mobile mode of life connected with preying, puts a mark of a fundamental nature upon the structure of the body. Apart from differences in many structural details, the predatory larvae of *Lestodiplosis* Kieffer, *Acaroletes* Kieffer, *Aphidoletes* Kieffer, and *Feltiella* Rübsaamen show a large external likeness. They all have long spindle-shaped bodies with a strongly narrowed anterior end. In comparison with them the first body segments of the phytophagous and mycetophagous larvae are broad and short, and their head capsule is not elongated, but broadly rounded at the anterior end.

The head of the larvae is always well-developed (fig. 6 and 7) and small compared to the body. The structure of the head capsule as well as that of the internal skeleton (the tentorium) are quite variable.

The most common structural types of the head capsule will now be first defined, and then each will be discussed in greater detail: 1. The simple head capsule without conspicuous internal skeleton (capsule relatively weakly sclerotized, tentorial arms not extending beyond the head capsule). 2. The head capsule with slender, elongated anterior and posterior tentorial arms (capsule relatively weakly sclerotized, as in the former type, but the posterior arms partly extending beyond the capsule). 3. The head capsule with robust elongated posterior tentorial arms (capsule strongly sclerotized, anterior tentorial arms not elongated, posterior tentorial arms half extending beyond the head capsule. This type of tentorial arms will from now on be called 'exocephalic' as distinct from the metacephalic arms of the brachycerous larvae, in which the posterior part of the head capsule is modified and mostly developed in the same way as the tentorial arms. 4. The head capsule with strongly elongated anterior and posterior tentorial arms (head capsule elongated, weakly sclerotized, especially at the posterior margin; robust anterior and posterior tentorial arms not extending beyond the head capsule).

1. Simple head capsule without conspicuous internal skeleton, relatively weakly sclerotized, brightly yellow or yellow, small, somewhat triangular, its length equalling or slightly surpassing its width at the base (fig. 6,1-3).

The dorsal and ventral side of the capsule are developed to nearly the same degree, the dorsal side is slightly longer than the ventral side. The posterior margin of the capsule is generally not very conspicuous. The dorsal side of the capsule is more or less uniformly sclerotized. The slightly less sclerotized spots of the cuticle around the base of the antennae have very weakly pronounced boundaries. On the ventral side of the capsule along the posterior margin sometimes a narrow more strongly sclerotized zone is found. The posterior margin of the capsule is smooth, without clear protrusions.

A head capsule of this type (fig. 6,1-3) is characteristic for a group of genera of the tribes Porricondylini (*Winnertzia* Rondani, *Camptomyia* Kieffer, *Rhipidoxylomyia* Mamaev, *Trichoxylomyia* Mamaev, *Solntsevia* Mamaev, etc.), for the representatives of some genera of the subfamily Lestremiinae (*Mycopriona* Mamaev, *Miastor* Meinert etc.), and is more rarely encountered in the tribes Oligotrophini (*Rhopalomyia* Rübsaamen) and Asphondyliini (*Asphondylia* H. Loew). The larvae of some Porricondylini (*Trichoxylomyia* Mamaev, *Asynapta* H. Loew. and *Dicerura* Kieffer) have an inconspicuous internal head skeleton, a feature that is strongly connected with the very weak sclerotization of the head capsule itself as well as with its component elements. In the more primitive larvae of *Camptomyia* Kieffer and in particular in those of *Winnertzia* Rondani (fig. 6,1) and *Rhipidoxylomyia* Mamaev, however, the sclerotization of the head capsule is more pronounced and in some cases small rods, running crosswise along the

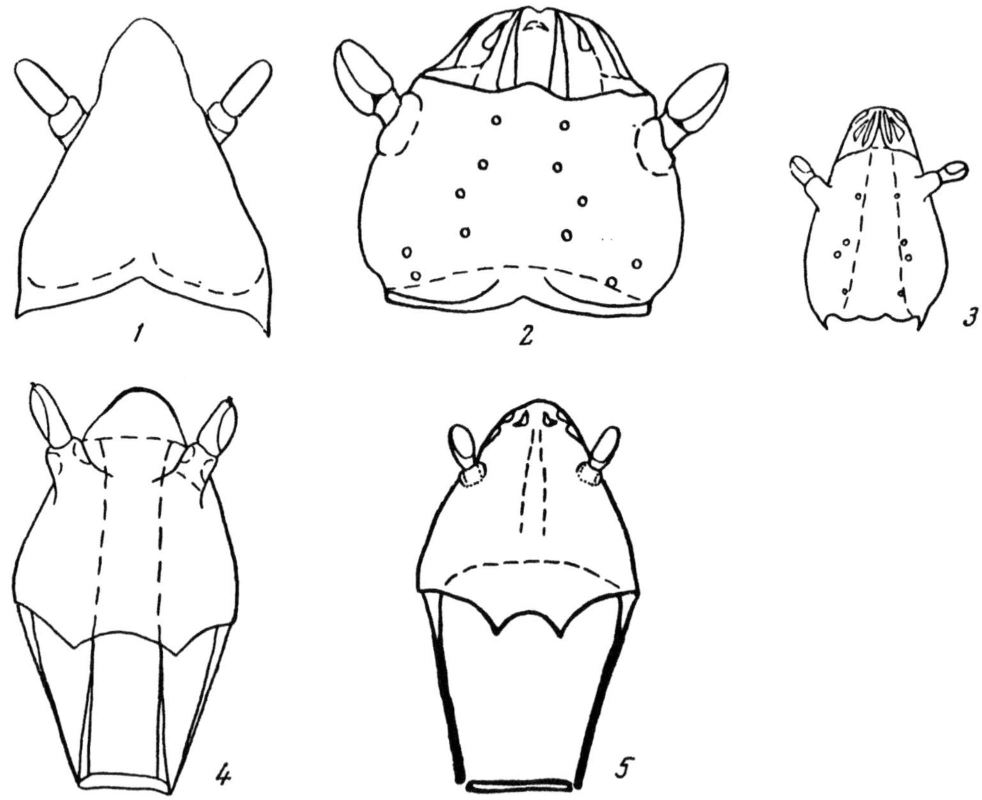


Figure 6. Dorsal view of head capsule (simple or with slender tentorial arms)
1 – *Winnertzia maxima* Mamaev; 2 – *Miastor metraloas* Meinert; 3 – *Mycopriona caucasica* Mamaev; 4 – *Aprionus similis* Mamaev; 5 – *Monardia longicauda* Mamaev (orig.)

posterior margin, can be discerned. At the latero-posterior corners of the head capsule, from where these rods originate, heavier, plate-like, thickenings can be distinguished. In this group of species the anterior tentorial arms are very weak and inconspicuous.

In this context the larvae of *Colomyia* Kieffer may be mentioned. Interestingly, they have a head capsule with highly developed, heavy anterior tentorial arms arising near the epistomal suture. The arms can be easily distinguished along their whole length. They are slightly curved towards the lateral parts of the head capsule to which they are attached by means of heavy plates that are, perhaps, parts of the posterior tentorial arms. A cross bar uniting the posterior ends of the anterior tentorial arms is hardly developed.

In general, the following features are characteristic for larvae with a simple head capsule. Irrespective of the weak sclerotization of the capsule, many of these larvae have preserved a rather well-developed internal skeleton which functions as a structural support in the head. The separate parts of the internal skeleton are differently developed. In some cases the anterior tentorial arms are more powerfully developed (in this context the genus *Colomyia* Kieffer may well be unique); in other cases the posterior tentorial arms are more developed and provide a sturdy support for the musculature of the posterior part of the head capsule. The capsule itself, its posterior margin in particular, is very weakly sclerotized.

In all investigated cases the tentorial arms are relatively short and do not extend beyond the head capsule, though the degree of development of the internal skeleton may vary.

2. The relatively large, mostly elongated-oval, weakly sclerotized, and yellowish head capsule with a rounded anterior margin and slender elongated anterior and posterior tentorial arms.

In this type of head construction the dorsal side is vaulted; whereas the almost flat ventral side is 1.5-2 times shorter than the dorsal one. The dorsal side of the posterior margin of the head capsule is more strongly sclerotized than its ventral side (fig. 6,4-5). A head capsule of this type is found in the genera *Campylomyza* Meigen, *Aprionus* Kieffer, and some other genera belonging to the subfamily *Lestremiinae* (fig. 6,4).

The more or less conspicuous, slender anterior tentorial arms are developed to various degrees. Their anterior tips are attached near the epistomal suture and their posterior ends extend far beyond the head capsule.

The posterior arms, which in general are slightly thicker and more conspicuous than the anterior arms, arise from the latero-posterior margins of the head capsule. Their free ends touch the tips of the anterior arms. The tips of the arms at the left and the right hand side of the head are united to form a bow shaped, curved cross bar. The posterior tentorial arms may be relatively long, as long as or longer than the width of the head capsule, as, for example, in the larvae of *Aprionus* Kieffer; or they may be very short and not longer than one half the width of the capsule as, for instance, is the case in the larvae of *Cordylomyia* Felt which have the anterior arms very weakly developed and hardly visible.

The peculiarity of the head capsule of the larvae belonging to this group consists of the fact that the well-developed, long but thin tentorial arms are not only situated inside the head capsule, but extend considerably beyond it as well. A similar construction of the internal skeleton of the head is found in the larvae of *Leptoconops* Skuse (Leptoconopidae, Diptera), and also in many non-nematoceran Diptera (Leptidae, Dolichopodidae).

3. The heavily sclerotized, dark brown, generally rather sturdy, comparatively short and broad head capsule with sturdy, elongated, exocephalic arms (fig. 7,1-4).

The length of the capsule is, as a rule, equal to or slightly surpassing its width at the base. The dorsal and ventral plates of the head capsule are highly developed, the ventral plate is only slightly shorter than the dorsal one. The unsclerotized rounded spots on the dorsal side of the capsule near the bases of the antennae are well-defined.

The larvae of *Parepidosis* Kieffer have a heavily sclerotized head capsule with an almost black posterior half. The posterior margin of the head capsule has a dark collar. The anterior part of the capsule is very weakly sclerotized on the dorsal side and clearly distinct from the posterior part. The capsule is ventrally closed as is the case in the majority of the gall midge genera. Two relatively thick, straight, parallel exocephalic arms arise from the latero-posterior margins of the capsule. Along their whole length the arms are equally thick or have clearly thickened posterior free ends. They are as long as the head capsule or extend considerably beyond it. The anterior arms lie within the capsule and are apparently weakly developed. In any case, they are not visible in the preparations. This type of head capsule is found in the larvae of *Porricondyla* Rondani, *Holoneurus* Kieffer etc.

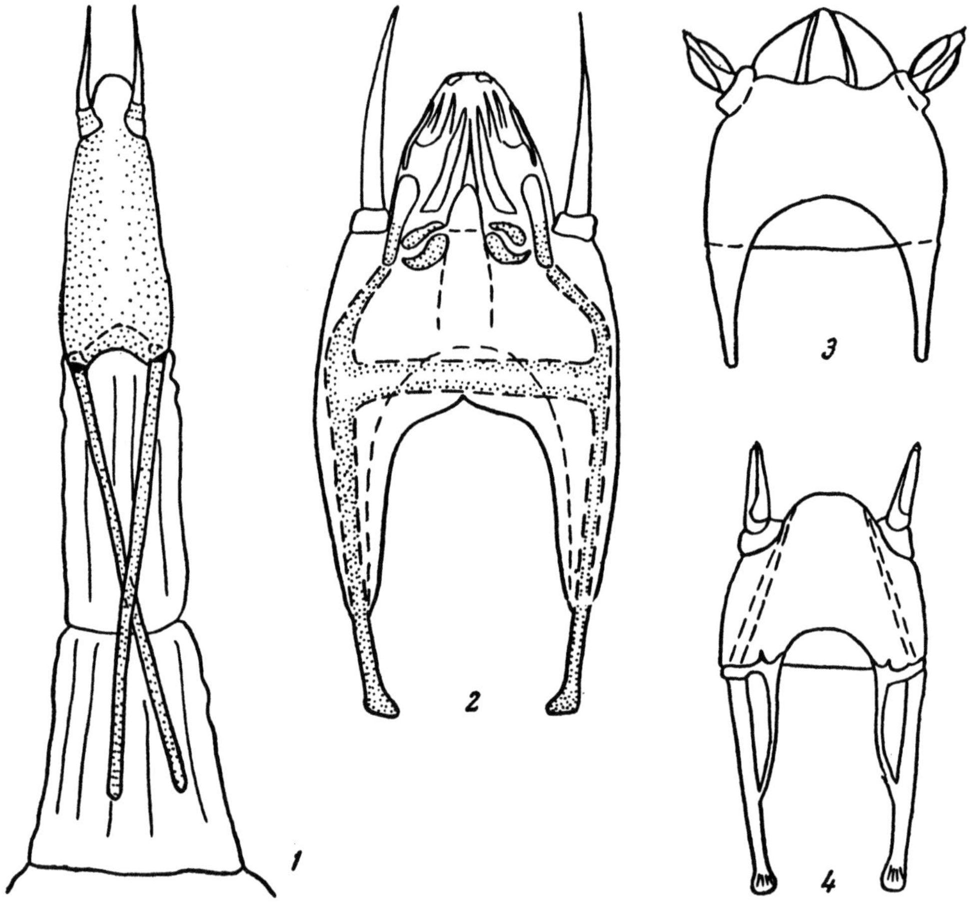

Figure 7. Head capsule with highly developed exocephalic (tentorial) arms
1 – *Arthrocnodax* sp.; 2 – *Lestodiplosis polypori* Rübsaamen; 3 – *Loewiola centaureae* (F. Loew); 4 – *Mycodiplosis pucciniae* Rübsaamen (1-2 – dorsal view, 3-4 – ventral view) (orig.)

The larvae of *Arthrocnodax* Rübsaamen (fig. 7,1) have a fundamentally similar structure. The head of these larvae is extremely long and narrow. The heavily sclerotized, dark brown head capsule measures up to three times as long as its width at the base. The posterior margin is dorsally and ventrally emarginate. Long (up to 1.5 times as long as the capsule) exocephalic arms are attached to the latero-posterior margins of the head capsule.

The head capsule of some other larvae of the tribe Cecidomyiini has a slightly different structure (fig. 7,2-4). The larvae of *Loewiola* Kieffer, *Macrodiplosis* Kieffer, and *Tricholaba* Rübsaamen have a relatively small capsule, whose length along the middle line is considerably less than its width. Dorsally the capsule is relatively short, with a smooth posterior margin. At the ventral side it is somewhat longer, with a small indentation up to the level of the posterior margin of the dorsal cover. From the latero-posterior margins broad, weakly sclerotized processes arise covering the exocephalic arms. The straight arms are not terminally broadened and are much longer than the dorsal cover of the capsule itself. Their structure closely resembles that of the exocephalic arms of the larvae of *Holoneurus* Kieffer.

The comparatively small head capsule of the larvae of *Mycodiplosis* Rübsaamen has a rounded anterior margin (fig. 7,4). The length of the capsule is equal to its width at the base. The posterior margin is fairly well-pronounced, but very weakly sclerotized. The margin is dorsally straight and smooth, ventrally it has a small, rounded emargination. The weakly developed anterior tentorial arms arise from the epistomal suture and run towards the latero-posterior margin of the capsule, with which they are fused by means of a small sclerotized plate. The long, rather thick, and massive exocephalic arms have a broadened tip and point backwards from this plate. Two small, weakly developed, sclerotized rods arise ventrally from the posterior margin of the head capsule. The anterior end of each of these rods is connected with the anterior end of an exocephalic arm, while the posterior end is fused with the middle part of the same arm.

A construction consisting of a relatively well-sclerotized head capsule and sturdy exocephalic arms, which in general are longer than the capsule, is characteristic for the larvae of this group. The arms are either completely free and lying wholly outside the capsule, or they are covered by its lateral processes. The posterior margin of the capsule is, in the former case, smooth and well-defined, whereas in the latter case it is weakly sclerotized and quite indistinct.

4. The large, relatively broad, though weakly sclerotized, long head capsule with dorsally as well as ventrally a deep emargination, and with sturdy anterior and posterior tentorial arms (fig. 7, 2).

The lateral sides of the capsule considerably surpass its dorsal and ventral sides. The short anterior tentorial arms diverge towards the lateral walls of the capsule. Their posterior ends are fused with a broad cross bar that divides the head capsule into an anterior and a posterior part. The posterior tentorial arms arise from this cross bar. Their length exceeds that of the anterior part of the capsule. The posterior arms have a widened tip and are as thick as the anterior arms but only half as thick as the cross bar. The lateral processes of the capsule cover the posterior arms but do not quite reach their tips. All internal arms and bars are rather sturdy and considerably more sclerotized than the head capsule itself. A head capsule of this type may be found in the larvae of *Lestodiplosis* Kieffer.

The construction of the head capsule of the larvae of *Rhizomyia* Kieffer has much in common with that of *Lestodiplosis* Kieffer. The head capsule itself is weakly sclerotized, but the sturdy internal rods are well-pronounced. The very broad cross bar, from whose margin the equally broad posterior arms run backward, is quite conspicuous. The posterior arms that are as long as the capsule anterior to the cross bar, are straight and not widened at the tip.

Thus, within the family of the gall midges a simple head capsule with an internal skeleton inside it occurs as well as a capsule beyond which the tentorial arms extend in various degrees. The problem of the genesis of the various types of head capsule of the gall midges is, morphologically seen, highly interesting.

In the larvae with a simple head capsule the tentorium lies within the capsule and has pronounced anterior, but small, often incompletely developed, posterior arms. The posterior margin of the head capsule of such larvae is smooth, without any emargination.

The larvae of *Mycopriona* Mamaev, for instance, have small, weakly pronounced rods that extend beyond the capsule and which arise from its latero-posterior corners. Comparably small posterior tentorial arms are also found inside the head capsule of some larvae (*Rhipidoxylomyia* Mamaev, *Winnertzia* Rondani). It seems very plausible that the small, short rods of *Mycopriona* Mamaev are nothing but the posterior tentorial arms extending beyond the capsule as a result of the reduction of its posterior parts. The internal skeleton in the head of the larvae of *Diallactes* Kieffer and *Aprionus* Kieffer is entirely of the same type: The anterior as well as the posterior tentorial arms are well-developed and partly extending beyond the head capsule. The difference between them is that in *Diallactes* Kieffer only the anterior arms are connected with the cross bar, while the posterior arms are free, whereas in *Aprionus* Kieffer the anterior as well as the posterior arms are interconnected. Though the position and the structure of the rods are entirely the same, it is by no means certain that in both cases we are concerned with tentorial rods, seeing that in the larvae of *Diallactes* Kieffer they happen to be completely independent, without connection with the other rod-like formations.

Some larvae of the tribe Porricondylini as well as the majority of the larvae belonging to the tribe Oligotrophini and Cecidomyiini characteristically have a well-developed head capsule to which sturdy exocephalic arms, lying entirely outside the capsule, are attached at the posterior margin. The lines along which the evolution of such structures, which are characteristic only for the larvae of gall midges, has proceeded may be more comprehensible after the analysis of some intermediate types of head structure.

The sturdy arms of the larvae of *Rhizomyia* Kieffer lie almost completely within the head capsule and are covered from above by its lateral processes. According to their position they are typical, strongly developed tentorial arms, not outgrowths of the posterior margin of the head capsule. Though a considerable part of the exocephalic arms in the larvae of many genera of the tribe Cecidomyiini is situated outside the head capsule, they are doubtlessly derived from the tentorium. As a matter of fact, in some species they are attached to the inside of the capsule and partly covered by its posterior margin.

Some peculiarities of the sturdy, long, exocephalic arms (as are found in the above mentioned larvae of *Parepidosis* Kieffer, *Porricondyla* Rondani, *Holoneurus* Kieffer and other genera of the tribe Porricondylini as well as in most species of the tribes Oligotrophini and Cecidomyiini) – above all the exocephalic position of the arms that are only connected to the head capsule with their anterior tip – led some authors to consider them as metacephalic arms that have nothing to do with the internal skeleton of the head (Hennig 1948). The comparative study of the structural idiosyncrasies of the head capsule shows, however, that these arms are, in fact, modified posterior tentorial arms. As it wouldn't be correct to call these arms 'tentorial' as far as they are located outside the head capsule, we will in the further discussion term them 'exocephalic' arms. Thus, the exocephalic arms differ from metacephalic ones by the fact that they do not stem from wings of the head capsule, but from the tentorial arms. We may suppose that during the evolutionary process of transformation on the one hand the posterior parts of the head capsule were being reduced, while on the other hand, the arms were simultaneously being lengthened and thickened. Thus, a simple head with an internal supporting skeleton became the starting-point from which a head capsule developed with sturdy exocephalic arms that are located outside the capsule.

The mouth apparatus of the gall midges is well-developed, though its dimensions are comparatively small. It consists of a labrum and a labium, mandibulae and maxillae (fig. 8-10). The labrum is a weakly sclerotized, variously shaped, generally short, plate. The mandibulae are situated laterally to the labrum which slightly covers them from above. They look like heavily sclerotized, relatively long, sickle-shaped plates. Next to the mandibulae the highly developed, but very weakly sclerotized maxillae are found. They consist of broad, elongated plates with a rounded tip. The one-segmented palpi are located on the upper half of the maxillae, near the outer margin. The labium, which covers the mouth apparatus from the ventral side, is weakly sclerotized, as are the maxillae. The labium is a broad lobe with a slightly bifurcated apex. The parts of the mouth apparatus, in particular the mandibulae, are connected to the head capsule by means of various sclerites that are outgrowths of the capsule. Apart from these a group of bow-shaped, curved sclerites – possibly representing the epi- and hypopharynx running down into the head capsule, as recorded from Ceratopogonidae (Lawson 1951) – is found in the anterior part of the throat. Study of the mouth apparatus shows that there are fixed modifications in the structure of its single parts, peculiar to the groups of larvae having different modes of feeding.

The larvae of *Campylomyza ormerodi* (Kieffer), which are mycetophages (fig. 8,1), have a broad and short labrum with a rounded tip, long but fairly sturdy, blunt, distally rounded mandibulae and a considerably widened labium with a weakly sclerotized anterior margin. The mouth apparatus of the larvae of *Lestremia cinerea* Macquart, in which the anterior half of the labrum is clearly divided into two lobes, has a similar construction (fig. 8,2-3).

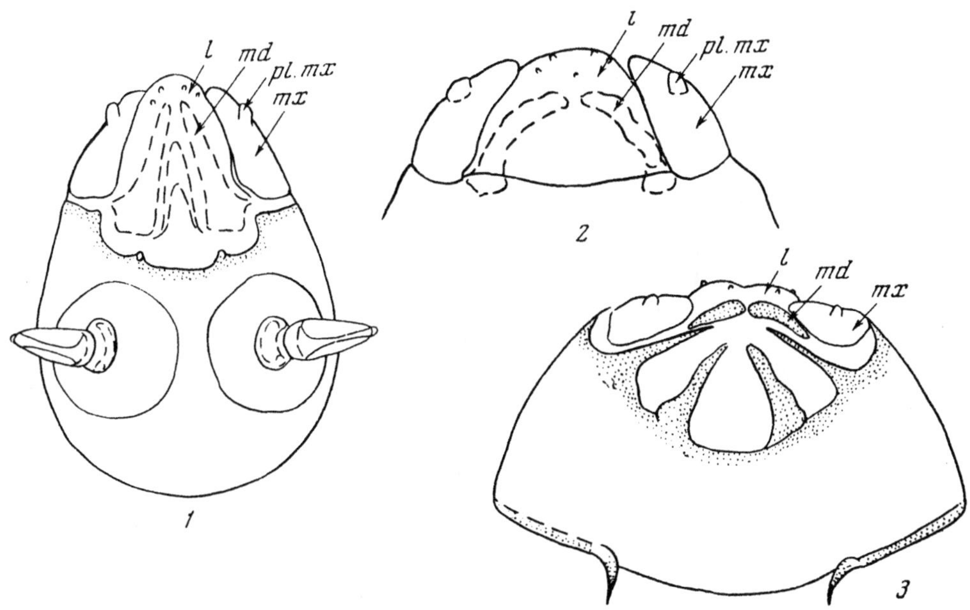


Figure 8. Mouth apparatus of larvae belonging to the subfamily Lestremiinae
1 – *Campylomyza ormerodi* (Kieffer), dorsal view; 2 – *Lestremia cinerea* Macquart, dorsal view; 3 – the same, ventral view (orig.); pl.mx. – maxillary palpi; other abbreviations as in figure 2

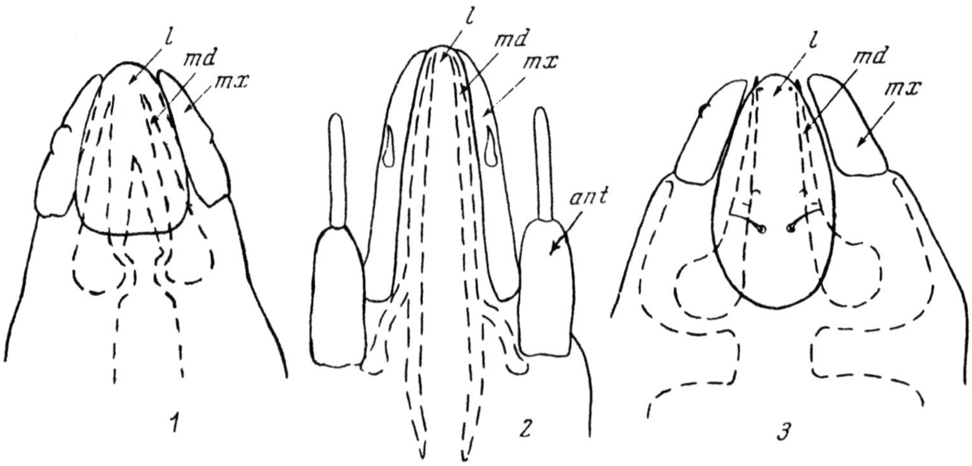


Figure 9. Mouth apparatus of larvae belonging to the subfamily Cecidomyiinae
1 – *Parepidosis arcuata* Mamaev, dorsal view; 2 – *Rhizomyia* sp., dorsal view; 3 – *Lestodiplosis polypori* Rübsaamen, dorsal view (orig.), abbreviations as in figure 2

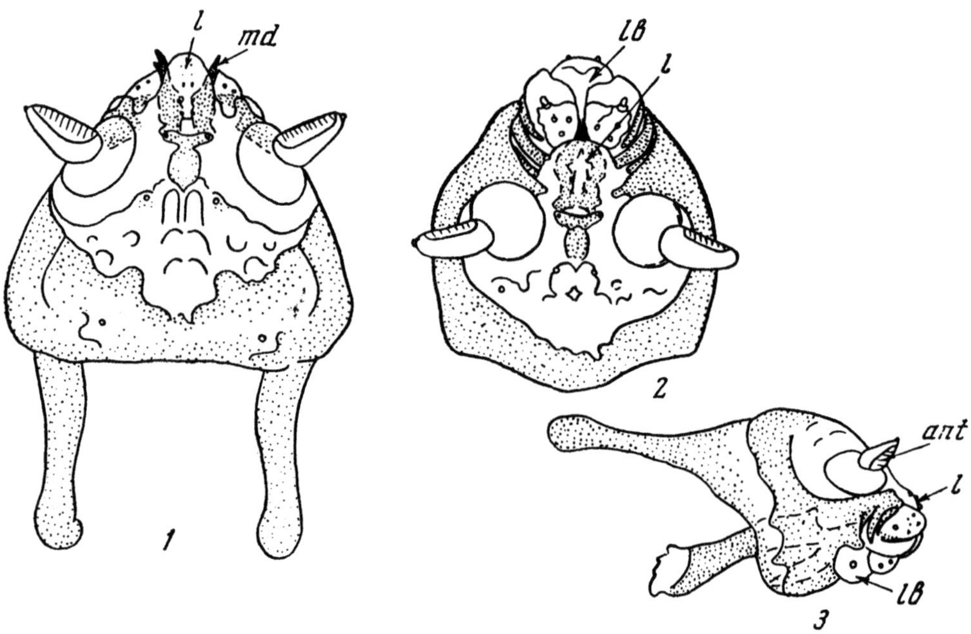


Figure 10. Mouth apparatus of the larva of *Putoniella marsupialis* (F. Loew)
1-2 – dorsal view; 3 – lateral view (after Principi 1958); lb. – labium; other abbreviations as in figure 2

The mouth apparatus of larvae that live on the surface of their substrate is somewhat differently built. In the larvae of *Parepidosis venusta* (Winnertz), and especially in those of *Rhizomyia* Kieffer, all parts of the mouth apparatus are strongly elongated and show some resemblance to a small snout (fig. 9,1-2). On the dorsal side strongly curved mandibulae forming sturdy pointed hooks are characteristic for the predatory larvae of *Lestodiplosis* Kieffer. The mouthparts of the phytophagous larvae that develop inside galls are relatively short and small. They, nevertheless, preserve the same type of construction that also prevails in other groups of larvae with a different mode of feeding. Thus, the larvae of *Putoniella marsupialis* (F. Loew) (fig. 10,1-3), for example, have a very small and short labrum, pointed mandibulae, short maxillae with clear-cut one-segmented palpi, and a lobe-shaped labium.

If we compare the constructions of the mouth apparatus of predators and mycetophages, in particular inhabitants of litter, they clearly show a strong resemblance (for example in the species of *Rhizomyia* Kieffer). Therefore, the conclusion is thrust upon us that, in spite of the different types of food, the manner of feeding is apparently the same. The predatory larvae suck their prey; mycetophagous and phytophagous larvae suck the cell contents of tissue, hyphae etc.

The antennae of gall midge larvae are two-segmented. The first segment is relatively short and broad, the second is relatively long and carries a rod-like sensorial appendage at its tip. In the larvae of *Lestremia* Macquart the second segment is again divided.

The spatula, being one of the most characteristic morphological structures, allows us to distinguish gall midge larvae from the larvae of other Diptera. The shape of the main parts of the spatula, the head and the manubrium, varies greatly among the various groups of gall midges. A three-toothed spatula (fig. 11,1) is characteristic for most larvae of the subfamily Lestremiinae (*Campylomyza* Meigen (fig. 11,1) excepted), and also for the species of *Holoneurus* Kieffer and *Diallactes* Kieffer of the Porricondylini. No larva belonging to the subfamily Cecidomyiinae has a three-toothed spatula.

For the subfamily Cecidomyiinae a two-toothed spatula is most characteristic (fig. 11,4-6). The large tribe Porricondylini is exceptional as it contains free living species whose larvae are distinguished by a one-toothed spatula (with the exception of the larvae of the genus *Colomyia* Kieffer where the spatula is two-toothed). A one-toothed spatula is also found in the larvae of some species belonging to the tribe Asphondyliini, (*Kiefferia* Mik) and Cecidomyiini (*Arthrocnodax* Rübsaamen). A four-toothed spatula arises as a result of a secondary division of the teeth as, for example, in the larvae of *Asphondylia* H. Loew. In some genera the head of the spatula has lateral processes that look like a third and fourth tooth (*Rhabdophaga triandraperda* Barnes). Though a spatula with a widened head, but without teeth, is of rare occurrence, it is, for instance, characteristic for the larvae of *Xylodiplosis* Kieffer, *Campylomyza* Meigen, *Itonida* Meigen, and some other genera (fig. 11,2). In a number of cases the spatula is completely reduced (*Rhizomyia* Kieffer, *Trichopteromyia* Williston, *Lestodiplosis* Kieffer, *Kaltenbachiola* Hedicke, *Feltiella* Rübsaamen, etc.).

Mostly, the base and the head of the spatula are both well-developed. In a number of cases, however, the base is strongly reduced. Thus, the spatula of the larvae of, for instance, *Rhabdophaga rosaria* H. Loew and *Planetella tuberifera* (Rübsaamen) is a broad plate with a toothed tip (fig. 11,4). *Dryomyia circinnans* Giraud also has a plate-shaped spatula. The spatula of the genus *Asphondylia* H. Loew with its strikingly widened head has a peculiar shape. Sometimes, however, it is not the head but the base of the manubrium that is widened, as is particularly characteristic for the large species of *Winnertzia* Rondani and *Rhabdophaga* Westwood.

The papillae, which formations are specific for the larvae of gall midges, are found on

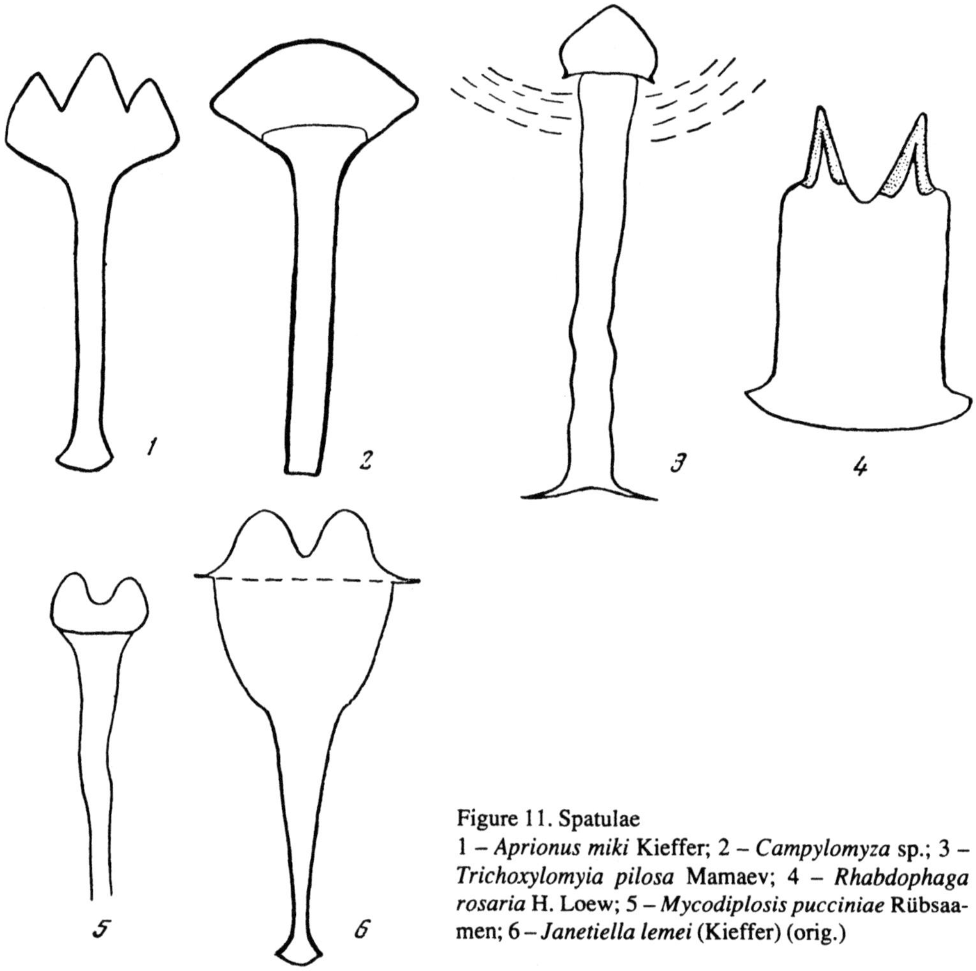

Figure 11. Spatulae
1 – *Aprionus miki* Kieffer; 2 – *Campylomyza* sp.; 3 – *Trichoxylomyia pilosa* Mamaev; 4 – *Rhabdophaga rosaria* H. Loew; 5 – *Mycodiplosis pucciniae* Rübsaamen; 6 – *Janetiella lemei* (Kieffer) (orig.)

all body segments. The distribution of the papillae is rather uniform in the various subfamilies, tribes, and genera; and may be summarized in a single scheme from which a few trivial deviations are possible.

On the dorsal side of the neck, the three thoracic, and the first seven abdominal segments, six dorsal papillae per segment are found (fig. 1,3-5). The larvae of the closely related genera *Dichrona* Rübsaamen, *Planetella* Westwood, and a few others are exceptional as they have six to ten dorsal papillae on the thoracic and the first seven abdominal segments. All dorsal papillae stand in a transversal row on the central or posterior part of the segment. On the eighth abdominal segment only two or four dorsal papillae stand in one or two transversal rows between the stigmata. The number of papillae on the last segment is quite variable.

Generally, three lateral papillae are found on each lateral side of the thoracic and first eight abdominal segments. One of these stand near the stigma, the other two are more ventrally located. As a rule, there are no deviations from this scheme. Only the larvae of

*Holoneurus* Kieffer have four lateral papillae on the thoracic and two on the abdominal segments. Lateral papillae are absent from the last segment.

Because the papillae are not uniformly distributed on the ventral side of the body various names have been assigned to them (fig. 1). On the thoracic segments we find the sternal, intercalary, and sublateral papillae; on the abdominal segments the ventral papillae are found. On the anterior part of each thoracic segment two sternal papillae stand near the middle line. On the prothorax the sternal papillae face the teeth of the spatula. Four groups of intercalary papillae are situated slightly more backward and nearer to the sides of the body. Each group consists of three, in some cases two, papillae. The two sublateral (one on each side) papillae are found between the intercalary and the lateral papillae. The larvae of *Diallactes* Kieffer have only four intercalary and two sublateral papillae on each of the three thoracic segments. The larvae of *Trichoxylomyia* Mamaev have a few solitary intercalary papillae. On the ventral side of the first seven abdominal segments are two transversal rows of papillae (fig. 1). The four anterior papillae form a single transversal row with about equal distances between the papillae. Sometimes the two central ones are slightly closer to each other. In a few cases they have been reduced. Behind the anterior ventral papillae is a row formed by the four posterior ventral papillae. The two central ones of these are considerably closer to each other than they are to the corresponding lateral ones. The anterior ventral papillae are found immediately behind or within fields of spines. In the porricondyline genera *Ulmomyia* Mamaev, *Synepidosis* Mamaev, and *Porricondyla* Rondani they are located within spinule fields. In *Asynapta* H. Loew they are found on spherical processes (prolegs) which are also located within spinule fields. On the eighth abdominal segment only one row of ventral papillae is found.

The degree of development is highest and the scheme of distribution of the papillae most complicated on the last segment. The six to eight anal papillae lie in a checkerboard arrangement around the anal slit on the ventral side of the segment. The terminal papillae are located at the end of the segment, mostly on its outgrowths and processes.*

The papillae on gall midge larvae may have very different shapes. The dorsal, lateral and marginal pairs of the posterior ventral papillae are the most variable ones. The shape of the papillae on the ventral side of the thoracic segments and the anterior ventral papillae of the abdominal segments have, within the various groups of gall midges, a relatively constant shape. The papillae may be divided into various groups, according to their shape and size.

First we must separate the simple papillae from the multiform composite papillae. The simple papillae are distinguishable (though, at times, hardly visible) as small, rounded, weakly sclerotized spots on the surface of the segments (fig. 12,1). *Winnertzia* Rondani, *Rhipidoxylomyia* Mamaev, and some other larvae of Porricondylini as well as the larvae of some species of the subfamily Lestremiinae (*Aprionus* Kieffer, *Monardia* Kieffer, *Trichopteromyia* Williston, *Mycopriona* Mamaev etc.) may serve as examples for which simple papillae are characteristic. In the majority of cases the sternal, the lateral and a part of the intercalary papillae on the thoracic segments, as well as the anterior and the central

*Note from the editor: The terminology of larval papillae as used by Mamaev differs in some respects from that of Möhn (1955b). Since Möhn's system is widely used in the western literature a comparism of both systems has been added to the caption of figure 1.

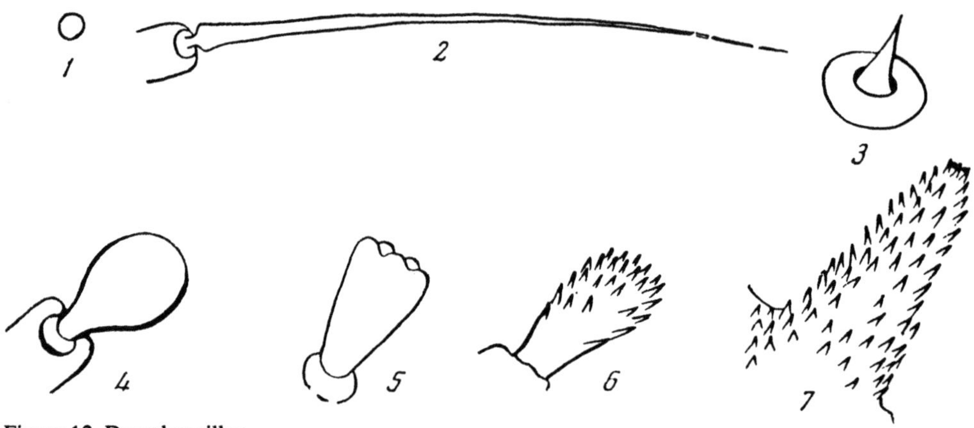


Figure 12. Dorsal papillae
1 – simple papilla of *Winnertzia rubricola* Mamaev; 2 – complex setaceous papilla of *Parepidosis arcuata* Mamaev; 3 – complex setaceous papilla of *Dasyneura pustulans* (Rübsaamen); 4 – complex pinhead-shaped papilla of *Dicroneurus montanus* Mamaev; 5 – complex squamiform papilla of *Holoneurus fulvus* Kieffer; 6 – complex brush shaped papilla of *Holoneurus corticalis* Mamaev; 7 – complex conical papilla of *Peromyia perpusilla* (Kieffer) (orig.)

pair or both pairs of posterior ventral papillae on the abdominal segments are simple.

The composite papillae, as distinct from the simple ones, are made up of two parts; a basal part and an apical appendage. The basal part consists of a bulge, a plate, or a conical protuberance (fig. 12,2-7). The apical part consists of a bristle-like appendage (for brevity's sake called 'bristle'), a rod-shaped, conical, tubular, or dendroid outgrowth. Composite papillae comprise generally: the dorsal papillae, the lateral one closest to the stigma or all lateral ones, the sublateral papillae, and the marginal pair of the posterior ventral papillae.

The apical part of a composite papilla is, as a rule, the most developed one. The basal part is only in a few cases substantially enlarged, as, for example, in the larvae of *Peromyia perpusilla* (Winnertz). In this species the papillae consist of a broad conical or spherical process which carries on its tip a very short, nearly invisible, bristle. In accordance with the shape of the terminal appendage the papillae are divided into bristle-, rod-, scale-, tree-, pinhead-, brush-shaped, and conical papillae.

Bristle-like papillae occur quite often (fig. 12,2-3). In some groups of larvae the papillae of this type consist of a small, flat, rounded plate from whose centre a bristle rises. In particular the dorsal, intercalary, and all or a part of the terminal papillae tend to be bristle-shaped. The length of the bristle is highly variable; in some cases it does not surpass the diameter of its basal plate, as, for instance, in the larvae of *Rhabdophaga triandraperda* Barnes. Not uncommonly, however, it is more than twice as long as the diameter of its basal plate (*Herbomyia* Möhn, many species of *Dasyneura* Rondani, *Janetiella* Kieffer, etc.). Bristle-like papillae are also of fairly common occurrence among the larvae of the tribe Porricondylini and some representatives of the subfamily Lestremiinae. Their shape, however, is different from that found in the larvae of Cecidomyiini. The dorsal papillae of the larvae of *Peromyia* Kieffer are small, rounded elevations with a bristle at their tip. The bristles themselves are also small, and about of equal length as the elevations. The terminal papillae of many species are of this type. Sometimes the basal elevation is very small or only represented by a small ring, in which case the papilla may look like a true bristle. The intercalary papillae of most species of the subfamily Lestremiinae and the tribe Porricondylini, as well as a number of species of the tribe Cecidomyiini are partly of this shape. Comparable papillae are sometimes found on the ventral side of the abdominal segments (for instance, a part of the papillae of *Ulmomyia* Mamaev). The papillae of the larvae of *Parepidosis* Kieffer have a bristle-like, yet very peculiar shape. The base of these papillae consists of a large, well-developed, conical protuberance displaying on its tip a hair-like bristle, whose length surpasses the width of the body segments.

The pinhead-shaped papillae consist of a large, conical base and a swollen apical part (fig. 12,4). In the larvae of the genus *Dicroneurus* Kieffer the dorsal, lateral, terminal, sublateral, and a part of the posterior ventral papillae are pinhead-shaped.

Scale-like papillae are characteristic for the genus *Holoneurus* Kieffer (fig. 12,5). Each papilla consists of a small elevation carrying a flattened outgrowth that is widened to the tip. In the larvae of *H. fulvus* Kieffer the dorsal papillae on the meso- and metathoracic, as well as on the abdominal segments are scale-like.

The larvae of *Holoneurus* Kieffer might also exhibit rod-shaped papillae. They consist of a small basal elevation and a stick-like, fairly long, apical part. These papillae are located laterally along the segments.

We have seen brush-like papillae (fig. 12,6) only on the larvae of *Holoneurus corticalis* Mamaev. These papillae are made up of a small basal elevation with a thick, brush-like bunch of hairs at the tip. In the larvae of *H. corticalis* Mamaev the central pair of dorsal papillae on the prothorax, the dorsal papillae on the other segments, as well as a part of the lateral papillae on the meso- and metathoracic and on the first eight abdominal segments are of this type.

Conical or spherical papillae that look like processes are characteristic for some species of the genera *Peromyia* Kieffer and *Skuhraviana* Mamaev (fig. 12,7). The processes are thickly covered with hair-, scale-, or bristle-like formations. On the top of the processes short bristles, representing the weakly developed apical parts of the papillae, are found. The dorsal papillae as well as a part of the lateral papillae of the investigated species of *Peromyia* Kieffer and *Skuhraviana* Mamaev belong to this type.

The integument of gall midge larvae is never completely smooth. A microscopic inspection of the larvae with a sufficiently high magnification will easily convince us of this fact. Often the entire surface of the larval cuticle is clearly sculptured, while fields of the covering on various parts of the segments have, as a rule, a different sculpturing. We can divide all these cuticular differences into macro- and microstructures. With 'macrostructures' we mean the cuticular plates, protuberances, teeth, spine- and hair-shaped formations that are generally located on the dorsal and lateral surfaces of the segments. We consider as 'microstructures' the crowded, though clearly discernable, rows of spines running transversely across the segments. The microstructures form separate fields on the anterior ventral, and sometimes also dorsal, third part of the segment. In many cases macrostructures as well as spinule fields may be developed, as, for example, in *Peromyia* Kieffer and *Skuhraviana* Mamaev (Lestremiinae), *Dicroneurus* Kieffer, *Parepidosis* Kieffer, *Porricondyla* Rondani (Porricondylini), *Dasyneura* Rondani, *Geocrypta*

Kieffer, *Macrolabis* Kieffer (Oligotrophini), *Contarinia* Rondani, *Clinodiplosis* Kieffer, and other Cecidomyiini.

In larvae that have ventrally as well as dorsally a heavily sculptured cuticle, spinule fields may be absent. Though the anterior third part of the ventral side of the segments will consequently carry spinule-like formations, these are relatively sturdy and broad, and not clearly separated from the cuticular structures on the remaining part of the segment. Such formations are only characteristic for the larvae of higher gall midges: *Dasyneura* Rondani, *Rhabdophaga* Westwood, *Iteomyia* Kieffer, *Herbomyia* Möhn, *Janetiella* Kieffer (Oligotrophini), *Loewiola* Kieffer, *Xylodiplosis* Kieffer (Cecidomyiini).

Finally, though the various macrostructures may be absent; clearly separate spinule fields occur on the dorsal and ventral sides (in some cases only on the ventral side) of the body. This situation is characteristic for the more primitive species of the subfamily Lestremiinae and the tribe Porricondylini.

If there are no manifest macrostructures the larvae are described as 'naked'. This, however, does not mean that they are wholly devoid of cuticular formations since, when macrostructures are lacking, spine fields are always present.

The macrostructures may be divided into the following groups: plate-, spine-, bristle-, and hair-shaped formations, and swellings. Plate- and spine-shaped structures are most commonly found. They are characteristic for the majority of the species belonging to the tribes Cecidomyiini and Oligotrophini.

The plate-like structures (fig. 13,1) may be oval or polygonal. The interstices between them are smaller than the diameters of the plates themselves (*Rhabdophaga* Westwood, *Iteomyia* Kieffer, *Janetiella* Kieffer, *Dasyneura* Rondani, *Macrolabis* Kieffer), or the plates are closely adjoining (*Wachtliella rosarum* (Hardy), some species of *Dasyneura* Rondani). The plates often carry small protuberances which look like toothlets, as, for example, in the larvae of *Wachtliella* Rübsaamen. The plates, which are generally located close to the middle line of the body, gradually change into the more elaborate processes at the lateral sides of the segments. The increasing size of the plates and, on the other hand, the increasing size of the processes lead to the development of spine-like structures.

Spine-like structures are especially characteristic for the larvae of the Cecidomyiini and the Oligotrophini, and are distributed on the ventral side of the body segments (fig. 13,2). These structures are either long and pointed (*Herbomyia* Möhn etc.), or relatively short and widened at the base (*Cystiphora* Kieffer, *Janetiella* Kieffer, etc.). Similar spine-like structures are also characteristic for some larvae of the tribe Porricondylini and the subfamily Lestremiinae. Their shape, however, (in particular that on the ventral side of the segments) reminds us of tooth-shaped processes (*Acoenonia* Pritchard, *Peromyia* Kieffer, *Holoneurus* Kieffer).

The bristle-like structures resemble well-developed bristles (fig. 13,3). Such structures cover the ventral surface of the last segment, and in some cases also the entire dorsal body surface of the larvae of *Parepidosis* Kieffer; a part of the surface of the segments of the larvae of *Gynapteromyia* Mamaev; and especially the ventral side of the body of *Dicroneurus montanus* Mamaev. Such structures are also found on some larvae belonging to the subfamily Lestremiinae, for instance, in the genus *Acoenonia* Pritchard.

The hair-shaped structures very strongly resemble the bristle-shaped ones, but the former are slenderer and longer (fig. 13,4). The dorsal surface of the body of the larvae of

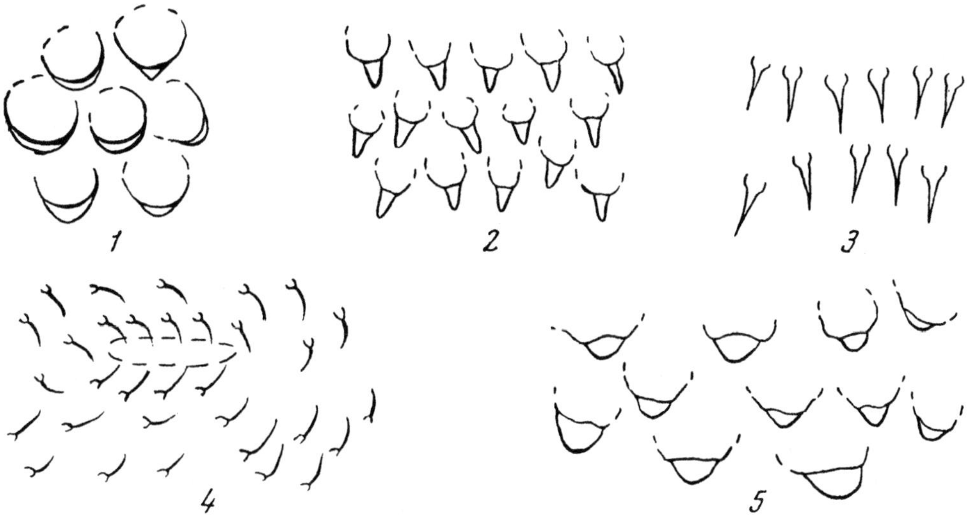


Figure 13. Cuticular structures
1 – platelets of *Rhabdophaga triandraperda* Barnes; 2 – spinule-shaped structures of *Herbomyia scirpicola* (Kh. Mamaeva); 3 – bristle-shaped structures of *Parepidosis venusta* (Winnertz); 4 – hair-shaped structures of *Bryomyia* sp.; 5 – knob-shaped structures of *Peromyia minutissima* Mamaev (orig.)

*Bryomyia* Kieffer and the ventral surface of the last segment of some species of *Parepidosis* Kieffer etc. are densily covered by such 'hairs'.

The cuticular verrucae (fig. 13,5) are, as a rule, small and oval. They often densily cover the surface of the body, for instance, in the larvae of *Peromyia* Kieffer. Such structures generally represent intermediate stages between plate- and spine-shaped formations.

The above mentioned structures do not exhaust the whole gamut of cuticular formations in gall midge larvae. Thus, for example, highly interesting, branched, needle-shaped processes, not found in other gall midge larvae, occur on the larvae of *Gynapteromyia carpatica* Mamaev. They remind us somewhat of the smaller and generally unbranched processes of the larvae of *Acoenonia europaea* Mamaev. We should point out that the various macrostructures are in particular characteristic for larvae living in litter, under bark, or in rotting wood.

The spinule fields, which we consider as microstructures, are located on the anterior third part of the meso- and metathoracic, and of seven to nine abdominal segments. They are found on small bolster-shaped elevations and remind us of analogous structures occurring in the larvae of the higher non-nematoceran flies. Though these fields are, among the Lestremiinae, pronounced on the dorsal as well as on the ventral side only on the larvae of *Miastor* Meinert and *Brittenia* Edwards, they are so pronounced on the larvae of many Porricondylini: *Colomyia* Kieffer, *Asynapta* H. Loew, *Dicerura* Kieffer, *Winnertzia* Rondani, *Camptomyia* Kieffer, *Rhipidoxylomyia* Mamaev, *Trichoxylomyia* Mamaev, and *Ulmomyia* Mamaev. On most larvae belonging to the subfamily Lestremiinae, on some representatives of the tribe Porricondylini (*Synepidosis* Mamaev, *Parepi-*

*dosis* Kieffer, *Diallactes* Kieffer, and *Dicroneurus* Kieffer), as well as on the larvae of the tribe Cecidomyiini only the ventral side carries spinule fields.

If the spinule fields occur dorsally as well as ventrally, they are located on the meso- and metathoracic and on seven abdominal segments. Only rarely are they also found on the eighth abdominal segment (for instance in *Colomyia* Kieffer). If the spinule fields occur only ventrally they are found on the meso- and metathoracic and on seven to nine abdominal segments. Thus, the larvae of the tribes Cecidomyiini and Oligotrophini have spinule fields on the thoracic and on seven to nine abdominal segments, as has the majority of the larvae belonging to the subfamily Lestremiinae; whereas in a few of the latter the spinule fields are confined to the thoracic and seven to eight abdominal segments.

The fields are oval, narrowing to the lateral margins, and consist of five to forty rows of spinules. On the segments halfway the body the fields are no less than four times as wide as long. If the fields consist of more or less separate rows of spinules, or if there are shiny, oval spots within them, the fields are called 'complex'. The spots represent sites where muscles are attached, as can be observed in larvae that have been slightly cleared in alkali. The bare, oval spots form a single row running from one margin of the field to the other.

Complex spinule fields occur in many larvae of the subfamily Lestremiinae (*Lestremia* Macquart, *Acoenonia* Pritchard, *Campylomyza* Meigen, *Aprionus* Kieffer, *Peromyia* Kieffer, *Skuhraviana* Mamaev, *Monardia* Kieffer), and in some larvae of the tribe Porricondylini (*Parepidosis* Kieffer, *Porricondyla* Rondani, *Dicroneurus* Kieffer), the tribe Cecidomyiini (*Ametrodiplosis* Rübsaamen, *Clinodiplosis* Kieffer, *Aphidoletes* Kieffer, *Monobremia* Kieffer), and the tribe Oligotrophini (*Dasyneura* Rondani, *Macrolabis* Kieffer, *Wachtliella* Rübsaamen). The larvae of some species belonging to the subfamily Lestremiinae have bare oval spots (*Micropteromyia* Mamaev, *Mycopriona* Mamaev, *Henria* Wyatt). In some cases rounded protuberances ('prolegs'), which carry the anterior ventral papillae, are found within the spinule fields (fig. 15,2). These protuberances are not covered with spinules and they appear as bright spots in the fields (*Asynapta* H. Loew). The ventral papillae on larvae of species belonging to the tribes Cecidomyiini and Oligotrophini are sometimes positioned in this way.

In general, the spinules in the fields are long and relatively slender; in some cases they are short, broad and blunt, as, for instance, in *Dicerura* Kieffer (fig. 14,3); or they appear as short needles with a widened base, as, for example, in *Henria psalliotae* Wyatt, *Miastor metraloas* Meinert and a number of related genera (fig. 14,4). The larvae belonging to the tribes Cecidomyiini and Oligotrophini have short, but pointed, spinules (fig. 14,5).

Gall midge larvae, in particular those of the free living species, are able to energetically move about in or on their substrate. This is made possible by the presence of specific locomotive structures. The spinule fields (discussed above) function as a grip, preventing the body from sliding backwards. Apart from the spinule fields, in some species the ventral side of the segments may be provided with sickle-shaped cuticular hooks (uncinuli), with various rounded protuberances, and also with long, conical processes with a tuft of hairs at their tip.

These sickle-shaped hooks are only characteristic for the larvae of some species of the tribe Porricondylini (*Holoneurus* Kieffer) (fig. 15,1). They are heavily sclerotized

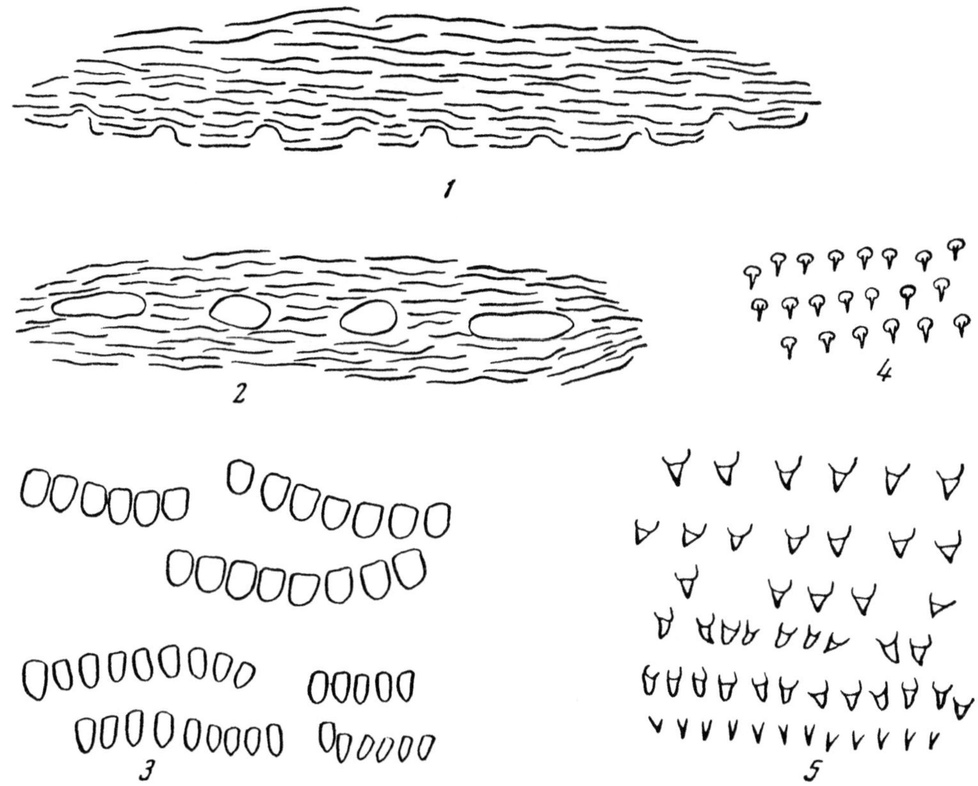


Figure 14. Fields of spinules on the abdominal segments
1 – complex of spinules of *Campylomyza* sp.; 2 – spinule fields with bare oval spots of *Asynapta* sp.; 3 – blunt spinules of *Dicerura* sp.; 4 – needle-shaped spinules of *Henria psalliotae* Wyatt; 5 – pointed spinules of *Profeltiella meridionalis* Mamaev (1-3, 5 – orig.; 4 – after Wyatt 1961)

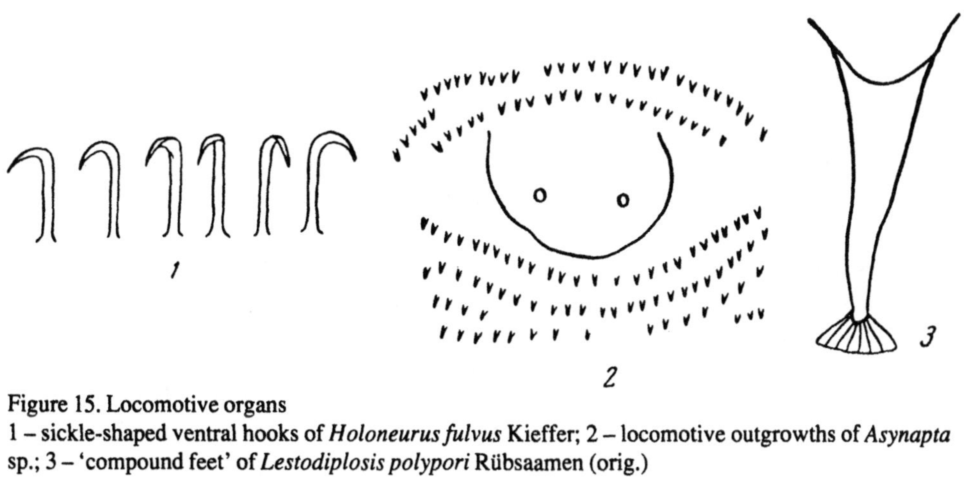


Figure 15. Locomotive organs
1 – sickle-shaped ventral hooks of *Holoneurus fulvus* Kieffer; 2 – locomotive outgrowths of *Asynapta* sp.; 3 – 'compound feet' of *Lestodiplosis polypori* Rübsaamen (orig.)

cuticular formations, which are considerably sturdier than the spinules. They are bent and pointed in such a way that their upper part resembles the end of a backwardly directed sickle. The hooks may be located on the ventral side of the thoracic as well as of the abdominal segments. On each segment they stand in a single transversal row, which is often divided into two parts by the middle line of the body. The rows, which consist of 6-16 hooks, are located on the anterior part of the segments in the same region where in most cases the spinule fields are found. The larvae of *Holoneurus* Kieffer have massive hooks in whose vicinity only a few rows of very small spinules are situated. In cases where the sickle-shaped hooks are highly developed, the spinule fields are strongly reduced.

Rounded, spherical or slightly elongated processes resembling incompletely developed prolegs of other insect larvae, are clearly pronounced in the representatives of the genus *Asynapta* H. Loew (fig. 15,2). These processes are located within the spinule fields on the thoracic and abdominal segments. As these processes are not covered with spinules they show up as bare spots amidst the spinule fields. On the thoracic segments they carry the sternal papillae on their tips; on the abdominal segments – the paired anterior ventral papillae.

The long, pointed processes with a tuft of hairs on their tips (fig. 15,3) slightly resemble the prolegs of other insects. They are found on the ventral surface of the body: in twos on the meso- and metathoracic, and in threes on the first seven abdominal segments. Such formations are well pronounced in the larvae of *Lestodiplosis* Kieffer.

Finally, the construction of the last two segments deserves special attention. These segments carry at their ends various appendages, processes, hooks, and spines (fig. 16,1-6). If the penultimate (eighth) segment has the usual shape (more or less equal to that of the other segments), it is separated from the last segment by a straight suture, which is dorsally as well as ventrally well-pronounced. In some larvae, however, the penultimate segment differs fundamentally from the other segments as it is laterally provided with highly developed outgrowths. These carry the stigmata and enclose the sides of the last segment. In some cases these outgrowths are small, not reaching beyond the end of the last segment, as, for instance, in the larvae of *Camptomyia* Kieffer (fig. 16,1), which have outgrowths like broad, rounded plates. In the larvae of *Ulmomyia* Mamaev (fig. 35,7) the outgrowths are still smaller and conical, with stigmata on the tip. The second to last segment of the larvae of *Profeltiella* Kieffer is of this type (fig. 16,2). Small, broad outgrowths are also found on the penultimate segment of the larvae of *Gynapteromyia carpatica* Mamaev (fig. 41,2). The bluntly truncated outgrowths of the larvae of this species are slightly shifted to the dorsal side of the segment. It is a characteristic feature of this species that similar outgrowths, which carry the stigmata, occur on the other abdominal segments as well.

In some cases the outgrowths of the eighth segment are very strongly developed, so that they may equal or surpass the length of the last segment (fig. 16,3). The stigmata are situated on the tip of the outgrowths (*Dicerura* Kieffer). The development of outgrowths carrying the stigmata is associated with a semi-aquatic mode of life.

The last segment is in most cases well-developed, albeit somewhat narrower, and, as a rule, shorter than the last but one (its length generally being only 2/3 of that of the preceding segment). Sometimes, though, the last segment is very long; in which case its length may surpass its width by 1.5-2 times, as, for example, in some species of

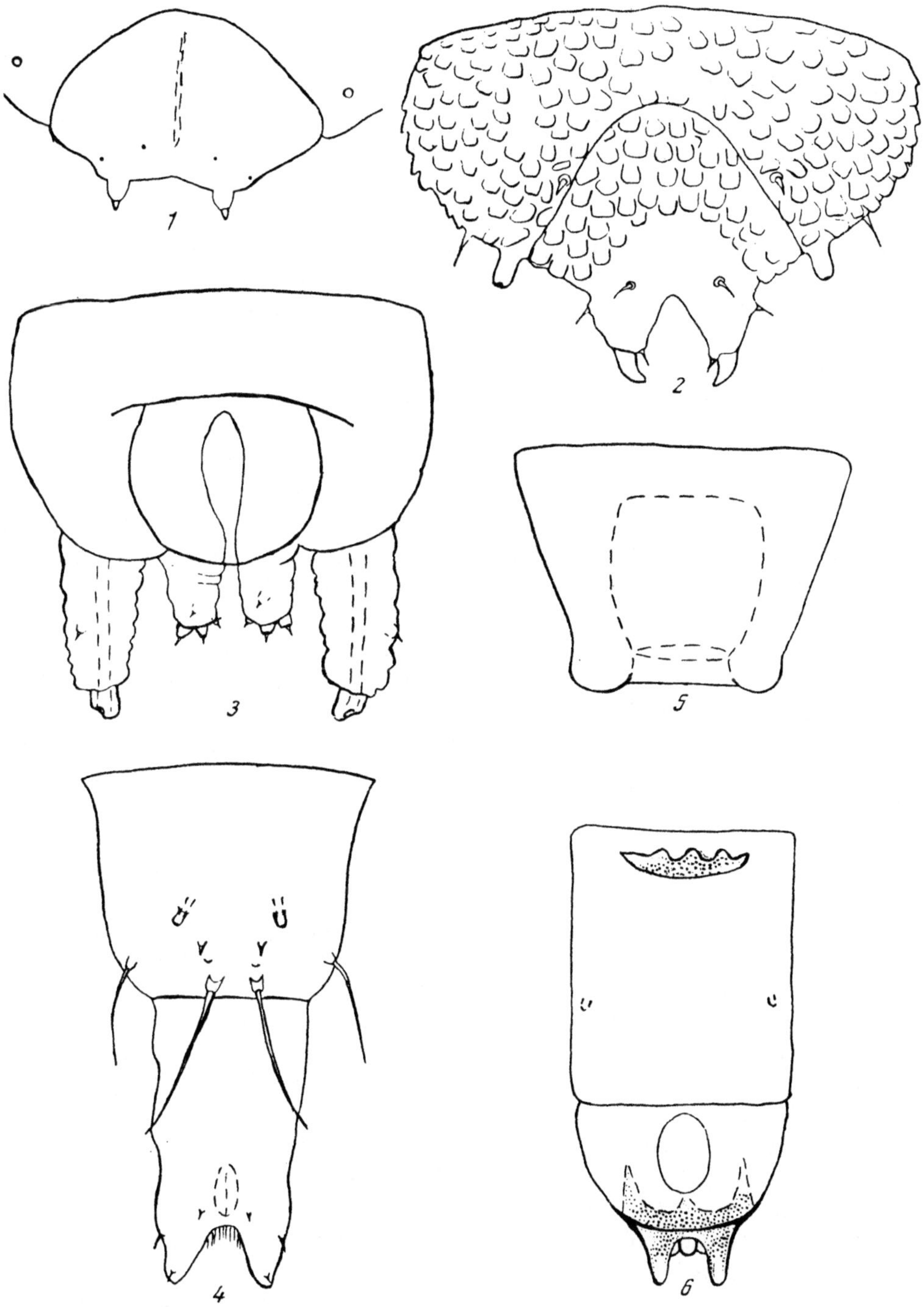


Figure 16. The body end
1 – *Campylomyza* sp.; 2 – *Profeltiella meridionalis* Mamaev; 3 – *Dicerura* sp.; 4 – *Parepidosis venusta* (Winnertz); 5 – *Aprionus similis* Mamaev; 6 – *Trichopteromyia modesta* Williston (1-2, 4-5 – dorsal view; 3, 6 – ventral view) (orig.)

*Parepidosis* Kieffer (fig. 16,4). In *Aprionus* Kieffer, *Peromyia* Kieffer, and *Skuhraviana* Mamaev (Lestremiinae) the last segment is slightly longer than the last but one. In a comparatively small group of larvae the terminal segment is very short and, whereas its dorsal side is hardly developed at all, its ventral side is clearly recognizable. This is characteristic for larvae in which the penultimate segment has highly developed lateral outgrowths.

The posterior margin of the terminal segment is rounded or provided with various appendages, processes, and sturdy papillae. An ordinary segment with a rounded posterior margin is characteristic for the larvae of *Lestremia* Macquart, *Campylomyza* Meigen, *Heteropeza* Winnertz, and *Heteropezina* Pritchard (Lestremiinae); as well as for many larvae of *Holoneurus* Kieffer, *Diallactes* Kieffer and some other Porricondylini. The terminal segment of many larvae of Oligotrophini (*Oligotrophus* Latreille, *Mayetiola* Kieffer, *Mikomyia* Kieffer, etc.) is rounded or has a small central notch. The last segment of the larvae of *Parepidosis* Kieffer is divided into two parts by a deep incision in the posterior margin along the middle line of the segment. As a result two, generally well-developed, conical processes are formed at the end of the segment (fig. 16,4). A terminal segment of this shape is also characteristic for the larvae of *Porricondyla* Rondani, *Dicroneurus* Kieffer, etc. The generally sturdy and variously shaped terminal papillae determine the shape of the last segment on whose end they are situated. Often the terminal papillae are just like the dorsal ones (for example the conical papillae of some species of *Peromyia* Kieffer, the bristle-like papillae of *Mikiola* Kieffer, *Rhabdophaga* Westwood, *Dasyneura* Rondani, *Rhizomyia* Kieffer, and many other genera). A whole group of naked larvae with weakly developed, simple papillae has well-pronounced bristle-like terminal papillae: *Miastor metraloas* Meinert, *Henria psalliotae* Wyatt, and *Peromyia* Kieffer. In some groups of larvae the terminal papillae have a different shape than the dorsal ones.

The terminal papillae of the larvae of *Micropteromyia ghilarovi* Mamaev (fig. 17,1) are represented by small protuberances, each with a single well-developed bristle at its tip. A part of the papillae on the larvae of *Loewiola* Kieffer, *Arthrocnodax* Rübsaamen, *Lestodiplosis* Kieffer, and *Aphidoletes* Kieffer has a similar shape. In many cases the bristle at the tip of the protuberance is short and broad, resembling a sclerotized spine (the larvae of *Tricholaba* Rübsaamen fig. 17,2). In the larvae of *Putoniella* Kieffer and *Macrodiplosis* Kieffer the papillae have at their tip partly a blunt sclerotized spine, partly a bristle. Sometimes the processes are elongated in which case they may have a considerable length and resemble sturdy conical processes. The terminal papillae of *Xenodiplosis* Kieffer and *Clinodiplosis* Kieffer etc. are of this type. The papillae on the larvae of *Harmandia* Kieffer look like large, broad protuberances with a sclerotized tip.

The terminal papillae of the larvae of *Henria* Wyatt and *Hygrodiplosis* Kieffer resemble nipples (fig. 17,3). The terminal papillae on the larvae of *Polystepha* Kieffer have the shape of broad, conical protuberances with small rods on their tip (fig. 17,4). The larvae of *Arthrocnodax* Rübsaamen carry at the end of the last segment two long, stick-like papillae with a broad protuberance at the base. Thus, they give an impression of two small protruding tubes. The terminal papillae of the larvae of *Trichodiplosis* Kieffer look like small, rounded processes with some long, bristle-shaped formations, which make a veritable tuft.

Finally, the last segment may be provided with small, conical, pointed (or broad and

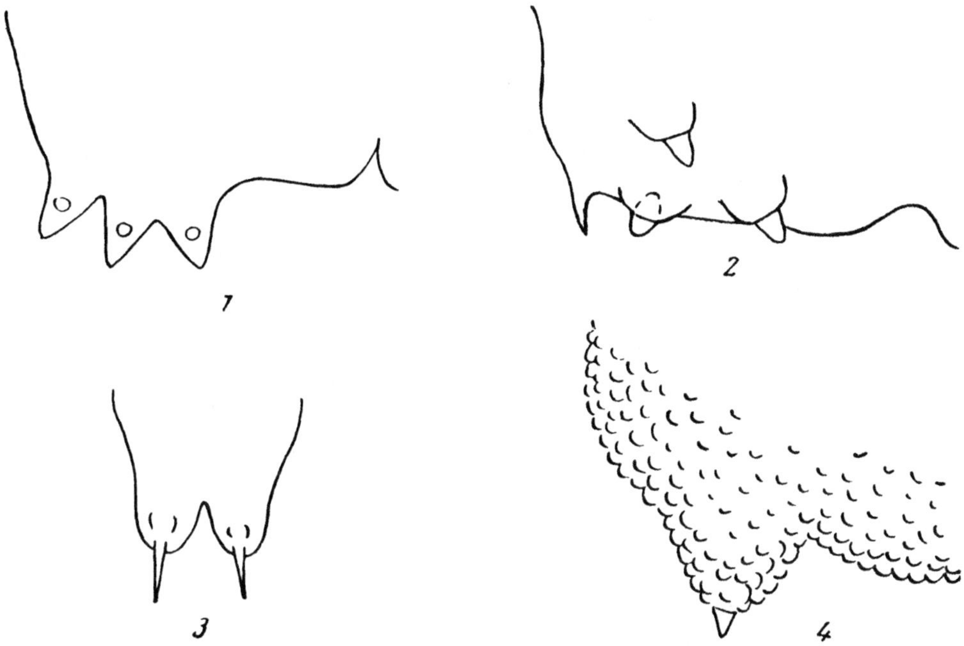

Figure 17. Terminal papillae
1 – *Micropteromyia ghilarovi* Mamaev; 2 – *Tricholaba trifolii* Rübsaamen; 3 – *Henria psalliotae* Wyatt; 4 – *Polystepha quercus* Kieffer (1-2, 4 – orig.; 3 – after Wyatt 1961)

blunt) sclerotized processes; a feature which is characteristic for the genera *Winnertzia* Rondani, *Rhipidoxylomyia* Mamaev, *Trichoxylomyia* Mamaev, etc. In the larvae of *Winnertzia* Rondani and *Camptomyia* Kieffer (fig. 16,1) these processes carry small, simple papillae, or bristles of various lengths. The shape and the location of these processes remind us of the papillae of many larvae of Cecidomyiini. It seems perfectly possible that the sclerotized processes on the end of the body of the larvae of *Winnertzia* Rondani and related genera are modifications of terminal papillae.

There is a definite link between the presence of the various cuticular macrostructures on the surface of the last segment and the degree of development of the terminal papillae. In smooth, naked larvae the terminal papillae or the supporting sclerotized processes replacing them tend to be highly developed. If the cuticular macrostructures are, however, massive, the terminal papillae are small and not strikingly different from the lateral ones in size.

All gall midge larvae do have stigmata on the prothorax (the anterior and only pair of thoracic stigmata) and on the first eight abdominal segments. Stigmata are absent from the meso- and metathoracic as well as from the terminal (ninth) abdominal segment. In the larvae of *Stelechodiplosis* Möhn, which develop in resin, the posterior pair of stigmata is located on strikingly lengthened processes of the second to last segment. In general, the stigmata are found on the lateral sides of the segments, only in some groups of larvae the posterior pair (and sometimes the anterior pair as well) has been shifted to the dorsal side of the segment. On by far most larvae belonging to the subfamily

Lestremiinae and to the tribe Porricondylini the stigmata are located on the lateral sides of the segments; a shift of the posterior pair to the dorsal side is, however, characteristic for some genera (*Peromyia* Kieffer, *Skuhraviana* Mamaev, *Holoneurus* Kieffer, *Parepidosis* Kieffer). On the larvae of the tribes Cecidomyiini and Oligotrophini, on the other hand, the posterior pair is located, as a rule, on the dorsal side of the eighth abdominal segment; only on the larvae of a few groups (*Tricholaba* Rübsaamen, *Profeltiella* Kieffer) is the posterior pair found on the lateral side.

Generally speaking, the stigmata are situated directly on the surface of the segments, only rarely is a part or all of the stigmata located on the end of processes of variable length and size. Among the porricondyline genera *Dicerura* Kieffer and *Ulmomyia* Mamaev the stigmata are located on special processes of the eighth abdominal segment. In the former the processes are very long, and considerably surpass the length of the anal segment (fig. 16,3). In the tribe Cecidomyiini the posterior pair, which is shifted far to the dorsal side of the segment, may be found on short processes (for instance in *Itonida pini* (De Geer)). In the larvae of *Gynapteromyia* Mamaev (Porricondylini) and *Acoenonia* Pritchard (Lestremiinae) all stigmata are situated on short, broad, more or less plate-like, lateral processes.

The shape of the stigmata is quite variable. Among the gall midge larvae, we can find all transitions from small oval protuberances to well-developed tubular cylindrical or conical formations. A considerable number of species belonging to the subfamily Lestremiinae and to the tribe Porricondylini has small stigmata which look like oval protuberances that are slightly elevated above the surface of the segment. In most cases all of them carry two to three respiratory openings. Some groups of larvae belonging to the subfamily Lestremiinae, like the genera *Acoenonia* Pritchard, *Peromyia* Kieffer (fig. 18,1), *Skuhraviana* Mamaev, and *Campylomyza* Meigen characteristically have sturdy stigmata which are shaped like short, cylindrical protuberances with a rounded tip. These stigmata are equal to or 1.5 times as long as their width at the base. Stigmata of the same type occur in *Dicerura* Kieffer, *Ulmomyia* Mamaev, *Gynapteromyia* Mamaev, *Holoneurus* Kieffer (Porricondylini) and, finally, in all larvae belonging to the Cecidomyiini and Oligotrophini that we know of. In comparison to these genera the stigmata of the larvae of *Itonida pini* (De Geer) (fig. 18,2) and of *Stelechodiplosis magna* Möhn (fig. 18,3) are

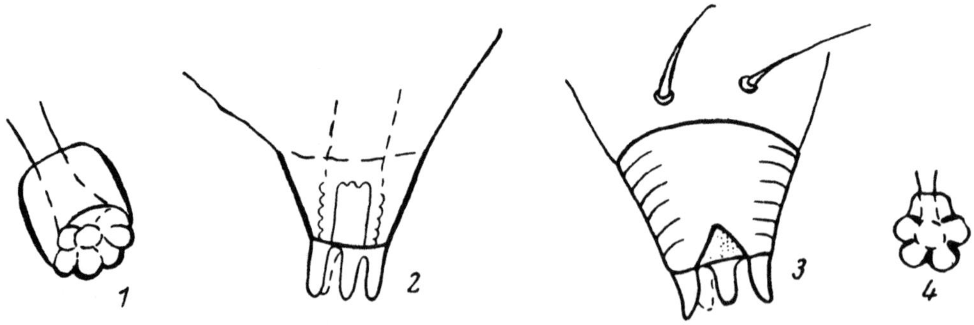


Figure 18. The stigmata on the eighth abdominal segment
1 – *Peromyia perpusilla* (Kieffer); 2 – *Itonida pini* (De Geer); 3 – *Stelechodiplosis magna* Möhn; 4 – *Trichoxylomyia pilosa* Mamaev (orig.)

longer, and their tip is incised, not rounded. The respiratory openings are in all cases located on the tip of the stigmata and are sometimes surrounded by short hairs.

A comparison of the stigmata of the various groups of larvae shows a definite interdependence between the shape, size, and distribution of the stigmata, and the thickness of the cuticle. All naked larvae, which do not have any massive cuticular formation, have very small stigmata which are located on the lateral sides of the segments. Such stigmata are characteristic for a part of the larvae of the subfamily Lestremiinae and the tribe Porricondylini, but they are not typical for the representatives of the tribes Cecidomyiini and Oligotrophini. Weakly developed stigmata are characteristic for some larvae living in very damp habitats (*Trichoxylomyia* Mamaev etc., fig. 18,4).

Larvae that are densely covered with various cuticular structures have sturdy stigmata. Moreover, the posterior pair is not located on the lateral sides of the segments, but has been shifted to the dorsal side or is found on the tip of processes of the eighth abdominal segment. Sturdy stigmata and various cuticular structures are especially characteristic for the larvae of the tribes Cecidomyiini and Oligotrophini, as well as for some Lestremiinae and Porricondylini.

In most cases the anal opening is slit-like and situated on the ventral side of the terminal segment (Porricondylini, Cecidomyiini, Oligotrophini). In the larvae of *Winnertzia* Rondani, *Camptomyia* Kieffer, *Trichoxylomyia* Mamaev, *Rhipidoxylomyia* Mamaev and some other Porricondylini the elongated-oval anal opening lies close to the anterior margin of the segment. The anal opening of the larvae of *Synepidosis* Mamaev is shifted to the end of the segment; though still on the ventral side, its posterior margin is not closed. The anal opening of some larvae belonging to the tribe Oligotrophini (*Dasyneura* Rondani, *Wachtliella* Rübsaamen, etc.) is found in the same place.

The anal opening is either smooth or accompanied by clearly distinct, longitudinal rows of tiny spinules. Only in the larvae of *Synepidosis* Mamaev do the spinules stand in transversal rows near the second half of the anal opening rather than in rows running lengthwise. In the larvae of *Porricondyla* Rondani and related genera the anal opening is completely covered by a dense tuft of hairs. A rounded anal opening located on the end of a simple, rounded terminal segment without distinct processes is characteristic for most Lestremiinae and for some species of the Porricondylini (fig. 16,5). The larvae of *Leptosyna* Kieffer, *Trichopteromyia* Williston, and *Micropteromyia* Mamaev are exceptional. Among these genera the end of the last segment is provided with various sclerotized processes, whereas the anal opening is shifted to the ventral side (fig. 16,6).

CHAPTER 3

# Collecting, preserving, and studying the material

A detailed investigation of any species of gall midges necessarily includes the study of the morphology of its larvae and their mode of life, and especially of the damage they cause. Larvae can only be successfully collected, in particular if we aim at the investigation of a whole complex of species living in a certain biotope, if we use multiple methods; for we must not only collect larvae from galls, but also from soil, litter, wood, and fungi.

At the very beginning the investigator encounters the following questions: How do we find gall midge larvae? How can we quickly tell them from the larvae of other Diptera? Their colour may help in recognizing gall midge larvae. The vast majority of dipteran larvae living in plant tissue, soil, and comparable substrates is white. Gall midge larvae, however, as a rule, are orange or bright yellow. White gall midge larvae are found considerably less often. For an accurate diagnosis of white larvae we should remember that gall midge larvae have neither a well-developed, large head (a feature that distinguishes them from most other nematoceran larvae) nor an oral-pharyngial apparatus formed by the integument of the anterior end of the head (as is characteristic for the larvae of the higher flies). With only a few exceptions gall midge larvae have on the ventral side of the thorax a special formation: the heavily sclerotized 'spatula' or 'breastbone'. This structure is not found in any other family of Diptera. Every entomologist acquainted with these characteristics and equipped with an average loupe will be able to immediately sort out his collected material on the spot.

We may recommend some elementary methods of observing and collecting gall midge larvae. The investigation of a region by means of an excursion aims at getting an impression of the habitats where gall midge larvae occur, and at revealing the damage done to plants. At the same time it serves to collect larval material which can be preserved in alcohol or some other fixative. During the excursion the presence of gall midge larvae should be verified in: galls, flowers, fruits, soil, litter, decaying wood, fruiting bodies of fungi, as well as in colonies of: plant lice, pine and spruce aphids, scales, spider mites, and galls of mites. In many cases it is possible, already during the excursion, to collect sufficiently large series of larvae of some species to render other collecting methods superfluous. This concerns, above all, larvae that develop within galls; first since galls are, as a rule, found in sufficient quantity; second since we may expect to find 10-20 larvae per gall.

A painstaking inspection of the plants making it impossible to miss some odd species, is a fundamental requisite for a thorough collection of phytophagous larvae. A fixed sequence is recommended for the inspection of trees and shrubs. We should first pay attention to the tips of the twigs. If these are affected, the young leaves do not unfold, they shrivel, or may be thickened, and there is often a rosette or spindle of leaves at the tip of the twig.

The inspection of the leaves comes next. We should open and examine folded leaf margins and fleshy galls which are developing on the leaf blades and stalks. Galls of plant mites are often found in large numbers on leaf blades. These galls, which present themselves as invaginations of the leaf blades, often have a fluted, hairy surface, contrary to the naked, smooth galls of gall midges. On the internal wall these galls form a special felt (erineum). Within mite galls larvae of gall midges, in particular of predatory species, may also be found. A method to collect these larvae will be described below. Gall midge larvae fairly often cause a pod-like folding of the leaf or the formation of a local, thick-walled, smooth invagination.

The inspection of the stems of twigs – the third stage of the examination – demands our special attention. Apart from inducing easily visible fleshy or blister-like swellings, gall midge larvae may also simply occur beneath the bark. The larvae of some species, for instance, live in, and cause the death of the cambial zone. Therefore, parts of a twig with dead or crackling bark should certainly be opened for inspection. Gall midges invade a stem where the bark has been damaged. When examining conifers it is necessary to painstakingly search for gall midge larvae between the needles, underneath the bark of young twigs, and in the resin oozing out of wounds. The inspection of trees and shrubs is completed by opening flowers and fruits. An efficient method for the collection of larvae from flowers and fruits will be described below.

Gall midge larvae that develop in herbs may also cause malformations of branch tips, leaf blades, stems, flowers, and fruits. They can therefore be collected during the examination, as recommended above. *Scirpus*, sedges, and grasses probably present a special case since gall midge larvae often develop within the leaf sheaths and hollow stems of these plants. It is, thus, absolutely inevitable to open stems and to examine leaf sheaths when inspecting plants of this group.

When collecting larvae from galls we should always be aware that larvae of two or even more species may develop together in a gall at the same time. One of these species is the host species, whereas the others are co-inhabitants (inquilines). Sometimes the inquilines keep on developing after the host larvae have abandoned the gall to pupate in the soil. We should, therefore, establish with certainty the actual gall maker among the various species living in a gall. If possible we should only use young, recently formed galls.

The collection of material in the field is probably the only way to procure the necessary series of larvae that develop in rotting wood. Since the breaking-up of wood by hand is very laborious, the ability to identify by external characteristics the parts of a decaying log or trunk that are inhabited by gall midge larvae is of extreme importance.

Already in the second year after felling, the trunk of a tree may be inhabited by gall midge larvae. They settle in the pores and underneath the bark at the butt-end. It is thus necessary to break apart the wood at the butt-end in order to establish the potential presence of gall midge larvae. When decay advances, the wood is destroyed by various fungi; it gradually becomes permeated with mycelium and loses its firmness. Gall midge larvae occur particularly under the bark and in the wood of trunks that show a pale coloured mouldering (as opposed to a brown coloured mouldering). Moreover, they are found only in wood that is not permeated with the mycelium of Polyporaceae and that is only weakly affected by other insect larvae. Gall midge larvae are concentrated near the cracks and clefts in the bark as well as near the butt-end. Thus, the mature larvae pupate in the immediate vicinity of the surface which facilitates the emergence of the adults. Gall midge larvae do not occur in porous, strongly decayed wood, and especially not in dried out wood and mould.

When examining stubs, our attention should be especially directed to the deep cracks and clefts that are caused by axe blows. The favourable dampness within the cracks allows the development of fungi on whose mycelium gall midge larve feed. When logs and stubs are considerably weakened, they are invaded by the larvae of some species that develop on the surface of remainders of wood. These larvae differ from the other gall midge larvae in their flattened, wide bodies.

The direct collection of gall midge larvae is often possible either when colonies of aphids are found, in which predatory larvae may develop en masse, or from samples of litter, where five to ten specimens per $dm^2$ may occur. If our examination of colonies of aphids, litter, etc. during an excursion has only yielded some odd larvae, we are forced to resort to other, more efficient methods of procuring large series of these larvae.

One of these methods – the use of transparent plastic boxes – is not only useful to collect larvae developing under conditions that are unfavourable for direct observation, but it can also be used to rear mature larvae from immature ones that during a preliminary inspection escaped observation. Thus, it is interesting to note that the observation of single, big larvae in an aphid colony indicates that the colony is more heavily infested with larvae of younger moults. If we put such colonies, together with the twigs on which they were found, for a few days in plastic boxes the young larvae will grow to maturity, thereby offering the opportunity to collect a large series of larvae of a certain species.

The use of plastic boxes is the only way to efficiently collect larvae that are endoparasitic in aphids. When examining a colony of lice a skilful investigator can see the larvae shining through the body integument of some individuals. If we place a colony that is infested with endoparasites in a plastic box the larvae will eventually leave their hosts and subsequently appear on the walls of the box, allowing their quick and efficient collection.

The tiny predatory larvae that live in the galls of plant mites can also be collected in sufficient quantity only by means of plastic boxes. To that end, leaves with mite galls are put into a box. After a few days the larvae, which by then have left the galls, may be collected from the walls of the box. Larvae which prey on colonies of scale insects, inside galls of pine and spruce aphids, etc. are collected in a similar way.

Flower inhabiting gall midge larvae may be collected directly by opening flowers. Collecting is, however, considerably more efficient when we make use of plastic boxes. Fairly often, the larvae of some (occasionally also rare) species develop together in the flowers of the same plant species, especially in the flower heads of composites. The use of plastic boxes doubtlessly allows the build-up of large collections of gall midge larvae from flowers with the least expenditure of time and effort. Larvae living in fruits may be collected in the same way.

The collector who puts the fruiting body of a fungus in a plastic box may well get a series comprising several hundreds of larvae. Gall midge larvae are, interestingly, only very rarely found in the fruiting bodies of Polyporaceae. Apart from economizing time and effort, the use of plastic boxes has the advantage of collecting fully mature larvae which have completed their development and which have left the flowers, fruits, fungi, etc. to pupate in the soil. If we open flowers, fruits, and galls in the field we often get in our collection tiny, immature larvae; a feature which subsequently hampers the study. Therefore, the use of plastic boxes is also recommended for the gathering of larvae from galls, but, of course, only in those cases where the larvae are known to pupate in the soil, and not inside their galls. In the latter case the opening of the galls can not be circumvented.

Gall midge larvae that develop in the soil and litter demand the most time-consuming and complicated method of collecting. As we already remarked, these larvae may be collected by manually searching through samples of soil and litter. In most cases, however, the low number of larvae renders this method inefficient. We recommend the drying of samples of the substrate in a set of thermo-eclectors as a reliable way of gathering sufficiently large series of larvae from soil and litter. This is a method used in paedo-zoological research which is concerned with the study of tiny arthropods. Experience has taught that in collecting the larvae from any single biotope it is necessary to use 40-50 funnels. Each funnel is made of tinplate or thick cardboard. It should measure 30-40 cm in height and 20-25 cm in diameter. In the upper third part of the funnel a gauze or small-mesh wire netting is fitted on which the sample is spread out. A test-tube is fastened to the neck of the funnel. Over each funnel a 15-40 watt light bulb is placed. The heat of the bulb gradually dries out the soil and litter samples. At first the larvae migrate to the lower layers of the sample, then they fall into the funnel and through it into the tube. The larvae that are thus collected belong, as a rule, to various species; therefore it may be desirable to sort them out prior to fixation.

Larvae that develop in fine-grained material (sandy soil, finely divided litter, stored flour, etc.) are conveniently collected by means of a number of soil sieves with various meshes. On sifting, the larvae will accumulate in one of the sieves from where they are easily collected.

In the funnel of the soil sieves a fraction of soil and litter particles containing gall midge larvae may also accumulate. The washing of soil samples being, however, both time-consuming and relatively inefficient we might consider using an eclector instead. For the study of paedogenic species we must necessarily have at our disposal larvae that will metamorphose into imagines, as well as reproducing larvae. During an excursion a collector generally gathers from rotting wood only paedogenic (i.e. parthenogenically reproducing) larvae. In order to breed imaginal larvae he should either keep the entire series in a plastic box for a period of time or resort to the breeding of paedogenic larvae on agar plates with the mycelium of various fungi (Nikolei 1961).

The simplest fixative for larvae is 70% alcohol. Larvae, fixed in alcohol swell slightly and their internal organs tend to become indistinct, to tear, and occasionally to macerate. Thus the use of material that has been fixed in alcohol is not feasible for anatomical and histological investigations. Such material is, however, perfectly suited to the study of cuticular structures, as is necessary for the identification of the larvae. The solutions of Carnoy and of Bouin are milder fixatives for gall midge larvae. These fixatives should therefore be chosen for the preparation of larvae for anatomo-histological investigations (for the composition of these fixatives see Romeys (1953) or any other histological textbook).

We used to prepare larvae for taxonomic studies as follows. After four to eight hours (depending on the size) the larvae were transferred from the 70% alcohol into a cold 10% solution of alkali. After the maceration of the intestines the larvae were taken from the alkali and cut open to remove the intestines. Then the envelopes of the larvae were washed in water and stained in a solution of fuchsin in water (30-60 min.). Other stains can be used as well. After staining the unused fuchsin was washed out with water and the larvae were transferred, first into 70% alcohol, and subsequently into absolute alcohol for no less than two hours.

After dehydration in absolute alcohol the larvae were placed onto a slide, washed in a few drops of xylol, and mounted in Canada or fir balsam. Three to four larvae were placed under one cover-glass, in such a way that the dorsal, lateral, and ventral sides of their bodies could be studied.

The scientific value of the material will only be guaranteed if all collections are accompanied by clearly written and accurate labels, and relevant notes in a diary. To achieve this it is necessary to observe the following rules. A serial number shall be assigned to each collection. On the label pertaining to the material that is to be fixed in alcohol the following information shall be indicated: the serial number, the date and the locality of collection (county, site), the biotope, the damaged plant species or habitat, and the surname of the collector. For example: No 5, 17 August 1961, Voronezhskaya oblast, Borisoglebsk, oak association, in pale mouldering oak wood, Yu. Ivanov. An identical label shall be fixed to the preparations mounted in balsam as well. An accurate description containing precise characteristics of the biotope, the habitat, the number, the colour, and the external appearance of the larvae must be entered in the diary under the same serial number. A label pertaining to the collection is also inserted into the diary.

Material of a poorly studied group, like the gall midges, is of great research value. We, therefore, emphatically recommend that the plant parts which have been damaged by gall midges be collected as herbarium specimens together with the collection of the larvae in alcohol. Moreover, it should be attempted to rear larvae in order to obtain the adult insects.

The simplest way to make a herbarium is the usual method of drying plants with galls. The keeping of galls in 70% alcohol or in 4% formaline (accompanied by a description of their original colour) is also recommended. Some manuals recommend the drying of plants in heated sand. This method, however, not only takes too much effort but, above all, it does not have any recognizable advantages over the usual way of drying. Plants with galls should be collected and dried with all organs needed for the subsequent identification of the plant species.

The rearing of adults from larvae does not only afford an opportunity to check on the correctness of the identification of the larvae, but it also allows the continual adding of new, reliably identified material to our gall midge collection. In general, rearing is only successful if the collected larvae are not too young. The experienced researcher can easily tell the mature larvae from the young ones, as the younger stage larvae are always pale and transparent, whereas mature larvae, as a result of the development of a fat-body, become opaque and are often brightly coloured.

Various methods may be recommended for the rearing of larvae, depending on their habitat. If their maturity is doubted, mycetophagous larvae must, after having been placed into a plastic box, be left for three to four days in the same substrate (fungus, wood, litter) in which they were found. Then they should be removed and put into a test-tube with sand. The test-tubes are prepared as follows. The preferably fine-grained quartz sand is first washed and sterilized. It is best sterilized by boiling in water. The sand should not be calcined as noxious oxides may then be formed. The washed, damp sand is poured into a test-tube with a diameter of two to three cm. and rammed down as tightly as possible. The sand should fill one third (3-4 cm) of the test-tube. It is not recommended to make the sand layer deeper, since then the pupae, unable to reach the surface, may perish. It is advisable to make a small hole in the surface with a needle or tweezers to prepare the way for the larvae digging into the sand. A digging larva makes its own burrow which often is subsequently used by the pupa on its way to the surface. Therefore, we should, as far as possible, avoid destroying these burrows when watering the tubes.

The sand in the tube should always be moderately damp. Dryness or, on the contrary, excessive moisture leads to the death of the larvae. If drying out, the sand should be moistened by adding drops of water along the wall of the tube. The tube is stoppered with cotton wool to guarantee ventilation. The tubes should not be stored in direct sun light, as this may rapidly lead to the death of the larvae by overheating.

It is also desirable to gather phytophagous larvae developing inside galls, flowers, fruits, or below bark together with the affected twigs, and to keep all of it in a plastic box for a few days before putting the larvae into a test-tube with sand. Mature larvae can be easily collected from the walls of the box. In order to get imagines from larvae that pupate within their galls it suffices to gather the plant material with the galls and keep it in a glass container with water until, on opening of the galls, pupae are found. Then the galls are placed into a test-tube or plastic box, until the adults hatch. In case a mould develops on the galls they must be slightly dried.

The larvae of predatory gall midges living in colonies of aphids, scale insects, within mite galls, etc. also easily reach maturity if they are kept together with their host animals in plastic boxes. A fixation procedure for adult gall midges and a method to make preparations of them have been previously described (Mamaev 1962).

CHAPTER 4

# Systematic outline of the gall midge larvae

## Differences between the larvae of gall midges and the larvae of related nematoceran families

Quite often, tiny larvae of other Nematocera that are not readily distinguished from gall midge larvae are found in soil, decaying wood, and fruiting bodies of fungi. More rarely such larvae are also found together with gall midge larvae in galls. In order to distinguish these larvae a short key to nematoceran larvae has been included. A complete key by Krivosheina was published in 1964.

## Key to the larvae of the gall midges and the larvae that morphologically resemble them

1(6) Head capsule absent or very small, weakly sclerotized, semi-transparent, generally with slender sclerotized rods extending beyond the head capsule. In case of a small head capsule without clear rods extending beyond the capsule, there is a well-developed special segment connecting the head with the thoracic segments. Larvae peri- or apneustic.

2(3) Larvae apneustic. Body naked, worm-shaped, without any cuticular formations. Head capsule with, in its central part, a group of crest-shaped sclerites constituting the mouth-throat apparatus. Larvae living in damp river sandbanks. **Leptoconopidae**

3(2) Larvae peripneustic. Body generally provided with various cuticular structures, particularly on the terminal segment. Mouth-throat apparatus absent.

4(5) Head capsule, tentorial arms, and mouth apparatus completely reduced. Body clearly consisting of twelve segments. . . . . . . . . . . . . . . . . . **Hyperoscelidae**

5(4) Head capsule, tentorial arms, and mouth apparatus present. Body clearly consisting of thirteen segments (a supernumerary segment is intercalated between head and thoracic segments). . . . . . . . . . . . . . . . . . . . . . . . . . . . . **Cecidomyiidae**

6(1) Head completely developed or with partly reduced ventral side; heavily sclerotized. Sclerotized rods not extending beyond the head capsule. Body consisting of three thoracic and ten abdominal segments or secondarily divided (Ryphidae). If prothorax secondarily divided (Bibionidae), then metathoracic stigmata present. Larvae holopneustic. . . . . . . . . . . . . . . . . . . . . . . . . . **Bibionidae, Scatopsidae, Fungivoridae, Ceratopogonidae, Chironomidae, Psychodidae, Rhyphidae**, etc.

## Division of the gall midge family into subfamilies

Up to the present no consensus on the system of the gall midges has yet been reached. The specialists have divided the family into three subfamilies: Lestremiinae, Heteropezinae, and Cecidomyiinae (Kieffer 1913, Felt 1925, Rübsaamen and Hedicke 1926-1938). Some leading systematical dipterologists consider these as three independent families, which are united into a superfamily (Hennig 1954, Rohdendorf 1964).

Comparatively recently the German entomologist Möhn (1955b) proposed to raise the tribe Porricondylini to the independent subfamily Porricondylinae. Pritchard (1960) showed the inappropriateness of the independent subfamily Heteropezinae; but he committed some errors in defining the systematic position of the groups that belonged there (Mamaev 1964b). In view of the fact that the question of the correct taxonomic rank of the various groups of gall midges is at present being thoroughly discussed, the authors think it appropriate, awaiting the accumulation of conclusive material, to maintain the division of the Cecidomyiidae into two subfamilies: Lestremiinae and Cecidomyiinae (= Itonidinae).

## Key to the subfamilies of Cecidomyiidae

1(2) Anal opening rounded, located at the end of the terminal segment. Head capsule simple, or with thin tentorial arms, never with long and robust exocephalic rods. If the anal opening is shifted to the ventral side of the segment (*Mycopriona* Mamaev, *Micropteromyia* Mamaev, *Leptosyna* Kieffer, *Trichopteromyia* Williston), then cuticular supporting hooks are developed at the end of the body, spinule fields at the dorsal side of the body absent, papillae simple. . . . . . . . **Lestremiinae** (p. 80)

2(1) Anal opening slit-like, always at the ventral side of the terminal segment. Head, as a rule, with robust exocephalic rods, occasionally simple. If cuticular hooks are present at the end of the body, then spinule fields are present at the ventral as well as at the dorsal side of the body, or the papillae are setiform (*Solntsevia* Mamaev). **Cecidomyiinae** (p. 125)

## SUBFAMILY LESTREMIINAE

The larvae belonging to this subfamily vary strikingly in both shape and appearance. Their body may be elongated-oval as well as widened; it may be bare or, on the contrary, densely covered with various protuberances and outgrowths. The head capsule may be conspicuously sclerotized and provided with well-developed tentorial arms, or weakly sclerotized and devoid of tentorial arms. The anal opening is, as a rule, located at the end of the terminal segment, where it may be found either directly on the surface or on a special, slightly retractible tube. If the end of the ninth abdominal segment bears supporting structures or large composite terminal papillae the anal pore is shifted to the ventral side (*Trichopteromyia* Williston).

The larvae of this subfamily may be divided into several groups according to the particularities of the structure of the head capsule and the covering of the body. We place the larvae of the genus *Lestremia* Macquart into the first group. Their head capsule is very short, semi-spherical, heavily sclerotized, and dark brown. Two short, thin, heavily sclerotized rods, which apparently represent the posterior tentorial arms found in the larvae of the other groups, originate on the posterior margin of the ventral side of the capsule. The anterior tentorial arms are not conspicuous. The antennae are well-developed, long, and strongly widened at the base – strikingly narrowed and acuminate at the tip. The one-toothed spatula has a strikingly widened, spindle-shaped head.

The body of the larvae is bare, all papillae are simple. The distribution of the papillae is only specific for the type of larvae under discussion. The dorsal papillae stand in a single transversal row of six papillae, whereas another two simple papillae are located closer to the anterior margin of the segment. On the eighth abdominal segment only two simple papillae are found near the stigmata. On the ventral side of the first seven abdominal segments eight simple ventral papillae are found. The two central ones are shifted in the direction of the anterior margin of the segment, whereas the other six make a single transversal row. Compact spinule fields without bare oval spots occur on the ventral side of the metathoracic and all abdominal segments. The broadly rounded terminal segment bears no appendages. The posterior pair of stigmata is located on the latero-posterior parts of the penultimate segment and is only slightly shifted to the dorsal side.

The larvae of the genus *Acoenonia* Pritchard form the second group. Their body is covered with specific cuticular structures, whereas spinule fields are absent. This combination is not found in any other group of larvae belonging to the subfamly in question. The platelets that densely cover the body at the dorsal as well as at the ventral side bear one to three teeth in various degrees of development. The papillae are complex: verruca- or bristle-like. The intercalary papillae are setiform and occur in twos, instead of in threes as is generally the case. An additional characteristic feature of the investigated larvae is the increase of the number of ventral papillae on the first seven to ten abdominal segments; besides the four anterior papillae six posterior ones occur.

The posterior pair of stigmata is found on small, broad processes of the eighth abdominal segment.

The heavily sclerotized head capsule bears at its latero-posterior parts two small, thick processes. There are no clearly pronounced anterior tentorial arms.

The clearly defined third group consists of the representatives of the genera *Peromyia* Kieffer, *Skuhraviana* Mamaev, and *Bryomyia* Kieffer. Their elongated head capsule resembles a truncated cone. The length surpasses, as a rule, the width at the base. The capsule is relatively well-sclerotized and completely developed at the dorsal as well as at the ventral side. The rounded posterior margin is clearly defined. Well-pronounced unsclerotized spots surround the bases of the antennae. We have not found anterior tentorial arms in our preparations. The posterior arms, which become gradually narrowed from the base to the tip, originate on the latero-posterior margins of the head capsule.

In the larvae of *Peromyia* Kieffer the arms gradually narrow into a free ending needle-shaped, hardly visible tip. Their length does not, or only slightly, surpass the width of the head capsule. The posterior tentorial arms of the larvae of *Skuhraviana* Mamaev are rather pronounced. Their length is equal to or even somewhat longer than the width of the head capsule. The posterior ends of the tentorial arms are fused with a fairly well-developed cross bar.

The anterior half of the spatula, which is in all cases well-developed, is three-toothed; its base is often inconspicuous.

The scheme of the distribution of the papillae departs somewhat from the rule. The thoracic and the first seven abdominal segments each bear, as always, six dorsal papillae. On the eighth abdominal segment only two conspicuous dorsal papillae, situated between the stigmata, are found. The other two papillae are very small and located very closely to the stigmata, so as to be nearly fused with them. At least in *Peromyia* Kieffer the three lateral papillae at each side of the segment are prominent. The thoracic segments each bear at the ventral side two sternal, eight doubled intercalary, and two sublateral papillae. The first seven abdominal segments are each provided with two (*Peromyia* Kieffer) or four (*Skuhraviana* Mamaev) anterior ventral papillae and with a row of six posterior ones. On the eighth segment the usual four simple ventral papillae occur. The dorsal and a part of the lateral papillae are conical, rounded, or sometimes verruca-like outgrowths with a sort bristle at the tip. In some species of *Peromyia* Kieffer the two central dorsal papillae are fused to form a robust conical outgrowth that is quite conspicuous among the other papillae. The papillae at the ventral side of the body are simple. Only the doubled intercalary papillae bear a short bristle. The anterior ventral papillae are located behind spinule fields.

The body is not smooth. Its dorsal and lateral surfaces are densely covered with various pointed protuberances, oval platelets, rounded verrucae, hair- and bristle-shaped formations. The metathoracic and all abdominal segments bear on the anterior third part of the ventral side dense fields of tiny spinules, as is the case in many other genera of this subfamily. The terminal segment has a truncated conical shape and is, as well as the other segments, densely covered with various cuticular formations. The anal slit, which is located at the very end of the body, is not very conspicuous.

The well-developed stigmata look like rather large, sclerotized verrucae. The first eight pairs are found directly along the lateral sides of the segments. They are mostly situated near the middle of the segment, or on its anterior half. The posterior pair is considerably larger than the other ones. It is shifted far to the dorsal side and does not lie at the same level as the other stigmata. The other investigated groups of larvae of the Lestremiinae characteristically have a bare, white body devoid of all cuticular formations and complex papillae. They only differ in the structure of the head capsule and in the spinule fields.

The fourth, large, apparently heterogeneous group comprises the larvae of the genera *Campylomyza* Meigen, *Aprionus* Kieffer, and *Cordylomyia* Felt. These can be linked up with many of the characteristics of the larvae of *Monardia* Kieffer and *Xylopriona* Kieffer. A well-developed head capsule combined with fairly conspicuous anterior and posterior tentorial arms is characteristic for this group. The anterior tentorial arms are developed to various degrees: they are well-pronounced, for example, in the larvae of *Aprionus* Kieffer and *Campylomyza* Meigen. The posterior tentorial arms are also comparatively slender, though twice as thick as the anterior ones. The larvae of *Campylomyza* Meigen have very short posterior arms. Their length does not surpass half the width of the head capsule. The posterior arms of the larvae of *Aprionus* Kieffer and *Monardia* Kieffer, on the contrary, are long, being no shorter than the width of the head capsule at its base. The posterior ends of the anterior and posterior tentorial arms of each side are connected by a bow-shaped bar, which is as thick as the arms are. In the larvae of *Xylopriona* Kieffer the cross bar is somewhat more weakly developed than in the larvae of the other genera.

The spatula is well-developed in this group. It is three-toothed (toothless only in *Campylomyza*), with a massive rhomboid head. The scheme of the distribution of the papillae is typical for the family and shows the usual set of dorsal and lateral papillae. The anterior and posterior ventral papillae make two rows on the abdominal segments; the anterior ones are found behind the spinule fields. The papillae, including the four ventral papillae on the eighth abdominal segment, are on all segments simple and without bristles. Only the doubled intercalary papillae bear the usual very short bristles.

The dorsal and lateral sides of the body are smooth and devoid of cuticular formations. As usual, spinule fields are found on the anterior third part of the ventral side of the segments. All fields are compact, without bare oval spots. The larvae of *Campylomyza* Meigen and *Aprionus* Kieffer bear spinule fields on the metathoracic and all abdominal segments. The larvae of *Monardia* Kieffer and *Xylopriona* Kieffer do not have spinule fields on the two last segments. All larvae of this group have an elongated-oval, broadly rounded terminal segment without conspicuous outgrowths and appendages, or they have small bumps near the latero-posterior corners of the segment. The anal opening of the larvae of *Xylopriona* Kieffer, *Aprionus* Kieffer, and *Campylomyza* Meigen lies at the end of a well-pronounced, slightly retractable, tubular formation.

The stigmata are of the usual type and look like very small sclerotized verrucae. All stigmata are located along the lateral sides of the segments. The first eight pairs are found on the anterior half of the segments, the last pair, though at the same level as the others, is situated on the posterior half.

The last, and by far the largest, heterogeneous group is composed of the larvae of the genera *Mycopriona* Mamaev, *Trichopteromyia* Williston, *Leptosyna* Kieffer, *Micropteromyia* Mamaev, *Miastor* Meinert, *Henria* Wyatt, *Brittenia* Edwards, *Heteropeza* Winnertz, *Heteropezina* Pritchard.

The head capsule of all larvae belonging to this group is well-developed, rather heavily sclerotized, elongated-oval or nearly square and provided with a clearly defined straight or slightly bent posterior margin. The internal posterior arms are not pronounced. The larvae of *Mycopriona* Mamaev and *Trichopteromyia* Williston have a head capsule with very small and short arms whose position and orientation are comparable with those of the posterior tentorial arms in *Aprionus* Kieffer, *Monardia* Kieffer, etc. The head capsule of the representatives of the other genera of this group characteristically bear no trace of arms. The head capsule of the larvae of *Miastor* Meinert, for example, is rather massive and broad, its width being equal to or slightly surpassing its length. A well-pronounced, clearly sclerotized ledge that superficially resembles the similar structure in the larvae of the Culicidae runs along its posterior margin.

The shape of the one- or three-toothed spatula is extremely variable; in some cases the spatula is not conspicuous.

The distribution of the papillae follows the scheme that is typical for the Lestremiinae. There is a complex set of intercalary papillae, each group of wich is made up of three papillae. The first seven abdominal segments each bear four anterior as well as four posterior ventral papillae. On the eighth abdominal segment the usual set consisting of four papillae on the dorsal as well as on the ventral side is found. All papillae are simple, without bristles, with the restriction that two out of every three intercalary papillae are provided with a short bristle. On all abdominal segments the anterior ventral papillae stand in a single row behind the spinule fields. The anterior ventral papillae are situated within the spinule fields only in the larvae of *Henria* Wyatt.

The dorsal and lateral sides of the body are smooth, without any cuticular formations. In the majority of the larvae spinule fields are only found on the ventral side of the body. The larvae of *Miastor* Meinert and *Brittenia* Edwards are the only ones that bear spinule fields on the dorsal side as well. In the larvae of *Mycopriona* Mamaev, *Trichopteromyia* Williston, and *Henria* Wyatt spinule fields are found on the metathoracic and the first eight abdominal segments, whereas the spinule fields on the eighth abdominal segment are lacking in *Micropteromyia* Mamaev, *Heteropeza* Winnertz, *Leptosyna* Kieffer, and *Brittenia* Edwards. In *Mycopriona* Mamaev, *Trichopteromyia* Williston, *Leptosyna* Kieffer, *Brittenia* Edwards, and *Heteropeza* Winnertz all spinule fields are compact; in *Micropteromyia* Mamaev, *Miastor* Meinert, and *Henria* Wyatt they contain well-defined oval spots.

The terminal segment of the larvae of *Mycopriona* Mamaev, *Trichopteromyia* Williston, and *Leptosyna* Kieffer bears clearly pronounced, heavily sclerotized appendages. *Micropteromyia* Mamaev has two unsclerotized outgrowths, which bear three pointed, conical protuberances at their tip. In *Brittenia* Edwards the terminal segment has well-developed conical papillae on the end. The terminal segment of the other larvae is rounded or bears weakly pronounced plate-like appendages. In conformity with the rest of the body the terminal segment is bare, without any processes or other formations on its surface.

The small stigmata are located at the sides of the segments. The well-pronounced posterior pair is found on the posterior third part of the penultimate segment, where it is situated at the same level as the other pairs.

Key to the genera of Lestremiinae

1(2) Spatula well-developed, three-toothed, middle tooth clearly shorter than the marginal ones. . . . . . . . . . . . . . . . . . . . . . . . . *Catocha* Haliday (p. 88)

2(1) Spatula without teeth or one-toothed; if three-toothed, then middle tooth longer than the marginal ones. Spatula sometimes inconspicuous.

3(10) Body, at least dorsally, covered with platelets, bearing rounded or pointed tooth-like protuberances (fig. 19,3; 23,6). Dorsal and lateral papillae (sometimes the ventral ones as well) complex: conical, knob-, bristle-like, etc. (fig. 24,2). Spinule fields present or absent. Anterior and posterior tentorial arms protruding beyond the head capsule (fig. 27,1), or small rod-like processes present; head capsule never simple. Spatula three-toothed.

4(5) Body dorsally and ventrally densely covered with sclerotized, one- to two-toothed platelets (fig. 19,3), spinule fields absent. Posterior pair of stigmata located on small outgrowths of the latero-posterior parts of the penultimate segment (fig. 19,5), not shifted to the dorsal side. Antennae (fig. 19,1) well-developed, cylindrical. Head capsule rounded-oval with short conical processes. Terminal segment forked, ending in two broad, conical outgrowths.
*Acoenonia* Pritchard (p. 89)

5(4) Body dorsally covered with platelets which bear variously shaped protuberances (fig. 23,6). Well-defined spinule fields present on the ventral side. Posterior pair of stigmata located directly on the surface of the segment, shifted to the dorsal side (fig. 20,1). Antennae well-developed, elongated, gradually tapering off to the tip (fig. 21,1). Head capsule triangular, its anterior part truncate; posterior tentorial arms thin and long, with ends either free or connected by a cross bar.

6(7) End of the posterior tentorial arms very thin, hair-like, not connected by a cross bar (fig. 21,1). Arms as long as or longer than the basal width of the head capsule. On the ventral side of the first seven abdominal segments only two anterior ventral papillae behind the spinule fields and a single cross row of six posterior ventral papillae present. . . . . . . . . . . . . . . . *Peromyia* Kieffer (p. 91)

7(6) Posterior tentorial arms well-developed and along their whole length heavily sclerotized, the ends connected by a bow-shaped cross bar (fig. 27,1). Length of the arms equal to or less than the basal width of the head capsule. Ventral side of the first seven abdominal segments bearing four anterior papillae behind the spinule fields and a single row of four posterior ventral papillae.

8(9) Posterior tentorial arms comparatively long, as long as the basal width of the head capsule. The central two ventral papillae removed from the spinule fields, the two lateral ones close to the spinule fields. Distance between the central papillae more than twice the distance between the papillae and the spinule fields. Dorsal and lateral surfaces of the segments (protuberances and papillae not counted) bare or provided with sparse pointed spinules, in particular on the central parts (fig. 27,3). The large protuberances and papillae covered with flattened scale-, hair-, or spine-like formations. The distance between the central dorsal papillae not more than half the width of the papillae themselves (fig. 27,3). . . . . . . . . . . . . . . . . . . . . . *Skuhraviana* Mamaev (p. 101)

9(8) Posterior tentorial arms comparatively short, as long as the basal width of the head capsule. Both pairs of anterior ventral papillae removed from the spinule fields. Distance between the central papillae equal to the distance between the papillae and the spinule fields, surpassing the distance between the exterior papillae and the spinule fields by no more than one third. Dorsal and lateral sides of the whole body, including protuberances and papillae, densely covered with comparatively long, thin hairs (fig. 28,2). Distance between the central dorsal papillae equal to or somewhat larger than the width of the papillae themselves (fig. 28,2). Head of the spatula well-pronounced; manubrium short, as long as the head (fig. 28,4). . . . . . . . . . . . . . . . . . . . *Bryomyia* Kieffer (p. 105)

10(3) Body naked, dorsal and ventral sides without sclerotized platelets, protuberances, or other cuticular formations. Dorsal, lateral, and most ventral papillae simple. Spinule fields always well-developed. Head capsule simple, without conspicuous arms or provided with well-developed anterior and posterior tentorial arms extending beyond the capsule.

11(12) Head capsule semi-spherical, clearly shorter than its basal width (fig. 29,2). Posterior ventral margin of head capsule with two short, heavily sclerotized, free-ending arms. Second segment of antennae strongly tapering from the base to the acuminate tip (fig. 29,1). Head of the spatula strongly laterally stretched, with a single central, conical tooth (fig. 29,3). Spinule fields present on the ventral sides of the metathoracic and all abdominal segments.
*Lestremia* Macquart (p. 107)

12(11) Head capsule, as a rule, elongated-oval, simple, without arms, or with well-developed anterior and posterior tentorial arms that extend beyond the capsule and whose ends are connected by a short cross bar. Second antennal segment cylindrical or triangular, but never with an acuminate tip (fig. 30,2,4). Spatula variously shaped or absent. Spinule fields variously distributed.

13(20) Anterior and posterior tentorial arms long and thin, extending far beyond the head capsule (fig. 29,5). Spinule fields present on the ventral side of the metathoracic and seven or nine abdominal segments.

14(17) Spinule fields present on the ventral side of the metathoracic and all abdominal segments.

15(16) Posterior tentorial arms relatively short, as long as or only slightly longer than half the basal width of the head capsule (fig. 29,5). Spatula well-developed, heavily sclerotized; its head rounded, without teeth at the tip (fig. 11,2). Second antennal segment nearly triangular, gradually tapering from the base tot the tip (fig. 29,5). . . . . . . . . . . . . . . . . . . . . . . *Campylomyza* Meigen (p. 109)

16(15) Posterior tentorial arms relatively long, as long as or longer than the basal width of the head capsule (fig. 6,4). Spatula generally well-developed, heavily sclerotized, three-toothed (fig. 30,1,3). Second antennal segment elongated-oval, with rounded tip (fig. 6,4). . . . . . . . . . . . . . . . . . . *Aprionus* Kieffer* (p. 110)

17(14) Spinule fields only present on the ventral side of the metathoracic and the first seven abdominal segments.

18(19) Tip of the terminal segment with two conspicuous outgrowths, each bearing four simple papillae (fig. 30,7). Dorsal side of the head capsule with two small, pointed protuberances on the posterior margin (fig. 6,5). Only posterior tentorial arms well-pronounced (fig. 6,5). Spatula with three well-developed, pointed teeth (fig. 30,6). . . . . . . . . . . . . . . . . . . *Monardia* Kieffer (p. 113)

19(18) Terminal segment without conspicuous outgrowths (fig. 31,3). Posterior margin of the dorsal side of the head capsule smooth (fig. 31,1). Anterior as well as posterior tentorial arms well-developed. Spatula with three well-developed, rounded teeth (fig. 31,2). . . . . . . . . . . . . . . . *Xylopriona* Kieffer (p. 114)

20(13) Head capsule simple, no conspicuous long and thin anterior and posterior tentorial arms extending beyond the capsule (fig. 32,3). Spinule fields present on the ventral side of the metathoracic and seven or eight abdominal segments, in a few cases also on the dorsal side of some segments.

21(24) Spinule fields present on the metathoracic and the first eight abdominal segments; the field on the eighth segment consisting of no less than 20 rows of spinules. End of the terminal segment provided with a sclerotized part bearing dark hooks (fig. 32,2).

22(23) Terminal segment strongly narrowed as compared to the penultimate one (slightly more than half as wide and about two thirds as long as the penultimate segment) (fig. 32,2). On the end of the terminal segment a simple sclerotized part bearing two heavily sclerotized, bent hooks (fig. 32,2); distance between the hooks about equal to the width of the hooks. Width of the segments, from the

*According to their descriptions the larvae of *Mycophila* Felt and *Tekomyia* Möhn cannot be distinguished from those of *Aprionus* Kieffer. The more heavily sclerotized spatula distinguishes the larvae of *Cordylomyia* Felt from those of *Aprionus* Kieffer (Hennig 1948).

mesothoracic segment onward, twice (the terminal segment 1.5 times) as wide as long. Spatula well-pronounced (fig. 32,1). . . . . *Mycopriona* Mamaev (p. 116)

23(22) Terminal segment short, about half as long and as wide as the penultimate segment (fig. 16,6). The end of the terminal segment bearing two sturdy, sclerotized, straight, apically rounded teeth (fig. 16,6); dorsally to these teeth three relatively short, also heavily sclerotized, protuberances are found. Protuberances located on a sclerotized plate on which three more weakly sclerotized internal processes originate. Bases of the sturdy teeth touching one another. Segments, at least from the third abdominal segment onward, clearly longer than wide. Penultimate segment about 1.5 times as long as wide. Spatula not pronounced. . . . . . . . . . . . . . . . . . . . . *Trichopteromyia* Williston (p. 118)

24(21) Spinule fields present on the metathoracic and the first seven abdominal segments (if a spinule field is found on the eighth abdominal segment then it consists of only a few short rows). Tip of the terminal segment (not counting *Leptosyna* Kieffer), generally, rounded, without any protuberances or outgrowths (fig. 32,6).

25(26) Segments comparatively wide. Eighth abdominal segment two to three times as wide as long. The relatively robust, wide manubrium of the three-toothed spatula considerably thicker than the middle tooth (fig. 32,4). Posterior margin of the terminal segment bearing two short, broad outgrowths, each provided with three conical papillae (fig. 32,5). . . . . . . . . . . *Micropteromyia* Mamaev (p. 119)

26(25) Segments comparatively long and narrow. Eighth abdominal segment as long as or longer than wide. Spatula not pronounced, if well-developed, then one-toothed. Terminal segment variously shaped.

27(30) Spinule fields present on the ventral as well as on the dorsal side of the segments. Anterior ventral papillae located behind the spinule fields. First two or three spinule rows not separated from the other rows.

28(29) Spinule fields on the dorsal as well as on the ventral side of the segments with bare oval spots. Spinule fields occur on the dorsal side of the metathoracic and the first seven, sometimes eight, abdominal segments. Spinule fields on the ventral side of the metathoracic and the first seven abdominal segments well-developed, but on the eighth abdominal segment consisting of only a few rows of spinules. Terminal segment provided with small, spade-like, bristle bearing appendages (fig. 32,6). Imaginal larvae provided with a one-toothed spatula (fig. 32,7). . . . . . . . . . . . . . . . . . . . . . . . . *Miastor* Meinert (p. 120)

29(28) Spinule fields on the dorsal as well as on the ventral side of the segments compact, without bare oval spots. Spinule fields located on the dorsal side of the metathoracic and the first four abdominal segments, as well as on the ventral side of the metathoracic and the first seven abdominal segments. End of the terminal segment bearing four well-developed knobs with a bristle at the tip (fig. 33,4). *Brittenia* Edwards (p. 121)

30(27) Spinule fields only present on the ventral side of the segments. Anterior ventral papillae located behind or within the anterior spinule fields. Anterior two or three rows of spinules separated from the other rows by a narrow, spinule-free strip.

31(32) Spinule fields on the first seven abdominal segments with eight bare, oval spots (fig. 33,5). Spinule fields on the other segments compact. Anterior ventral

papillae of the first seven abdominal segments located within the spinule fields. Terminal segment provided with two small, unsclerotized, bristle bearing outgrowths, the tip of which is split into two parts (fig. 33,6).
*Henria* Wyatt (p. 123)

32(31) All spinule fields compact, without clear-cut bare oval spots. A single row of anterior ventral papillae behind the spinule fields. Terminal segment (fig. 33,7) rounded or its tip provided with well-developed teeth.

33(34) Cuticle of the first five abdominal segments transversely striated. Tip of the terminal segment provided with two well-developed, dorsally bent hooks (fig. 33,7) in front of which two short, blunt processes are located. Two heavily sclerotized platelets, the pointed tip of each is directed to the anterior margin of the segment, are found ventrally to the base of the hooks.
*Leptosyna* Kieffer (p. 124)

34(33) Segments not transversely striated. Terminal segment rounded, without sclerotized hooks at the tip.

35(36) Spinule fields consisting of ten rows on the metathoracic, and of 14-18 rows (including three separate anterior rows) on the abdominal segments. All spinules of the same size. . . . . . . . . . . . . . . . . . . . *Heteropeza* Winnertz (p. 124)

36(35) Spinule fields consisting of six rows on the metathoracic, of eight rows on the first, and of thirteen rows on the other six abdominal segments. Spinules of the last three rows much smaller than those of the preceding rows.
*Heteropezina* Pritchard (p. 125)

## 1. Genus **Catocha** Haliday

*Morphology*. The second antennal segment is as wide as long.

The spatula is well-pronounced and three-toothed. Its lateral teeth are divergent and considerably longer than the central one.

The yellow-orange body is bare and cylindrical.

The papillae are, apart from some lateral ones, simple. The thoracic and the first seven abdominal segments each bear four (the eighth abdominal segment only two) dorsal papillae. The four anterior ventral papillae are located within spinule fields. Two out of the three lateral papillae are provided with short bristles.

Spinule fields are found on the ventral side of the metathoracic and all abdominal segments. The spinules are setiform. Ten to fifteen rows of very small spinules are located on the posterior half of the last two metathoracic and the first seven abdominal segments.

The terminal segment is tubular. A row of four papillae occurs on the dorsal side in front of its middle. Another row of four papillae is found on the tip of the body. Near the middle of the ventral side six papillae are found.

*Biology*. The mycetophagous larvae live in moss and forest litter.

### **Catocha muscicola** Kieffer

*Morphology*. The spinule fields consist of 38-40 rows of spinules. On the first seven abdominal segments they contain eight oval, bare spots.

*Biology*. The larvae are found in moss.

*Comment*. The description of the larvae of *C. muscicola* Kieffer, as used by the authors, has been given by Kieffer (1913). Edwards (1938) considers this species synonymous with *C. latipes* Haliday.

2. Genus **Acoenonia** Pritchard (fig. 19)

*Morphology*. The larvae of this genus are dirty white, short and broad (length-width ratio = 2.5:1). They bear two pointed processes on the terminal segment.

The head capsule is small and short. Its length is equal to or slightly surpassing the width at the base (fig. 19,1). Short and thick outgrowths originate on the ventral margin of the capsule. The tentorial arms are not pronounced. The antennae are well-developed.

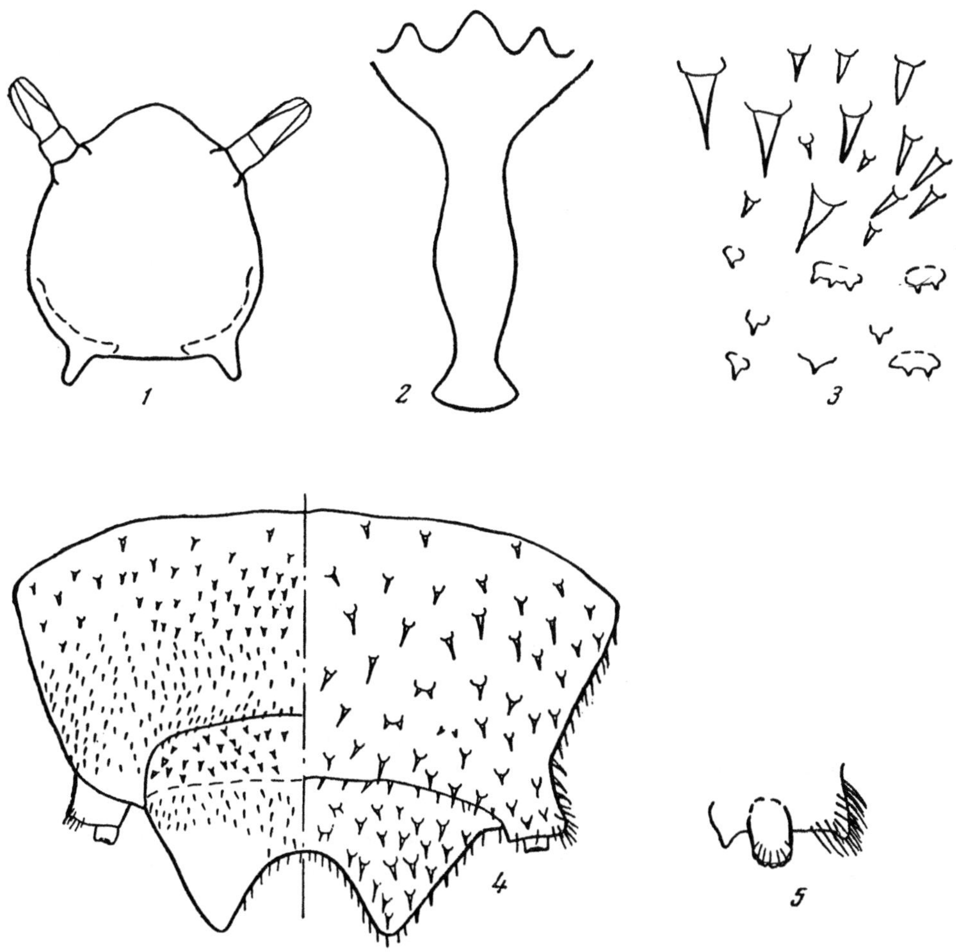


Figure 19. Larva of *Acoenonia europaea* Mamaev
1 – head; 2 – spatula; 3 – third abdominal segment (dorsal view), cuticular structures; 4 – last two segments, dorsal (right) and ventral (left) view; 5 – eighth abdominal segment, stigmata (orig.)

The first segment is only slightly wider than long. The second segment is 3-3.5 times as long as its base is wide. Its upper half is somewhat broadened.

The spatula is three-toothed and has a well-developed manubrium (fig. 19,2).

The body is densely covered with variously shaped platelets and bristle-like formations (fig. 19,3). The knob-shaped dorsal papillae are only weakly discernable against the background of other cuticular formations. The thoracic and the first seven abdominal segments each bear six small knob-like dorsal papillae. The eighth abdominal segment has four dorsal papillae, which are located somewhat in front of the stigmata. The ventral surface of the thoracic segments shows the usual set of papillae: two sternal and eight intercalary papillae per segment, the latter arranged in four groups of two papillae, each of which bears a short bristle. Apart from these, every thoracic segment bears two sublateral papillae. On the first seven abdominal segments four simple, weakly developed anterior ventral papillae are found. The posterior half of these segments bears, besides the usual four papillae, six posterior ventral papillae. Two of these are slightly shifted to the anterior side. On the eighth abdominal segment only four ventral papillae are found. The lateral surfaces of the segments bear two rather sturdy and one small knob-shaped papilla.

Spinule fields are lacking.

The terminal segment (fig. 19,4) is quite narrow (as compared to the other abdominal segments). It is small and short, and its anterior part is covered by the penultimate segment. It ends in two robust conical outgrowths which are covered with the same cuticular formations as are found on the whole body. The anal slit is located on the ventral side of the terminal segment.

The well-developed stigmata (fig. 19,5) look like cylindrical swellings. The posterior pair of stigmata is found on small, broad processes of the penultimate segment. Each of these processes also bears a robust knob-shaped papilla with a bristle. The stigmata occur on the lateral surfaces of the segments and are not shifted to the dorsal side.

*Biology*. The larvae live in decomposing detritus.

*Comment*. The diagnosis of the genus is given for the first time.

**Acoenonia europaea** Mamaev

*Morphology*. The spatula is provided with three conical teeth (fig. 19,2). The central tooth is slightly larger than the lateral teeth. The spindle-shaped head is transversely stretched. The broad manubrium is well-developed. It is broader than the maximal width of the central tooth and it has a widened base.

The papillae are knob-shaped and simple. The body is densely covered with variously shaped cuticular formations. The cuticular platelets are smaller and more densely distributed than the papillae. They tend to be grouped around the stigmata and the dorsal papillae. A large portion of the platelets, especially on the central part of the segments, bears pointed processes (fig. 19,3). The ventral surface of the segments is covered with similar, though smaller, formations which superficially resemble bristles. The platelets on the ventral edges of the segments bear no processes.

*Biology*. The larvae have been found in cracks on the lower side of a beech log in an advanced state of decay.

*Material*. Kvasy, Zakarpatskaya oblast, U.S.S.R., 11 April 1963, two larvae.

*Comment*. The diagnosis of the species is given for the first time.

3. Genus **Peromyia** Kieffer (fig. 5,2; 12,7; 13,5; 18,1; 20-25)

*Morphology.* The bright orange larvae are somewhat flattened. The terminal segment is conical and bears at its tip two terminal papillae which form small processes.

The conical head capsule (fig. 21,1) is well developed and as long as, or slightly longer than, its width at the base. The posterior pair of slightly concave, thin tentorial arms originate on the posterior lateral margins of the head capsule. The arms are only conspicuous near their base, and gradually narrow at the posterior end. Their length is slightly less than the width of the head capsule. The weakly developed anterior tentorial arms are hardly visible. The antennae are elongated. Their first segment is well-developed and no shorter than its width at the base. The second segment is long, gradually tapering to the tip. It is three to eight times as long as broad and bears a clearly visible stick-like appendage on its tip.

The spatula is three-toothed (fig. 21,2). The teeth are pointed and of unequal length, the central one being somewhat larger than the lateral ones. The manubrium is, as a rule, well-pronounced, long and broad. Its width is no less than that of the base of the central tooth. The base of the manubrium is, generally, not widened.

The dorsal and ventral surfaces of the body are covered with blunt or pointed oval protuberances of various shapes. The protuberances may be bare or covered with hairs, various scale-like formations etc. Though the ventral surface of the body is devoid of protuberances, compact spinule fields are found on the anterior parts of the metathoracic and all abdominal segments (fig. 25,4). The thoracic and the first seven abdominal segments each bear six papillae on the dorsal side. On the eighth abdominal segment only two well-developed dorsal papilllae occur. Two other, vestigial ones, are located near the stigmata. Six dorsal papillae of the same shape as those on the other segments are found on the central part of the terminal segment. Two terminal papillae of the same shape are found on the tip of this segment.

The conical or tubercle-shaped dorsal papillae (fig. 24,2) bear a short bristle on the tip. The papillae on the neck, the prothoracic, and the last two abdominal segments are all about the same size; but on the other segments the central papillae are slightly larger. In some cases the central dorsal papillae (including those on the eighth abdominal segment) have the shape of conical outgrowths which are nearly as long as the segment. The tip of these outgrowths is split into two and bears two short bristles (fig. 22,2).

The lateral papillae are well-developed. Out of the three papillae that occur on each side of a segment, the one that is closest to the stigmata is largest and in most cases conical. The two other ones are smaller and look like small knobs or are simple. All complex papillae, i.e. the conical, knob-like, etc., as well as the dorsal papillae, bear a short bristle on the tip. The papillae on the ventral surface of the body are mostly simple. On every thoracic segment are found: two simple sternal papillae, two simple sublateral papillae, and four groups each consisting of a single pair of bristle-shaped intercalary papillae.

On each of the first seven abdominal segments are found: two simple anterior and six simple posterior ventral papillae. The distance separating the two exterior posterior papillae from the central ones is nearly two times as large as the distance between the central papillae. On the eighth abdominal segment only four, simple, papillae are found.

The conical terminal segment (fig. 20,1) has a protruding tubular end on whose tip the anal opening is located. The entire dorsal surface of the segment is covered with the same protuberances as are found on the other segments. The two terminal papillae are the only well-developed ones. The stigmata (fig. 20,1) are shaped like oval swellings and are no wider than long. The posterior pair of stigmata is shifted to the dorsal side.

*Biology.* The larvae of a considerable portion of the species of this genus develop in burrows below the strongly decaying bark of felled trunks. Inhabitants of litter are known as well.

*Comment.* The descriptions of the larvae of *Peromyia aurantiaca* (Kieffer), *P. intermedia* (Kieffer), *P. sanguinea* (Kieffer), and *P. leveillei* Kieffer (Kieffer 1895) that are presently available, are incomplete and could not be used in preparing the key.

Key to the species of *Peromyia* Kieffer

1(6) Second antennal segment (fig. 21,1) very long, tapering off to the tip, whip-like, 7-8 times as long as the width at the base. Dorsal and ventral papillae appearing as robust, easily recognizable, conical outgrowths. Central pair of dorsal papillae on the thoracic and the first eight abdominal segments (fig. 20,1) entirely or at least along their basal halves fused, forming a single, quite robust outgrowth, the size of which surpasses that of all other outgrowths several times; sometimes as long as the segment itself. The tip of the outgrowth split into two and bearing two small bristles.

2(5) Lateral papillae (fig. 20,3) located close to the stigmata, nearly twice as long as the surrounding cuticular protuberances. Cuticular protuberances on the lateral surfaces of the first seven abdominal segments especially concentrated on the anterior half of the segments, in front of the lateral papillae. If the lateral surface of the segments is densely covered with protuberances, then protuberances are diverse: pointed and tooth-like on the anterior half and blunt on the posterior half (behind the lateral papillae) of the segment.

3(4) Cuticular protuberances (fig. 20,3) tooth-like, concentrated only on the anterior half of the first seven abdominal segments, in front of the lateral papillae. Posterior half of the segment, behind the lateral papillae, bare or covered with weakly pronounced protuberances. Protuberances located on or near the dorsal papillae mostly smooth, tooth- or scale-like. Spinule fields on the metathoracic and the last two segments consisting of 7-10 rows, on the first seven abdominal segments of 13-15 rows. . . . . . . . . . . . . . . . . . . . . . . . *P. muscorum* (Kieffer)

4(3) Cuticular protuberances (fig. 21,4-6) of different shapes, forming an even and dense covering on the lateral surfaces of the first seven abdominal segments. Protuberances on the anterior half of the segments, in front of the lateral papillae, pointed and tooth-like, but on the posterior half, behind the lateral papillae, blunt. Spinule fields located on the metathoracic and first seven abdominal segments, compact, comprising 13-14 rows. . . . . . . . . . . . . . *P. monilis* Mamaev

5(2) Length of the lateral papillae (fig. 22,3) that are located near the stigmata, not surpassing the length of the surrounding cuticular protuberances. Cuticular protuberances resembling teeth, forming an even and dense covering on the entire lateral surface of the first seven abdominal segments, including the posterior part behind the lateral papillae. Protuberances (fig. 22,2) close to or on the dorsal papillae mostly branched, brush-like. Spinule fields on the metathoracic and last two abdominal segments consisting of 12-15 rows of spinules, on the first seven abdominal segments comprising 16-18 rows. . . . . . . *P. perpusilla* (Winnertz)

6(1) Second antennal segment (fig. 23,1) comparatively short, nearly cylindrical, only slightly narrowed at the tip, three to four times as long as its width at the base. Dorsal and lateral papillae (fig. 23,6) resembling small swellings. Central pair of dorsal papillae on the thoracic and first eight abdominal segments not fused, i.e. there is a pair of clearly separated central dorsal papillae. On a segment all dorsal papillae the same size, the central ones may be slightly enlarged.

7(10) Dorsal and lateral surfaces covered with small blunt or pointed protuberances (fig. 23,6). Protuberances on the prothoracic and last two segments more or less uniformly distributed, on the first seven abdominal segments arranged into four transverse rows which are most pronounced in front of the rows of dorsal papillae. Dorsal papillae appearing as small swellings with a single small bristle on the tip, and surrounded by smooth parts of the cuticle. Papillar swellings considerably larger than all surrounding cuticular protuberances.

8(9) Protuberances on the lateral surfaces of the segments (fig. 23,4) pointed, tooth-shaped. Papillae on the neck (fig. 23,5) appearing as well-developed, broad, conical outgrowths, at least as long as wide. Prothoracic segment densely covered with well-pronounced, blunt tubercles. Spinule fields on the metathoracic and last two segments consisting of 8-9 rows, on the first seven abdominal segments comprising 13-15 rows. . . . . . . . . . . . . . . . . . . *P. sanguinea* (Kieffer)

9(8) Protuberances (fig. 24,3) along the lateral surface of the segments blunt, broadly rounded. Papillae on the neck (fig. 24,1) very small, appearing like small swellings, considerably shorter than the width at the base. Prothorax bare or with very small, hardly visible protuberances. Spinule fields consisting of 11 rows on the metathoracic, of 4-5 rows on the last two, and of 9-14 rows on the first seven abdominal segments. . . . . . . . . . . . . . . . . . . . . *P. minutissima* Mamaev

10(7) Dorsal and lateral surfaces densely covered with small, rounded protuberances (fig. 25,3). Protuberances adjacent to each other, forming compact fields, not arranged in transverse rows. Protuberances more or less the same size, on the central part of the segment, however, somewhat smoothed and more weakly pronounced. Dorsal papillae comparatively small, appearing as small tubercles with a small bristle on the tip. Papillar tubercles densely surrounded by cuticular protuberances and smaller than these. . . . . . . . . . . . . . . . *Peromyia* sp.

**Peromyia muscorum** (Kieffer)

*Morphology*. The head capsule is relatively long, considerably longer than its width at the base. The antennae are very long and well-developed. The length of the first antennal segment is equal to or somewhat surpasses its width at the base. The second segment is acuminate and seven or eight times as long as the width at the base.

The spatula has a long, well-pronounced manubrium.

The very long central dorsal papillae (fig. 20,1) are fused and covered with blunt or pointed cuticular protuberances. The protuberances on the prothoracic and the last two abdominal segments are more or less uniformly distributed, whereas on the other segments they are arranged into three to four transverse rows. These rows are particularly

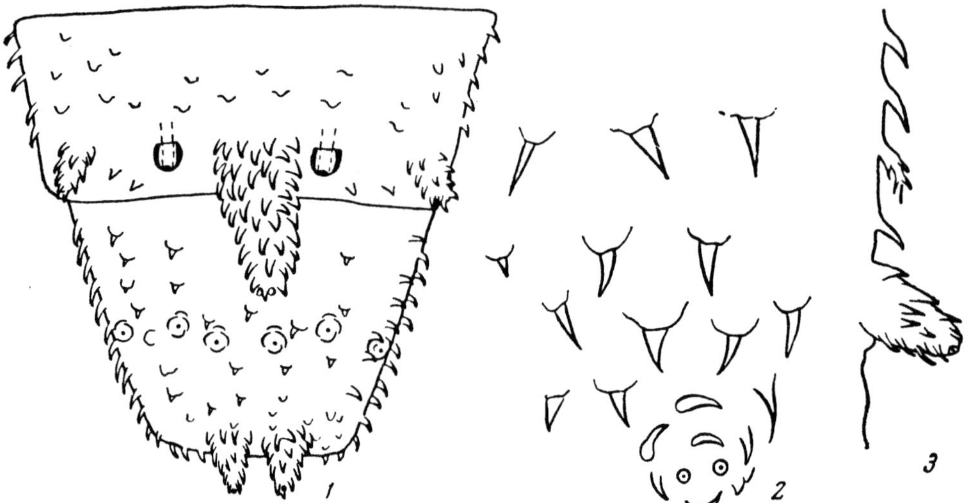


Figure 20. Larva of *Peromyia muscorum* (Kieffer)
1 – last two segments (dorsal view); 2 – abdominal segment, cuticular structures around central dorsal papilla; 3 – abdominal segment, cuticular structures and lateral papillae (orig.)

concentrated on the anterior half of the segment, in front of the papillae. Along the lateral surfaces of the segment the protuberances are more densely distributed. The anterior and posterior margins of a large portion of the first seven abdominal segments are, however, devoid of protuberances. The protuberances on the anterior part of the segments are branched and brush-like, but in the vicinity of the dorsal papillae most of them are tooth-shaped (fig. 20,2).

The dorsal papillae (fig. 20,2) are very robust outgrowths which are covered with squamuliform protuberances. The two central dorsal papillae on the thoracic and the first eight abdominal segments are fused. As a result, the thoracic and the first seven abdominal segments bear five dorsal papillae, and the eighth abdominal segment only a single one. On the terminal segment six central dorsal and two terminal papillae are found. All papillae on the terminal segment are small and conical, and of about equal size. The lateral papilla that is found on the abdominal segments close to the stigma is quite robust, no less than twice as long as the surrounding protuberances (fig. 20,3).

The spinule fields on the ventral surface are comparatively small. On the metathoracic and the last two segments they consist of 7-10 rows of spinules, on the first seven abdominal segments they comprise 13-15 rows.

*Biology.* The larvae have been found below the bark of a decaying oak log.

*Material.* Vicinity of Tula, 7 May 1958, below the bark of an oak log, one larva.

*Comment.* The larva was found together with larvae of *P. perpusilla* (Winnertz) and is described for the first time.

**Peromyia monilis** Mamaev

*Morphology.* The constructional details of the head capsule, the antennae, and the dorsal papillae of this species very strongly resemble those of the previous species.

The entire dorsal surface of the segments is densely covered with pointed tooth-shaped, or broad and rounded protuberances.

On the neck and the prothorax the protuberances appear as blunt verrucae. The protuberances on the meso- and metathoracic as well as on the first seven abdominal segments are heterogeneous (fig. 21,4-6). On the anterior half they are pointed and tooth-like, on the posterior half they appear as blunt, robust verrucae which in some cases are as wide as the lateral papillae. The protuberances on the last two segments are pointed and tooth-like (fig. 21,3). A considerable portion of the ventral body surface is bare. It is covered with platelets of an irregular shape only near the lateral sides.

The lateral papillae (fig. 21,5) are well-developed and robust. They are about twice as long as the surrounding protuberances. The spinule fields are compact and comprise on the metathoracic and the first seven abdominal segments 13 to 14 rows of spinules.

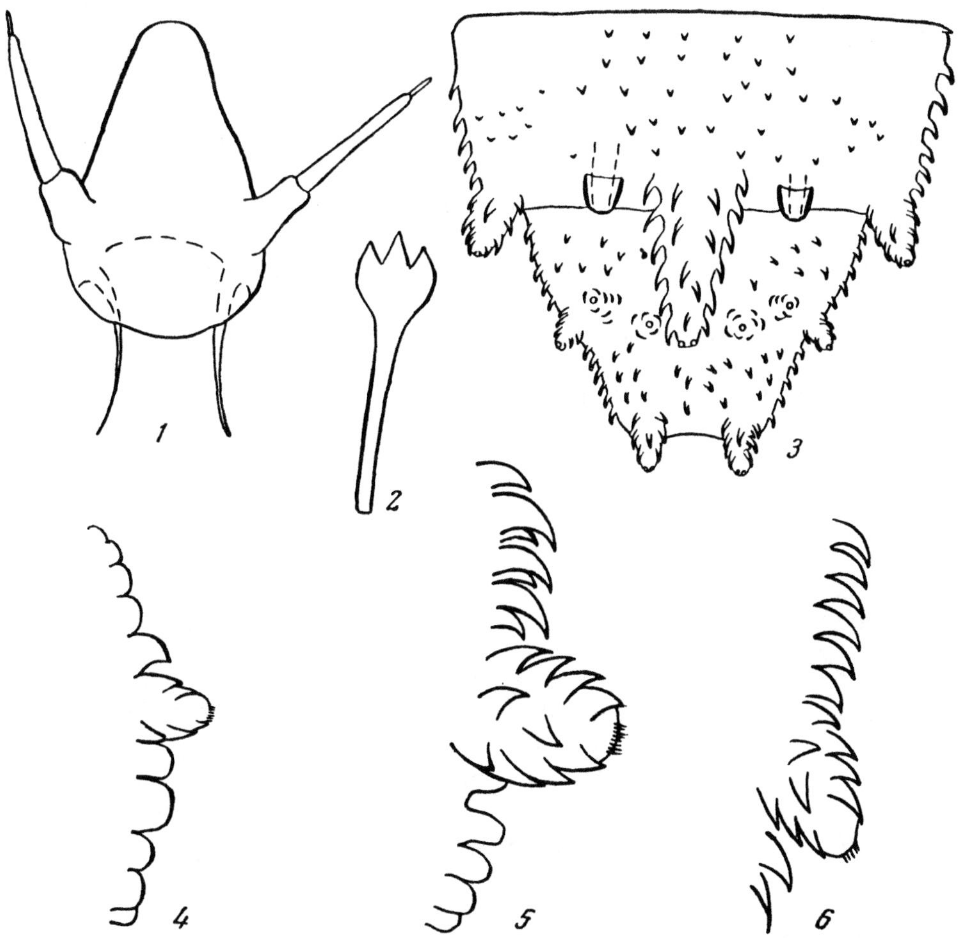


Figure 21. Larva of *Peromyia monilis* Mamaev

1 – head; 2 – spatula; 3 – last two segments (dorsal view); 4-6 – prothoracic, second and eighth abdominal segments, cuticular structures and lateral papillae (orig.)

*Biology.* The larvae live in holes in corroding fir wood. They have also been observed below bark that had come off from spruce logs.

*Material.* Vicinity of Tula, 9 May 1958, two larvae; Khust, Zakarpatskaya oblast, Ukr. S.S.R., 4 July 1963, one larva.

*Comment.* The larva is described for the first time.

**Peromyia perpusilla** (Winnertz)

*Morphology.* The larvae have characteristic unpaired central dorsal papillae (fig. 22,2). The dorsal surface is covered with blunt or pointed protuberances. The protuberances are more or less evenly distributed on the prothoracic and the last two segments.

On the other segments the protuberances are arranged into transverse rows which are mostly located in front of the dorsal papillae. The protuberances uniformly cover the entire lateral surfaces (fig. 22,3). On the anterior half of the dorsal surface of the segments they are comparatively short and provided with some teeth but near the dorsal papillae they are, as a rule, branched and brush-like.

The dorsal papillae (fig. 22,2) are covered with branched protuberances and hairs. Although the papilla that is closest to the stigma is the largest one of the three lateral papillae occurring on the abdominal segments, it is no longer than the surrounding protuberances. The spinule fields are comparatively small. On the metathoracic and the last two abdominal segments they consist of 12-15 rows, on the first seven abdominal segments of 16-18 rows. The other characters match those of the larvae of *P. muscorum* (Kieffer) (see above).

*Biology.* The larvae have been observed on films of mycelium in clefts in a rotting oak log.

*Material.* Novo-Prokhladnoe, Krasnodarskiy kray, 27 May 1959, one larva; idem, 8 May 1959, three larvae.

*Comment.* The larva is described for the first time.

Figure 22. Larva of *Peromyia perpusilla* (Winnertz)
1 – spatula; 2 – cuticular structures around central dorsal papilla; 3 – abdominal segment, cuticular structures and lateral papilla (orig.)

**Peromyia sanguinea** (Kieffer)

*Morphology*. The well-developed first segment of the antennae is no shorter than its width at the base (fig. 23,1). The second segment is shorter than in the preceding larvae, being 3.5-4 times as long as its width at the base.

The three-toothed spatula is well-developed (fig. 23,2). The central tooth is slightly longer than the other ones. All teeth have a pointed tip. The manubrium is well-pronounced, rather long and not widened at the base.

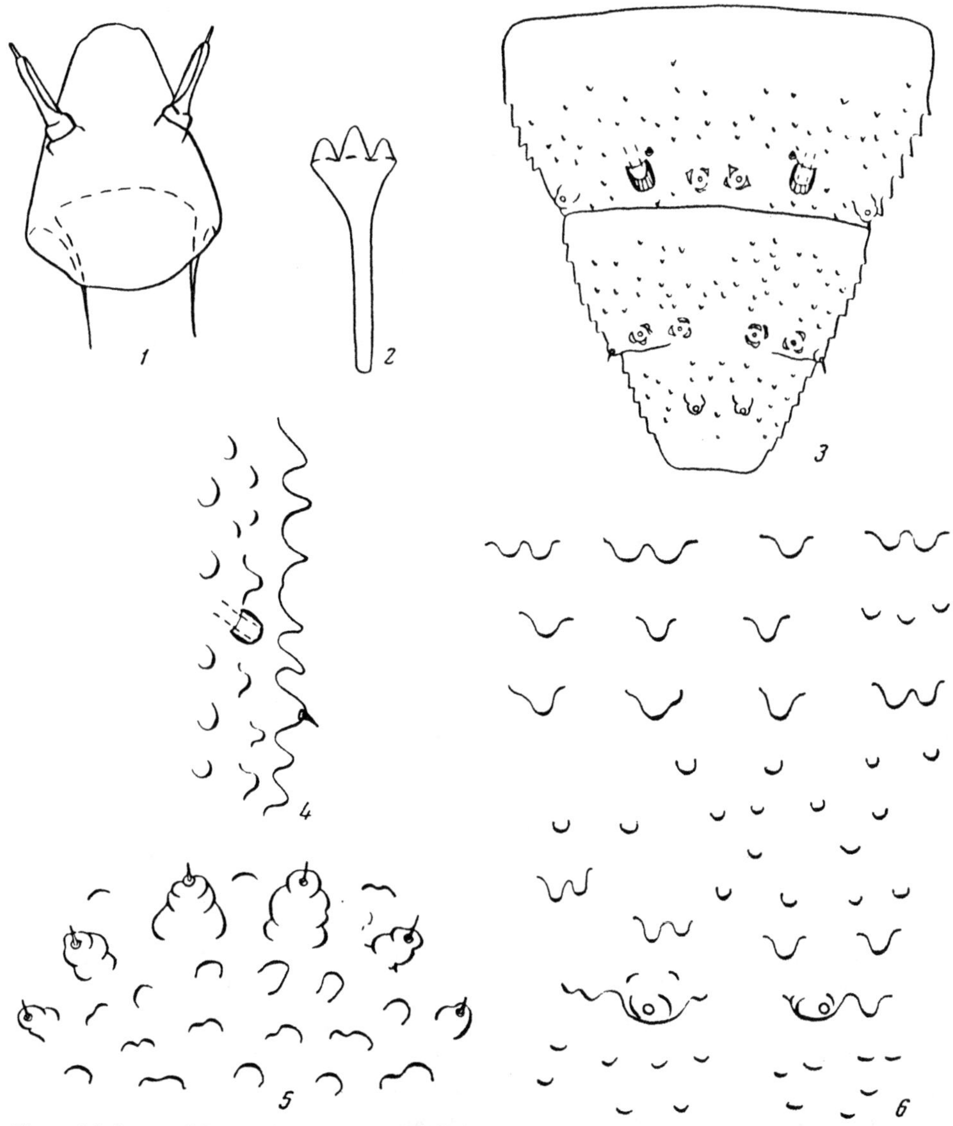


Figure 23. Larva of *Peromyia sanguinea* (Kieffer)

1 – head; 2 – spatula; 3 – last two segments (dorsal view); 4 – first abdominal segment, cuticular structures and lateral papilla; 5 – supernumerary segment, cuticular structures and dorsal papillae; 6 – abdominal segment, cuticular structures and dorsal papillae (orig.)

The dorsal and lateral surfaces of the body are covered with sparse, clearly observable, blunt or pointed, in some cases paired, protuberances of various sizes. The protuberances on the neck, the prothorax, and the terminal segment are more or less evenly distributed (fig. 23,5). On the other segments all protuberances are arranged into rather distinct rows. The anterior part of the segments, i.e. in front of the dorsal papillae, bears three to four rows of robust protuberances between which a few small ones are found (fig. 23,6). The surface behind the dorsal papillae is bare or carries very small protuberances. The protuberances on the lateral surfaces of the segments are pointed and longer than their width at the base (fig. 23,4). The surfaces of the body, protuberances, and papillae are smooth and not covered with hairs. Compact spinule fields occur on the ventral surface of the metathoracic and all abdominal segments. They comprise 13-15 rows of spinules on the first seven abdominal segments and 8-9 rows on the remaining ones.

Six verruca-shaped dorsal papillae are found on each of the neck, thoracic, and first seven abdominal segments (fig. 23,5-6). Especially those on the neck are conspicuous and rather robust, being as long as or longer than their width at the base. On the eighth abdominal segment only four papillae occur, two of which are located near the stigmata (fig. 23,3). The central dorsal papillae are close together, the distance between them being about half their width. The lateral papillae have various sizes, but all of them are comparatively small, no longer than the surrounding verrucae. The largest one is found closest to the stigma (fig. 23,4).

*Biology.* The larvae have been observed in alder stumps.

*Material.* Volkovo, Yaroslavskaya oblast, 27 June 1961, three larvae.

*Comment.* A fragmentary description of the larva was given by Kieffer (1895).

**Peromyia minutissima** Mamaev

*Morphology.* The antennae are comparatively short. Though the first segment is well-developed, it is somewhat shorter than its width at the base. The second segment is also relatively short (no more than 3.5 times as long as its width at the base), nearly cylindrical, and rounded at the tip.

The well-developed spatula has three pointed teeth, the central tooth being the longest. The broad manubrium is no narrower than the central tooth and has a thickened base.

The dorsal body surface is covered with small, hardly visible, pointed protuberances of various sizes (fig. 24,2). On the neck and prothorax small, weakly pronounced verrucae occur (fig. 24,1). The meso-, and metathoracic as well as the first eight abdominal segments bear well-developed cuticular protuberances on a considerable portion of their surface. The posterior part of the segments, i.e. behind the dorsal papillae, may be bare or covered with small, blunt verrucae (fig. 24,2). The robust protuberances are arranged into three rather distinct transverse rows on each segment: two rows on the anterior half and one in front of the row of dorsal papillae. Between the rows smaller protuberances are found. The protuberances on the lateral surfaces of the segments are small, and broadly rounded (fig. 24,3). They are no longer, and may even be considerably shorter, than their width at the base. The surfaces of the body, protuberances, and papillae are smooth, and devoid of hairs, scales, or other cuticular formations.

Compact, small spinule fields occur on the metathoracic and all abdominal segments. On the metathorax they consist of 11, on the last two segments of 4-5, and on the remaining segments of 9-14 rows of spinules.

Figure 24. Larva of *Peromyia minutissima* Mamaev
1 – supernumerary segment, cuticular structures and dorsal papillae; 2 – abdominal segment, cuticular structures and dorsal papillae; 3 – first abdominal segment, cuticular structures, lateral papillae, and stigma; 4 – terminal papillae (orig.)

The distances between all dorsal papillae on a segment are equal. The neck, thoracic and first seven abdominal segments each bear six small, verruca-like dorsal papillae (fig. 24,1,2). The papillae on the neck are very small, considerably shorter than the width at the base. The eighth abdominal segment bears the usual set of four dorsal papillae. The lateral papillae are small, and appear as blunt verrucae with a small bristle on the tip. The size of these papillae does not surpass that of the surrounding verrucae.

*Biology*. The larvae occur in layers of mouldering wood of broad-leaved trees. They have also been observed on dark films of mycelium on the surface of a linden log.

*Material*. Khust, Zakarpatskaya oblast, Ukr. S.S.R., 13 July 1963, three larvae.

*Comment*. The larva is described for the first time.

**Peromyia** sp.

*Morphology*. The head capsule is of the same type as that of the other larvae of *Peromyia* Kieffer. The antennae are comparatively short (fig. 25,1). The first antennal segment is well-developed and as long as it is wide at the base. The second segment is lengthened and tapers off to the tip, which bears a stick-like appendage. The segment is no more than three times as long as its width at the base.

The spatula is three-toothed (fig. 25,2). The base of the manubrium is weakly

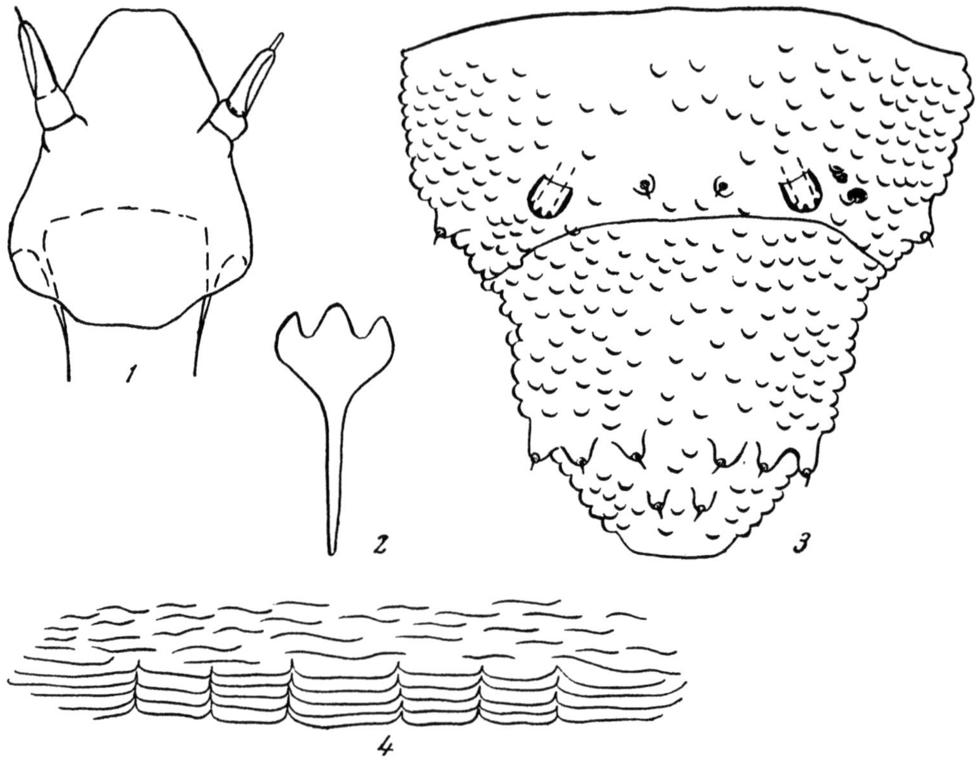


Figure 25. Larva of *Peromyia* sp.
1 – head ; 2 – spatula; 3 – last two segments (dorsal view); 4 – abdominal segment, spinule fields (ventral view) (orig.)

sclerotized. The teeth are of about equal size, the central one being slightly larger than the lateral ones. The dorsal and lateral body surfaces bear a uniform, dense covering of well-marked, oval, squamuliform protuberances. The covering is thinner at the borders between the segments, on their central parts, and near the lateral sides. The ventral body surface is bare. Spinule fields occur on the anterior part of the metathoracic and all abdominal segments. They consist of 18-21 rows on the thoracic and first seven abdominal segments; on the last two segments they comprise 6-8 rows. In some cases the rows of spinules are partly fused. The posterior part of the field is divided into separate parts by six longitudinal grooves.

The dorsal papillae and one of the three lateral papillae on each side of the body have the appearance of low verrucae with a short and small bristle on their tip. All ventral and the remaining two lateral papillae are simple and are directly situated on the surface of the segments.

The conical terminal segment bears, in addition to the six verruca-like dorsal papillae which it has in common with the other body segments, two similar terminal papillae. The six dorsal papillae are not located on the central part of the segment, as in the other species of this genus, but are shifted to the posterior end where they are found near the terminal papillae.

The conspicuous stigmata appear as oval verrucae and are as long as their width at the base. The posterior pair of stigmata is well-developed. It is elevated by protruding verrucae and shifted to the dorsal side of the segment.

*Biology*. The larvae have been found within the dark mould below the bark of a rotting aspen trunk.

*Material*. Volkovo, Yaroslavskaya oblast, 21 June 1961, three larvae.

*Comment*. The larva is described for the first time.

4. Genus **Skuhraviana** Mamaev (fig. 26, 27)

*Morphology*. The larvae have well-developed conical or tubercle-shaped dorsal papillae. Superficially, they strongly resemble the larvae of *Peromyia* Kieffer, but they are light coloured, and yellowish.

The head capsule is well-developed and heavily sclerotized (fig. 27,1). The posterior margin is rounded. The length somewhat surpasses the width at the base. The two posterior tentorial arms originate on the latero-posterior margin of the head capsule. Over their whole length they are of the same width and as long as or slightly shorter than the width of the capsule. Their posterior ends are connected by a small bow-shaped cross bar of the same thickness. The anterior tentorial arms are very weakly developed. The dorsal cover of the head capsule is provided with large membranous parts at the base of the antennae.

The well-developed antennae (fig. 27,1) quite strongly resemble those of *Peromyia* Kieffer. The first antennal segment is as long as or longer than its width at the base. The second segment is very long, narrower than the first one, and slightly tapers off at the tip. It is no less than five times as long as the width at the base. The tip bears a small, stick-like appendage.

The central tooth of the three-toothed spatula is considerably longer than the lateral teeth (fig. 26,6). The manubrium is not developed or very short.

The dorsal and lateral surfaces of the body are covered with reasonably long, pointed protuberances of various sizes (fig. 27,3). The dorsal surface of the neck, prothoracic, and last two segments is completely covered with large protuberances. On the remaining segments these protuberances are arranged into two or three transverse rows, which are located in front of the row of dorsal papillae. Apart from these, some rows of smaller protuberances are found on the segments halfway the body. Such small protuberances may also occur on the posterior part of the segments, behind the dorsal papillae. The lateral surfaces of the segments have a more uniform covering of protuberances. The protuberances, papillae, and sometimes the surface of the segments as well, may be covered with various cuticular formations resembling scales or small spinules. This covering is especially dense on the lateral parts of the segments. The ventral surface of the pro-, meso-, and metathoracic, as well as of all abdominal segments bears broad, compact spinule fields which cover the anterior half of the segments. The posterior part of the segments is bare.

Simple, bristle-shaped, and conical papillae may be found (fig. 26,3). The conical dorsal papillae are well-developed and covered with scale-shaped protuberances. The neck, thoracic, and first seven abdominal segments each bear six dorsal papillae; though these are all more or less about the same size, the marginal ones are slightly smaller. The length of the papillae is no less than, and in the majority of cases noticeably surpasses, their width at the base. The central papillae are close together; the distance between them is less than the width of the papillae themselves. On the first three abdominal segments the distance between the papillae 'b'-'c' is less than that between the papillae 'a'-'b' (fig. 1). On the last abdominal segments, including the seventh, the papillae are found at equal distances from each other.

On the eighth abdominal segment only two well-developed, conical dorsal papillae are found between the stigmata, whereas two other, simple ones, are located on small verrucae close to the stigmata, as is the case on the larvae of *Peromyia* Kieffer. The terminal segment bears six conical papillae on its central part and two on its tip (fig. 26,5). The largest of the conical lateral papillae is found closest to the stigma, whereas the two smaller ones are shifted to the ventral surface of the segment. On each of the thoracic segments two simple sternal and two sublateral papillae occur as well as four groups of intercalary ones, each group of which consists of two simple papillae. On each of the first seven abdominal segments eight ventral papillae are found. The four anterior papillae are located immediately behind the spinule rows; the four posterior ones are found closer to the posterior margin of the segment. The marginal posterior papillae stand on parts of the cuticle that are covered with hairs, the other papillae are situated on bare parts of the central region of the segment. On the eighth abdominal segment only four ventral papillae are found. The two marginal ones are located close to the spinule fields, from which the central ones are further removed. The distance between the central papillae is twice that between the papillae and the spinule fields.

The tip of the conical terminal segment is truncated (fig. 27,2). The segment is comparatively short, as long as it is wide at the base. As we have already mentioned, the dorsal surface of the segments is uniformly covered with pointed protuberances. The anal opening is located on the tip of a conspicuous, broad, retractable tube.

The stigmata are well-developed and robust. The posterior pair is shifted to the dorsal surface of the segment (fig. 26,5).

*Biology*. The larvae of the known species develop in rotting wood. Their mode of life resembles that of the larvae of *Peromyia* Kieffer.

*Comment*. The characterization of the genus is given for the first time.

Key to the species of *Skuhraviana* Mamaev

1(2) Dorsal papillae conical, in particular on the last segments longer than the width at the base (fig. 26,2-3). Robust protuberances on the lateral surfaces (fig. 26,4) as well as on the dorsal surface of the first seven abdominal segments arranged into 2-3 rows. Spinule fields on the ventral surface consisting of uniform, small spinules.
*S. triangulifera* Mamaev

2(1) Dorsal papillae short, oval, appearing like broad and blunt outgrowths, considerably shorter than the width at the base (fig. 27,3). Lateral surfaces of the first seven abdominal segments uniformly covered with robust protuberances (fig. 27,4). Apart from the spinules and scales which cover the large protuberances and papillae, also small protuberances are found sparsely distributed on the dorsal surface of the segments. Spinule fields on the ventral surface consisting of diverse small spinules. Rows of robust, broad spinules that are sharply different in size from the remaining ones present within the anterio-lateral parts of the fields.

*Skuhraviana* sp.

**Skuhraviana triangulifera** Mamaev

*Morphology*. The body of the larvae bears weakly developed protuberances. The spatula is extremely weakly developed; the head and manubrium are not developed (fig. 26,6).

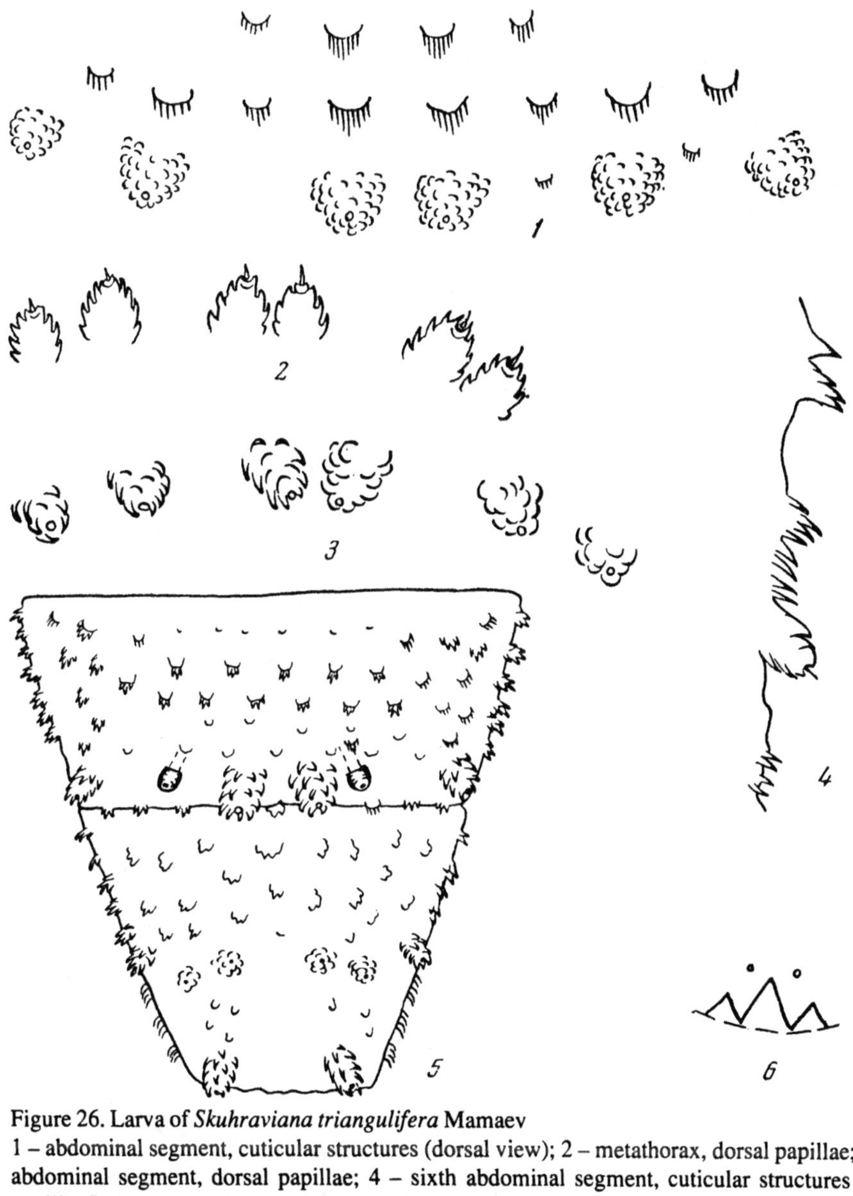


Figure 26. Larva of *Skuhraviana triangulifera* Mamaev
1 – abdominal segment, cuticular structures (dorsal view); 2 – metathorax, dorsal papillae; 3 – second abdominal segment, dorsal papillae; 4 – sixth abdominal segment, cuticular structures and lateral papilla; 5 – last two segments (dorsal view); 6 – spatula (orig.)

On the dorsal surface pointed, conical protuberances which are as long as, or longer than, the width at the base, are found. The protuberances are arranged into two or three clearly discernable transverse rows which are located in front of the papillae (fig. 26,1). The protuberances are covered with various cuticular formations resembling scales, hairs, etc. Apart from these protuberances the surface of the segments is smooth, though sometimes a few, hardly visible, small protuberances may be found. On the lateral surfaces of the first six abdominal segments the protuberances are also arranged into a few rows (fig. 26,4). The dorsal papillae are conical (fig. 26,5).

The spinule fields are uniform and consist of 10-14 rows on the metathoracic and the last two segments, and of 17-20 rows on the other segments.

*Biology*. The larvae develop beneath rotting bark and in wood of broad-leaved trees.

*Material*. Novo-Prokhladnoe, Krasnodarskiy kray, 20 May 1959, two larvae.

*Comment*. The larvae are described for the first time.

**Skuhraviana** sp.

*Morphology*. The larvae bear well-pronounced protuberances. The spatula is small; the central tooth is considerably more robust than the lateral ones; the head and the manubrium are neither pronounced nor heavily sclerotized.

Figure 27. Larva of *Skuhraviana* sp.
1 – head; 2 – terminal segment (dorsal view); 3 – first abdominal segment, cuticular structures and dorsal papillae; 4 – first abdominal segment, cuticular structures and lateral papilla (orig.)

The dorsal and lateral surfaces are covered with broad, pointed, short protuberances which are shorter than the width at the base (fig. 27,3). As in the preceding species, the protuberances bear hairs, scales, and spinules. These formations do not only cover the protuberances and papillae, but they cover the dorsal surface of the segments between the protuberances as well. The protuberances form two to three distinct transverse rows on the dorsal surface of the first seven abdominal segments. Along the lateral surfaces they are uniformly distributed, and do not form clear rows (fig. 27,4). The dorsal papillae are broadly oval (fig. 27,3).

The spinule fields are heterogeneous. In the anterio-lateral part of the fields a row of robust, broad spinules with strikingly different dimensions is found. On the first seven abdominal segments the spinule fields consist of 20-24 rows, but on the metathoracic and the last two segments of 14-17 rows.

*Biology.* The larvae develop below the bark and in the wood of rotting beech trunks.

*Material.* Kvasy, Zakarpatskaya oblast, Ukr. S.S.R., 12 June 1963, three larvae; idem, 12 July 1963, three larvae.

*Comment.* The larva is described for the first time.

### 5. Genus **Bryomyia** Kieffer (fig. 13,4; 28)

*Morphology.* Except for their white colour, the appearance of the larvae is very similar to that of the larvae of *Skuhraviana* Mamaev and *Peromyia* Kieffer.

The conical head capsule is well-developed, and noticeably longer than the width at the base. The well-developed posterior tentorial arms originate on the posterior margin of the head capsule. The tips of the arms are connected by a cross bar. The arms are comparatively short, no longer than half the width of the capsule at its base. The highly developed antennae are provided with elongated first and second segments, as well as with a well-developed stick-like appendage (fig. 28,1). The second antennal segment is five to ten times as long as its width at the base.

The spatula bears conical teeth, the central one of which is considerably larger than the lateral ones. The head of the spatula is well-developed, but the manubrium is very short (no longer than the head), clearly distinct, heavily sclerotized, and (like the head) almost black.

On the dorsal and lateral surfaces of the body broad, blunt verrucae are found (fig. 28,2). These verrucae occur in four to five rows on each segment; between the rows smaller verrucae may be interspersed. The ventral surface is bare, without verrucae or hairs. Compact spinule fields only occur on the anterior third part of the metathoracic and all abdominal segments.

On the dorsal surface of the thoracic and the first eight abdominal segments the usual set of six dorsal papillae is found. All papillae appear as reasonably robust, conical verrucae with a small bristle on the tip (fig. 28,2). On the eighth abdominal segment only four dorsal papillae are found, the two largest of which are located between the stigmata, and the two smaller ones near the stigmata. The distance between the papillae 'a'-'a' is equal to the width of the base of a papilla and about two third of the distance between the papillae 'b'-'c' (fig. 1).

On the ventral surface of each thoracic segment two simple sternal papillae, two

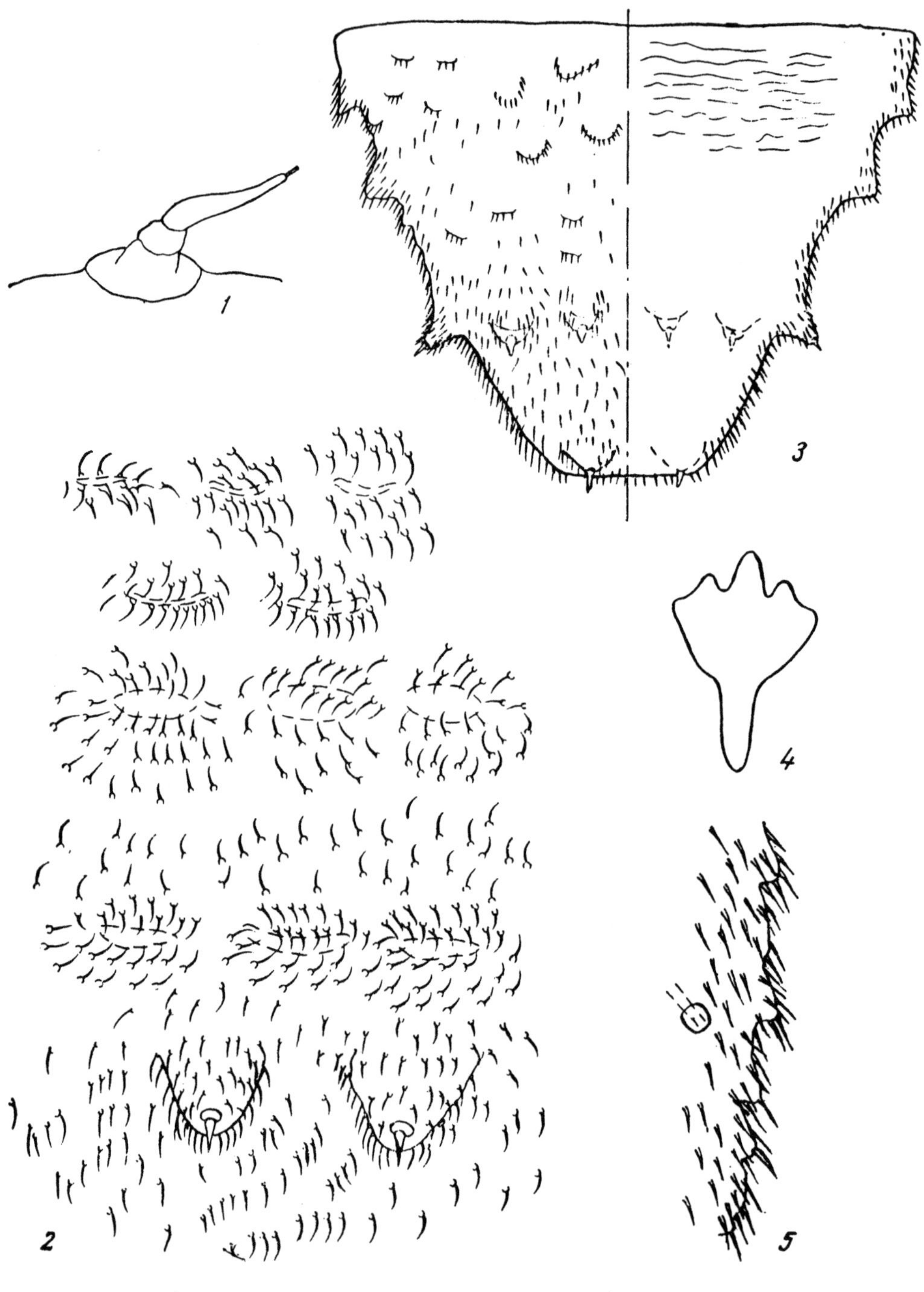


Figure 28. Larva of *Bryomyia* sp.
1 – antenna; 2 – abdominal segment, cuticular structures and dorsal papillae; 3 – terminal segment (left: dorsal view; right: ventral view); 4 – spatula; 5 – abdominal segment, cuticular structures on the lateral surface (orig.)

simple or setiform sublateral ones, and four groups, each consisting of two intercalary papillae, are found. Four anterior as well as four posterior ventral papillae occur on each of the first seven abdominal segments. The eighth abdominal segment bears only four simple ventral papillae. Both pairs of anterior ventral papillae are found at some distance from the spinule fields. The distance between the central papillae is equal to that between the papillae and the spinule fields, and surpasses the distance between the marginal papillae and the spinule fields by no more than one third. The lateral papillae are weakly developed. Both conical lateral papillae (one on each side) on every segment are conspicuous. Moreover, two other small papillae (situated directly on the surface of the body, and sometimes simple) are rather clearly visible on each side of the segment.

The terminal segment is typical for the genus (fig. 28,3). It is conical and bears six clearly pronounced conical dorsal papillae as well as two similar papillae on the tip of the segment. Eight verrucae, arranged into two rows, occur on the anterior part of the segment in front of the dorsal papillae. The stigmata are similar to those of *Skuhraviana* Mamaev.

*Biology*. The larvae develop in soil, litter, and rotting wood.

*Comment*. Up to this time only the larva of *Bryomyia bergrothi* Kieffer has been described (Kieffer 1895).

**Bryomyia** sp.

*Morphology*. The entire dorsal and lateral surfaces of the segments, and in particular the papillae and verrucae, are densely covered with inarticulate (simple) hairs (fig. 28,2,5). On the metathoracic and the last two segments the spinule fields comprise 10-15, on the first seven abdominal segments 20-25 rows.

*Biology*. The larvae have been found on the surface of a strongly rotting beech log beneath the fruiting body of a fungus.

*Material*. Kvasy, Zakarpatskaya oblast, 28 June 1963, one larva.

*Comment*. As a result of the incompleteness of its description the larva of *Bryomyia bergrothi* Kieffer (Kieffer 1895) can only be distinguished from the larva that we have diagnosed by a single character: according to Kieffer, the central dorsal papillae of the larva of *B. bergrothi* Kieffer are fused at the base; in *Bryomyia* sp. they are free.

### 6. Genus **Lestremia** Macquart (fig. 8,2-3; 29,1-4)

*Morphology*. The elongated larvae are cream-coloured and have a heavily sclerotized dark brown head capsule without distinct papillae.

The head capsule is very short (considerably shorter than its width at the base) and without emargination at the ventral side (fig. 29,2). The anterior margin of the capsule is broadly rounded, the dorsal side of the posterior margin bears two small, hardly visible protuberances. The ventral side bears two short arms which are as heavily sclerotized as the head capsule itself. The antennae have a curious shape; they are strongly widened at the base and pointed at the tip. The first antennal segment is very short and broad. The conical second segment strongly tapers off to the tip, where it passes into a flagelliform appendage which is considerably longer than the second segment itself. The base of the antennae is positioned at the centre of an oval, unsclerotized spot.

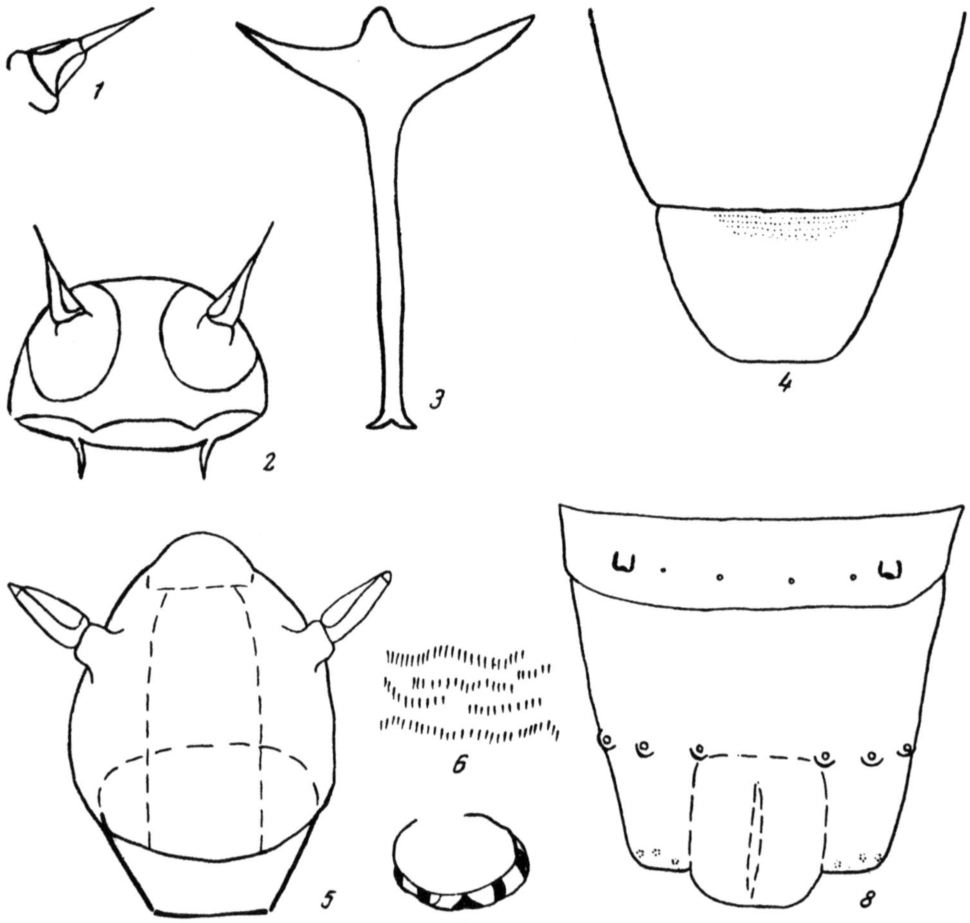


Figure 29. Larvae of *Lestremia* Macquart and *Campylomyza* Meigen
*L. cinerea* Macquart: 1 – antennae; 2 – head; 3 – spatula; 4 – last two segments (ventral view); *Campylomyza* sp.: 5 – head; 6 – ventral surface, rows of spinules; 7 – stigma; 8 – last two segments (dorsal view) (orig.)

The spatula (fig. 29,3) has a strikingly widened head, whose anterior margin bears a single, small, conical tooth with a rounded tip. At the base of the tooth originate long, pointed wings. The long and narrow manubrium has a slightly broadened base.

The body is smooth and provided with hardly visible simple papillae. On each of the thoracic and the first seven abdominal segments a transverse row of six simple dorsal papillae is found. Two other simple papillae are located in front of this row, nearer to the anterior margin. On each of the lateral surfaces of the thoracic and the first eight abdominal segments three simple lateral papillae (on the prothoracic and the eighth abdominal segment often only two of them are well-pronounced) occur. On the ventral surface of each thoracic segment are found: two simple sternal papillae, four groups of intercalary papillae, two simple sublateral papillae. Each group of intercalary papillae consists of two papillae close to each other. The first seven abdominal segments each bear six simple ventral papillae, arranged into a single transverse row. On some specimens two setiform papillae have been observed within the spinule fields. The ventral surface of the eighth abdominal segment is provided with four ventral papillae.

Spinule fields are absent from the dorsal surface. On the ventral surface compact spinule fields occur on the metathoracic and all abdominal segments. The spinules stand in a variable number of rows.

The terminal segment narrows somewhat at the tip and bears no outgrowths or protuberances at all (fig. 29,4).

All stigmata are located on the anterior half of the lateral surfaces of the segments. The last pair of stigmata is slightly shifted to the dorsal side of the penultimate segment.

*Biology.* The larvae develop on fungi in the soil, litter, and decaying wood, and have also been recorded from the mycelium of mushrooms.

*Comment.* Kieffer drafted a fragmentary diagnosis of this genus (Kieffer 1913, vide genus *Prionellus* Kieffer).

**Lestremia cinerea** Macquart

*Morphology.* See also the diagnosis of the genus. The spinule fields consist of 15-27 rows. On the metathoracic and the last two segments the spinule fields comprise 15-18 rows, on the first seven segments 20-27 (mostly 23-25) rows are found.

*Biology.* The larvae have been observed in rotting leaves that were covered with mould as well as in brown, decaying wood (corrosive type of decay).

*Material.* Vicinity of Tula, 3 June 1958, four larvae; Pass of Takhta-Karacha, Samarkandskaya oblast, Uzbekskaya S. S. R., 29 May 1962, five larvae.

*Comment.* Kieffer's description of the larva of *Prionellus* Kieffer (Kieffer 1913) pertains, in fact, to a species of the genus *Lestremia* Macquart. The diagnosis of the larvae of *Lestremia* Macquart in the same work is erroneous.

7. Genus **Campylomyza** Meigen (fig. 8,1; 11,2; 14,1; 29,5-8)

*Morphology.* The bright orange larvae are elongated and mobile. The head capsule (fig. 29,5) is elongated-oval, comparatively weakly sclerotized, though clearly discernable; its length somewhat surpasses the width at the base. The anterior and posterior tentorial arms are well-developed. The anterior arms are long and slender whereas the posterior arms are relatively short, no longer than the width of the head capsule. The ends of the posterior arms are connected by a short, slightly bow-shaped cross-bar. The cover of the head capsule has small, indistinctly bounded membranous parts at the base of the antennae. The antennae are short, but highly developed. The first antennal segment is broad and short; the second is comparatively well-developed, nearly triangular, and no less than 2.5 times as long as its width at the base.

The spatula (fig. 11,2) is quite often weakly sclerotized. Its head is shaped like a somewhat laterally stretched rhombus with a rounded tip.

The smooth body is provided with simple papillae. The dorsal papillae are distributed according to the typical scheme: six papillae on each of the thoracic and first seven abdominal segments. On the eighth abdominal segment only four, and on the central part of the terminal segment six simple papillae are found (fig. 29,8). The lateral surfaces of the thoracic and the first eight abdominal segments each bear three simple lateral papillae. On the ventral side of the thoracic segments the simple sternal and sublateral papillae occur. On each segment the intercalary papillae are arranged into four groups of three papillae, two of the three papillae bear a short bristle whereas the slightly longer third papilla lacks a bristle. The first seven abdominal segments are each provided with eight ventral papillae, the four anterior ones of which are found behind the spinule fields. On the ventral surface of the eighth abdominal segment only four simple papillae occur.

Compact spinule fields are located on the anterior part of the ventral surface of the metathoracic up to the ninth abdominal segment (fig. 29,6). The irregular posterior margin of the fields is subdivided into several parts by eight bare grooves.

The anal slit is found on the narrow posterior part of the conical terminal segment (fig. 29,8). The posterior pair of stigmata has the shape of rather robust verrucae which are located on the posterior third part of the penultimate segment. They are shifted slightly to the dorsal side. The other stigmata (fig. 29,7) appear as well-developed, sclerotized verrucae which are located on the anterior third part of the lateral surfaces of the segments.

*Biology.* The larvae develop in rotting vegetable remains.

*Comment.* Kieffer gave the most complete previous diagnosis of the larvae (Kieffer 1913).

**Campylomyza** sp.

*Morphology.* The head of the spatula is well-sclerotized. The comparatively narrow manubrium is parallel-sided, and the base is not widened (fig. 11,2).

The spinule fields consist of long rows which are very close together. Only on the last two segments the rows of spinules are very short and rather widely apart. The spinules on the metathoracic segment stand in 14 rows, on the seventh abdominal segment in 15, on the eighth in 11, and on the ninth in 8 rows. The spinule fields on the first six abdominal segments comprise 17-19 rows.

*Biology.* The larvae have been observed below a damp rotting log of *Populus nigra*.

*Material.* Pass of Takhta-Karacha, Samarkandskaya oblast, Uzbekskaya S.S.R., 16 May 1962, three larvae.

8. Genus **Aprionus** Kieffer (fig. 3,1; 6,4; 11,1; 16,5; 30,1-5)

*Morphology.* The bright orange, elongated larvae bear vestigial papillae. There are no distinct outgrowths at the end of the body.

The head capsule is big, but weakly sclerotized; its length somewhat surpasses its width; the ventral side is no more than two-thirds as long as the dorsal side. There are no membranous spots at the base of the antennae. The anterior tentorial arms are extremely long and narrow. The posterior arms originate on the latero-posterior margins of the head capsule, their ends are connected by a bow-shaped, small cross-bar. The posterior tentorial arms are about as long as the width of the base of the capsule. The antennae are well-developed, but comparatively short (fig. 30,2). As its length is no less than its width, the first antennal segment considerably protrudes above the surface of the head capsule. The elongated-oval second segment is 2-2.5 times as long as wide.

The central tooth of the three-toothed spatula is slightly larger than the lateral teeth (fig. 30,1). The manubrium is relatively narrow (half as broad as the base of the central tooth). It may be variously sclerotized.

The smooth body bears small, simple papillae. On each of the dorsal surfaces of the thoracic and the first seven abdominal segments six dorsal papillae are found. On the eighth abdominal segment only four dorsal papillae are located. On each lateral surface of the thoracic and the first eight abdominal segments three conspicuous, simple lateral papillae are found. Two simple sternal and sublateral papillae, as well as four groups, each comprising two papillae bearing a short bristle, are distributed on the ventral surface of the thoracic segments. Four anterior and four posterior ventral papillae occur on each of the first seven abdominal segments. The eighth abdominal segment bears only four simple ventral papillae.

Compact spinule fields without bare spots are located on the ventral surface of the metathoracic and all abdominal segments.

The rounded-oval terminal segment (fig. 16,5) is slightly narrowed as compared to the preceding one. On its latero-posterior margins it bears small, oval outgrowths, which are

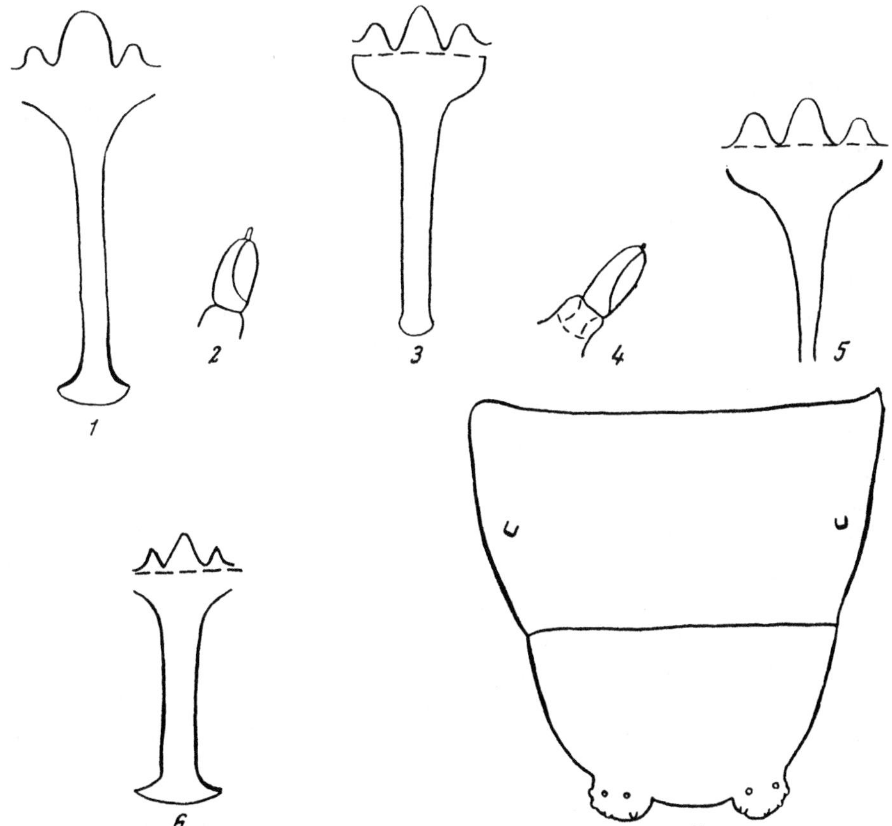


Figure 30. Larvae of *Aprionus* Kieffer and *Monardia* Kieffer
*Aprionus* sp.: 1 – spatula; 2 – antenna; *A. flavidus* (Winnertz): 3 – spatula; 4 – antenna; *A. similis* Mamaev: 5 – spatula; *M. longicauda* Mamaev: 6 – spatula; 7 – last two segments (dorsal view) (orig.)

not always conspicuous if the posterior edge of the segment is retracted. The anal pore is situated on the tip of a tubular formation which is mostly retracted into the terminal segment.

All stigmata are small and distributed along the lateral surfaces of the segments.

*Biology.* The larvae inhabit decaying wood, a number of species is found in soil and forest litter.

*Comment.* Some characters of the larvae of *Aprionus bidentatus* Kieffer, *A. miki* Kieffer, *A. praecox* Kieffer, and *A. spiniger* Kieffer are known (Kieffer 1895, 1913).

### Key to the species of *Aprionus* Kieffer

The larvae of the species of this genus differ almost exclusively in the shape of the spatula and the number of rows of spinules on the segments. As this character is dependent on the age of the larvae it is obligatory to use imaginal material obtained by rearing larvae up to the imaginal phase.

1(4) Spinule fields on the ventral surface of the middle abdominal segments comparatively small, consisting of 16-20 rows, on the terminal segment of no more than 10-12 rows.

2(3) All teeth of the spatula rounded and broad; central tooth big (considerably longer than the head of the spatula), broad (width no less than the length), and broadly rounded at the tip. Lateral teeth rather robust, more than half as long as the central tooth. . . . . . . . . . . . . . . . . . . . . . . . . . . . . . . . *A. terrestris* Mamaev

3(2) Central tooth of the spatula conical, rounded at the tip; as long as or shorter than the length of the head of the spatula, and longer than the width of the base of the tooth. Lateral teeth of the spatula small, considerably shorter than half the length of the central tooth (fig. 30,1). . . . . . . . . . . . . . . . . . . . . . . *Aprionus* sp.

4(1) Spinule fields on the ventral surface of the middle abdominal segments large, consisting of more than 20 rows, on the terminal segment of no less than 13-14 rows.

5(8) Spinule fields on the ventral surface of the middle abdominal segments consisting of 26-30 rows, on the last segments of 15-20 rows.

6(7) Central tooth of the spatula rather big, nearly twice as long as the lateral teeth, and slightly longer than the head of the spatula. Manubrium well-pronounced, with distinct and slightly broadened base (fig. 30,3). . . . . . . . *A. flavidus* (Winnertz)

7(6) Central tooth of the spatula somewhat smaller, less than twice as long as the lateral teeth, and as long as the head of the spatula. As a rule, only the anterior half of the manubrium well-pronounced (fig. 30,5). . . . . . . . . . . . . . *A. similis* Mamaev

8(5) Spinule fields on the last two segments consisting of no more than 13-14 rows.

9(10) Teeth of the spatula broadly rounded, blunt; the central tooth as long as, or shorter than, half the total length of the head of the spatula. Second antennal segment rather long, nearly three times as long as the width of its base. *A. dentifer* Mamaev

10(9) Teeth of the spatula pointed, triangular (fig. 11,1); the central tooth considerably longer than half the total length of the spatula. Second antennal segment broader and shorter, no more than 2.5 times as long as the width of its base. *A. miki* Kieffer

*Comment*. In view of the large similarity of the larvae the morphological diagnoses of the species are not given. All differentiating characters have already been used in the identification key.

**Aprionus terrestris** Mamaev
*Biology*. The larvae develop in the upper layers of the soil to a depth of 20 cm.
*Material*. Vicinity of Tula, 27 June 1958, seven larvae.
*Comment*. The diagnosis of the larva is given for the first time.

**Aprionus** sp.
*Biology*. The orange larvae have been found deep within beech wood (light coloured type of decay).
*Material*. Kvasy, Zakarpatskaya oblast, Ukr. S.S.R., 16 June 1963, four larvae.
*Comment*. The diagnosis of the larvae is given for the first time.

**Aprionus flavidus** (Winnertz)
*Biology*. The larvae have been found in brown, porous fir wood (corrosive type of decay).
*Material*. Khust, Zakarpatskaya oblast, Ukr. S.S.R., 5 July 1963, four larvae.
*Comment*. The diagnosis of the larva is given for the first time.

**Aprionus similis** Mamaev
*Biology*. The larvae develop on the mycelium below strongly rotting bark of spruce.
*Material*. Kvasy, Zakarpatskaya oblast, Ukr. S.S.R., 18 June 1963, three larvae; idem, 22 June 1963, four larvae.
*Comment*. The diagnosis of the larva is given for the first time.

**Aprionus dentifer** Mamaev
*Biology*. The larvae have been found in a pine forest, in deep litter permeated by mycelium.
*Material*. Pavlova Sloboda, Moskovskaya oblast, 21 May 1963, three larvae.
*Comment*. The diagnosis of the larva is given for the first time.

**Aprionus miki** Kieffer
*Biology*. The larvae have been found in spruce stumps.
*Material*. Pavlova Sloboda, Moskovskaya oblast, 14 May 1963, one larva.
*Comment*. Some data on the larvae of this species is given by Kieffer (1895).

9. Genus **Monardia** Kieffer (fig. 6,5; 30,6-7)

*Morphology*. The elongated-oval larvae have a short, weakly sclerotized head and small, rounded outgrowths at the body end. The nearly triangular head capsule gradually tapers off from the base to the tip (fig. 6,5). It is at most slightly longer than wide. The posterior margin of the head capsule bears at its dorsal side two small, pointed protuberances. The well-developed posterior pair of tentorial arms is considerably longer than the width of the base of the head capsule. The arms are nearly equally thick over their entire length; their ends are connected by a small bow-shaped cross-bar. The anterior tentorial arms are not pronounced. The antennae are relatively short. Their first segment is very short, the second is well-developed and at least three times as long as the first segment.

The three-toothed spatula is well-developed (fig. 30,6). The teeth are heavily sclerotized.

The body is smooth, without outgrowths or other cuticular formations. The thoracic and all abdominal segments, except the eighth on which four dorsal papillae are found, each bear six dorsal papillae. The lateral surfaces are each provided with the usual set of three lateral papillae. The number and the distribution of the papillae on the ventral surface of the thoracic segments do not deviate from the rule. Two out of the three intercalary papillae of each group bear a short bristle. The four anterior ventral papillae on the first seven abdominal segments are situated behind the spinule fields. The four posterior ventral papillae are found in the usual location. The eighth abdominal segment bears four simple ventral papillae.

Compact spinule fields are found on the ventral surface of the metathoracic and the first seven abdominal segments.

The terminal segment bears two broad, oval, soft outgrowths, each with four simple papillae (fig. 30,7).

All stigmata are very small and situated directly on the surface of the body; sclerotized tubercles are not developed. The terminal pair of stigmata, which is comparatively far removed from the posterior margin of the penultimate segment, is not shifted to the dorsal side (fig. 30,7).

*Biology*. The larvae of the known species develop in decaying wood.

*Comment*. The genus *Monardia* Kieffer (= *Peromyia* Kieffer) as conceived by Edwards (1938) and Pritchard (1947) is a heterogeneous array of species. We have retained only two species in the genus: *M. stirpium* Kieffer (= *Peromyia vanderwulpi* De Meijere) and *M. longicauda* Mamaev.

**Monardia longicauda** Mamaev

*Morphology*. The spatula is rather broad and short (fig. 30,6). Its head is not clearly outlined. The manubrium is broad (its width noticeably surpasses the width of the base of the central tooth) and has parallel sides; the base is widened.

The spinule fields on the metathorax comprise 13 rows, those on the first and second abdominal segments 15, and those on the other abdominal segments 19-21 rows.

*Biology*. The larvae develop in rotting wood of stumps of *Salix alba*.

*Material*. Pass of Takhta-Karacha, Samarkandskaya oblast, Uzbekskaya S.S.R., 8-13 May 1962, four larvae.

*Comment*. The larvae of *M. stirpium* Kieffer (Kieffer 1895, 1913) have a noticeably lower number of rows of spinules (only eight) on the segments as compared to the species we described.

### 10. Genus **Xylopriona** Kieffer (fig. 31)

*Morphology*. The larvae are white (orange just before the pupation), big, elongated-oval, and have a narrowed anterior and posterior end. The length-width ratio equals 4:1.

The light brown head capsule is comparatively small and slightly longer than the width at the base; it has an emargination at the ventral side. The tentorial arms are well-

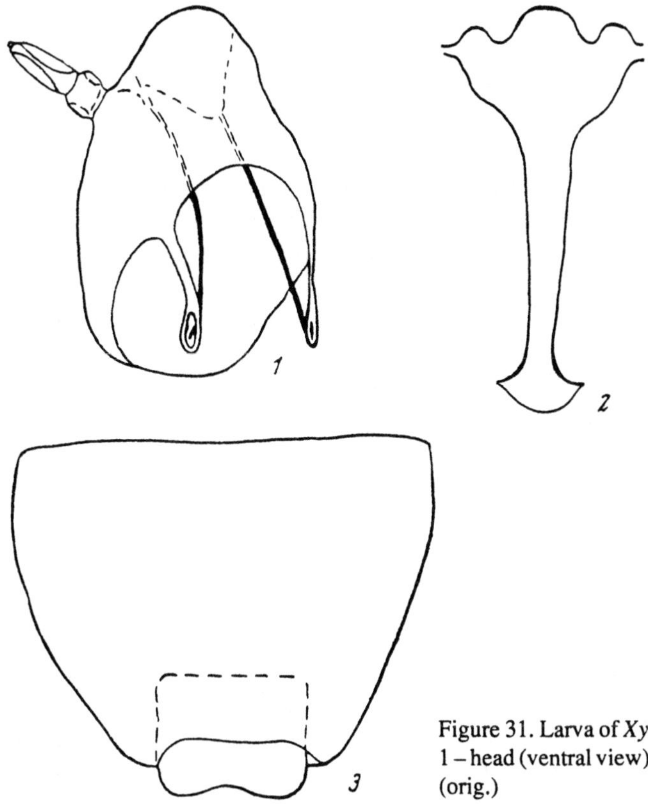


Figure 31. Larva of *Xylopriona atra* (Meigen)
1 – head (ventral view); 2 – spatula; 3 – terminal segment (orig.)

developed. The long and narrow anterior tentorial arms originate on the anterior third part of the head capsule. The posterior pair arises from the ventral side of the posterior margin of the head capsule. These arms are somewhat thicker and 2.5 times shorter than the anterior arms. The arms of both pairs are interconnected outside the capsule. The antennae are well-developed, the first segment is rather robust and as long as wide; the second segment is nearly cylindrical and slightly narrowed at the tip. It is nearly three times as long as the width at the base.

The three-toothed spatula has a highly sclerotized manubrium (fig. 31,2).

The smooth body lacks protuberances and is provided with simple, weakly developed papillae. Each of the three thoracic segments bears the usual set of papillae on its ventral surface: two simple sternal, two simple sublateral, and four groups of intercalary papillae, each group consisting of two setiform papillae.

Large, compact spinule fields are found on the ventral surface of the metathoracic and the first seven abdominal segments.

The terminal segment (fig. 31,3) is slightly narrowed compared to the penultimate segment. On its tip it has a relatively well-developed tube which bears the anal opening.

The small stigmata are located along the lateral sides of the segments; the terminal pair is not shifted to the dorsal side.

*Biology*. The larvae develop in soil and decaying wood.

*Comment*. The diagnosis of the genus is given for the first time.

**Xylopriona atra** (Meigen)
*Morphology.* The black spatula is heavily sclerotized and bears three well-developed, rounded teeth (fig. 31,2). The central tooth is considerably larger than the lateral ones. The long manubrium is well-pronounced and broadened at the base; its width is slightly less than that of the central tooth.

The papillae are simple. Compact spinule fields are found on the ventral surface of the metathoracic and the first seven abdominal segments. The field on the metathorax comprises 19 rows, that on the seventh abdominal segment 24 rows. The fields on the other segments consist of 28-34 rows.

*Biology.* The larvae develop in the upper layers of the soil (to a depth of 40 cm) of beech forests.

*Material.* Kvasy, Zakarpatskaya oblast, Ukr. S.S.R., 13 June 1963, five larvae.

*Comment.* The diagnosis of the larva is given for the first time.

11. Genus **Mycopriona** Mamaev (fig. 6,3; 32,1-2)

*Morphology.* The robust larvae are bright yellow-orange. The head capsule is elongated. The papillae are weakly pronounced.

The brown head capsule is sclerotized and elongated. It is no less than 1.5 times as long as the width of its base (fig. 6,3). The posterior tentorial arms are very thin. There are no unsclerotized spots developed around the bases of the antennae. The antennae are comparatively short; the second segment is approximately twice as long as its width, and 1.5 times as long as the first. The tip of the second segment is rounded instead of pointed.

The central tooth of the three-toothed spatula is larger than the lateral teeth (fig. 32,1).

The body is smooth and bears simple, small papillae. On each of the thoracic and the first seven abdominal segments six dorsal papillae are found, on the eighth abdominal segment only four. Three lateral papillae are located on each of the lateral surfaces of the thoracic and the first eight abdominal segments. On the ventral surface of the thoracic segments the usual simple sternal and sublateral papillae are found. Each lateral surface of the segments bears six intercalary papillae, four of which have a short bristle and occur in pairs, whereas the other two are simple and very badly pronounced. On each of the first seven abdominal segments four simple anterior ventral papillae and four posterior ones are located. All papillae are found behind the spinule fields. Only four simple central papillae occur on the eighth abdominal segment.

The ventral surfaces of the metathoracic and the first eight abdominal segments bear conspicuous spinule fields. The compact fields, which do not contain clearly delimited bare spots, are transversely oval (only 5-6 times as wide as long). The fields on the penultimate segment are no more than twice as wide as long.

The small terminal segment is noticeably shorter than the penultimate segment and somewhat broader than half the width of that segment (fig. 32,2). Two heavily sclerotized hooks are situated at the tip of the terminal segment. The surface of the segment around the bases of the hooks is also sclerotized. The distance between the hooks is approximately equal to the width of the base of a hook. There are no rows of spinules around the anal pore, which is situated on the ventral side of the terminal segment.

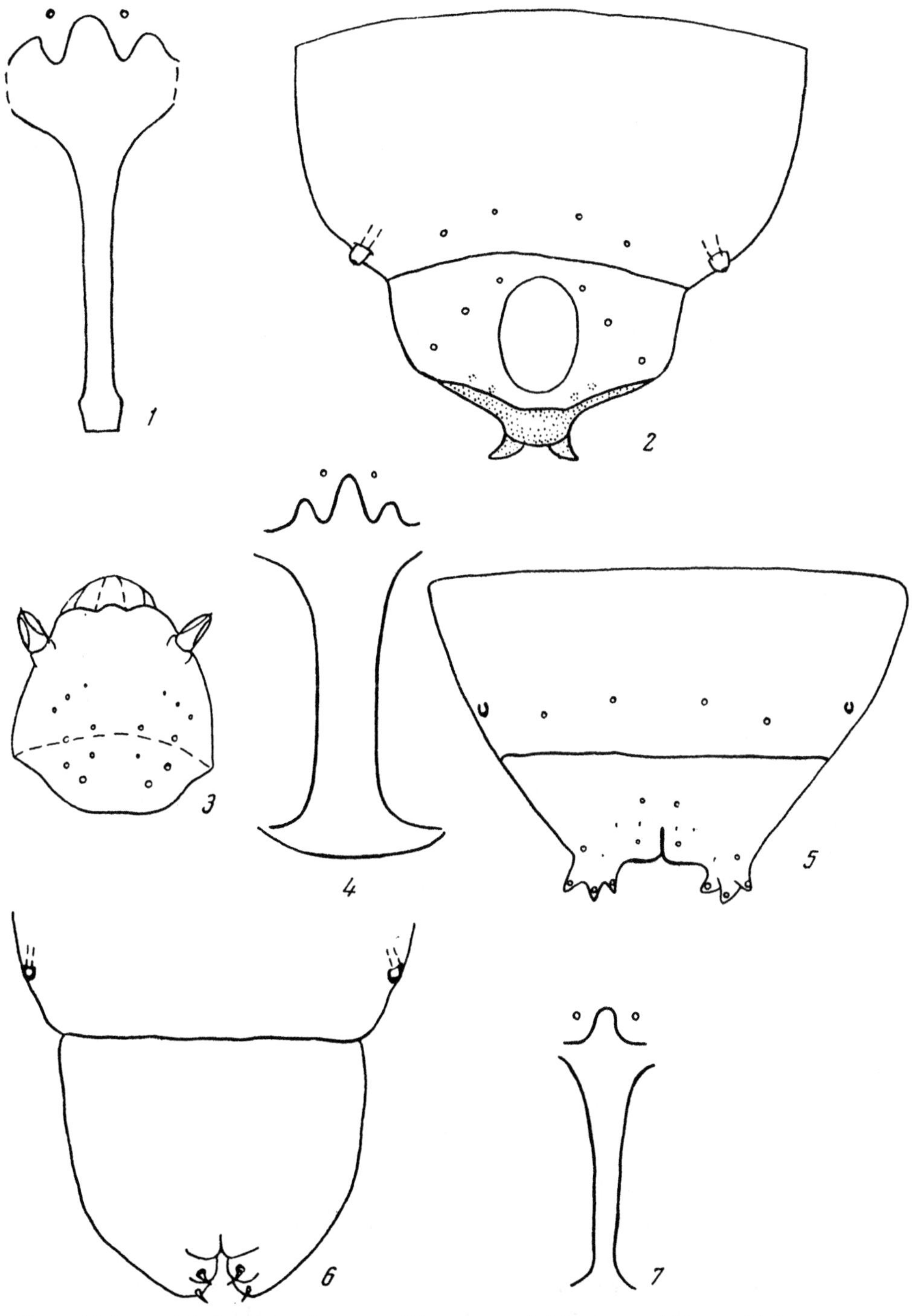


Figure 32. Larvae of *Mycopriona* Mamaev, *Micropteromyia* Mamaev, and *Miastor* Meinert
*Mycopriona caucasica* (Mamaev): 1 – spatula; 2 – last two segments (ventral view); *Micropteromyia ghilarovi* Mamaev: 3 – head; 4 – spatula; 5 – last two segments (dorsal view); *Miastor metraloas* Meinert: 6 – last two segments (dorsal view); 7 – spatula (orig.)

The terminal pair of stigmata is located on the latero-posterior corners of the penultimate segment; it is not shifted to the dorsal side. The other stigmata appear as small, though well-sclerotized tubercles. They are found on the anterior half of the lateral surfaces of the segments.

*Biology*. The majority of the presently known species develops in decaying wood.

*Comment*. The diagnosis of the genus is given for the first time.

**Mycopriona caucasica** (Mamaev)

*Morphology*. The dark brown teeth and manubrium of the spatula are heavily sclerotized. The base is not widened.

The compact spinule fields are large. On the metathoracic and the eighth abdominal segment they consist of 26-28 rows; on the other segments of 37-42 rows.

*Biology*. The larvae have been found in a very damp, strongly decomposing beech stump (light coloured type of decay).

*Material*. Novo-Prokhladnoe, Krasnodarskiy kray, 14 June 1959, two larvae.

*Comment*. The diagnosis of the larva is given for the first time.

## 12. Genus **Trichopteromyia** Williston (fig. 3,2; 16,6)

*Morphology*. The violet-orange larvae are strongly elongated and cylindrical. They have two robust, sclerotized processes at the end of the body.

The weakly sclerotized, nearly transparent head capsule is slightly longer than the width of its base. The hardly developed tentorial arms look like small sclerites and are located within the head capsule. The second segment of the relatively short antennae is nearly cylindrical and rounded at the tip; it is longer than the first antennal segment.

The spatula is not pronounced.

The papillae on the smooth body are mostly simple. The thoracic and the first seven abdominal segments each bear six simple dorsal papillae; the eighth abdominal segment bears four dorsal papillae. On each lateral surface of the thoracic and the first eight abdominal segments three simple lateral papillae are found. Two simple sternal, two simple sublateral and four groups of intercalary papillae are found on the ventral surface of each thoracic segment. Each group of the latter comprises two setiform and one simple papilla. The first seven abdominal segments each bear four anterior ventral papillae, which are located behind the spinule fields, and four posterior ventral papillae; on the eighth abdominal segment only four ventral papillae occur.

Conspicuous, compact spinule fields are found on the ventral surface of the metathoracic and the first eight abdominal segments (fig. 16,6). Apart from the terminal segment, the segments are considerably longer than wide, at least from the third abdominal segment onwards. The penultimate segment is about 1.5 times as long as wide.

The terminal segment is short, but relatively well-developed. It is as wide, or nearly as wide, as the penultimate segment. The terminal segment ends in two robust, sclerotized, straight processes with a pointed tip. Three comparatively short, though also highly sclerotized processes are located dorsally to these processes. The processes are situated on a sclerotized plate on which originate three more weakly sclerotized outgrowths which point to the anterior margin of the segment. The chink-shaped anal opening is situated on the ventral side of the segment; it is bare, i.e. not surrounded by rows of spinules running lengthwise (fig. 16,6).

The small stigmata of the terminal pair are located on the lateral sides of the penultimate segment, at a considerable distance from its posterior margin (fig. 16,6). The other stigmata are also small; they are found on the anterior half of the lateral sides of the segments.

*Biology*. The larvae develop in pieces of decomposing wood with luxuriant mycelium.

*Comment*. The diagnosis of the genus is given for the first time.

**Trichopteromyia modesta** Williston

*Morphology*. The transversely oval spinule fields consist of 20-29 rows. The fields on the first four abdominal segments are 3-3.5 times as wide as long, those on the metathoracic and the other abdominal segments are broader (4-4.5 times as wide as long).

*Biology*. The larvae have been found in strongly decaying and comparatively dry beech wood.

*Material*. Novo-Prokhladnoe, Krasnodarskiy kray, 2 June 1959, three larvae.

*Comment*. The diagnosis of the larva is given for the first time.

## 13. Genus **Micropteromyia** Mamaev (fig. 17,1; 32,3-5)

*Morphology*. The elongated-oval larvae have simple papillae and are provided with two outgrowths, each bearing three sclerotized teeth, on the terminal segment.

The relatively small and short head capsule is at most only slightly longer than its width at the base (fig. 32,3). The tentorial arms are not developed. There are no pronounced light-coloured, unsclerotized spots around the bases of the antennae. The antennae are comparatively short and somewhat pointed at the tip. The second antennal segment is nearly twice as long as its width at the base (and nearly twice as long as the first segment). The spatula is three-toothed (fig. 32,4).

The body is smooth and bears simple papillae. On the dorsal surface of the thoracic and the first seven abdominal segments the usual set of six papillae is found; on the eighth abdominal segment four papillae are located. The lateral surfaces of the segments each bear two well-developed simple lateral papillae. Two simple sternal and two simple sublateral papillae are found on the ventral surface of each thoracic segment. On each side of the segments occur five intercalary papillae. Four of them bear a short bristle and are grouped in pairs, between which the fifth, unpaired and simple, papilla is situated. On the ventral surface of the first seven abdominal segments occur four anterior ventral papillae (which are located behind the spinule fields) as well as four posterior ventral papillae. The ventral surface of the eighth abdominal segment bears only four simple ventral papillae. On the terminal segment are found: eight conspicuous simple papillae around the anal pore, a single terminal papilla on the base of each of the six teeth, and, finally, two simple papillae on the base of each outgrowth.

The spinule fields are located on the ventral surface of the metathoracic and the first seven abdominal segments. The spinule fields on the metathorax are compact, whereas those on the first six abdominal segments show eight, and on the seventh six bare oval spots.

On the posterior margin of the small terminal segment (which is noticeably shorter than the penultimate one) two short outgrowths are found (fig. 32,5). Each outgrowth bears three conical sclerotized teeth. A single simple papilla is found on the base of each tooth, as has already been mentioned above.

The stigmata are small and positioned directly on the surface of the segments (fig. 32,5). The terminal pair of stigmata, which is found on the posterior third part of the penultimate segment, is only slightly shifted to the dorsal side. The other stigmata occur on the anterior half of the segments.

*Biology.* The larvae develop in decaying wood (light coloured type of decay).

*Comment.* The diagnosis of the genus is given for the first time.

**Micropteromyia ghilarovi** Mamaev

*Morphology.* The central tooth of the spatula is approximately twice as long as the lateral ones (fig. 32,4). The well-pronounced manubrium is comparatively broad and has a widened base. The slightly broadened lateral parts of the head of the spatula are indistinct. The spinule fields comprise 14 rows of spinules on the metathorax, and 17-22 rows on the abdominal segments.

*Biology.* The larvae have been seen in rotting wood of: *Betula*, *Tilia*, *Ulmus*, and *Quercus*. They occur in connection with the mycelium of *Fomes fomentarius*.

*Material.* Vicinity of Tula, 18 May 1958, three larvae; Tellermanov forestry, Voronezhkaya oblast, 25 September 1960, five larvae.

*Comment.* A complete diagnosis of the larvae is given for the first time.

14. Genus **Miastor** Meinert fig. 6,2; 32,6-7)

*Morphology.* The larvae are white (the young ones transparent). They have a small head and a comparatively weakly pronounced spatula.

The clearly defined, relatively short head capsule is heavily sclerotized (fig. 6,2). A distinct, heavily sclerotized collar is found on the posterior edge of the capsule. The tentorial arms are not pronounced. The first segment of the well-developed antennae is very short, whereas the second is robust, broadened at the base and rounded at the tip. It is no less than three times as long as the width of its base.

The degree of sclerotization of the one-toothed spatula (fig. 32,7) varies.

The papillae on the smooth body are simple. The usual set of six simple dorsal papillae is found on the dorsal surface of the thoracic and all, except the eighth, abdominal segments. Only four simple dorsal papillae occur on the eighth abdominal segment. Two conspicuous lateral papillae are located on each of the lateral surfaces of the segments. The ventral surface of each thoracic segment bears: two simple sternal papillae, four groups of paired setiform and two single simple intercalary papillae, and two simple sublateral papillae. The ventral surface of the first seven abdominal segments is provided with: four anterior ventral papillae, situated behind the spinule fields, and four posterior ventral papillae. Four simple ventral papillae occur on the eighth abdominal segment. Eight simple papillae are found in front, and two at the bases, of the terminal outgrowths.

Spinule fields occur on the dorsal as well as on the ventral surface of the metathoracic and the first eight abdominal segments. The short spinules have a broadened base. The spinule fields on the dorsal surface contain bare, elongated-oval spots. Within the fields on the first six abdominal segments four oval spots are found, the lateral ones of which are, on the first two abdominal segments, situated at the very edge of the field. The fields on the seventh abdominal segment contain only two spots. On the eighth abdominal segment just some sparse rows of spinules are found, or spinules are lacking altogether. The fields on the ventral surface of the first six abdominal segments each contain six bare oval spots; the two lateral ones of which are shifted to the lower margin of the field, so that the posterior edge of these spots is not closed. The field on the seventh abdominal segment encloses four bare spots. On the eighth abdominal segment the field is only represented by a few rows of spinules. The dorsal as well as the ventral surface of the ninth abdominal segment lack spinule fields. The bare spots are distributed as follows: the four inner ones are arranged into pairs, which are located at some distance from each other.

The terminal segment (fig. 32,6) bears four lobe-shaped outgrowths, each bearing a bristle. The two stigmata (fig. 32,6) have the appearance of two small, sclerotized verrucae and are located on the lateral surfaces of the segments. The terminal pair of stigmata is shifted to the posterior margin of the eighth segment, the other pairs are shifted to the anterior margin of their segments.

*Biology.* The larvae occur in decaying detritus and feed by sucking mycelium.

*Comment.* Up to this time Springer gave the most accurate description (Springer 1915). The drawing of the last two segments of the larva of *Miastor metraloas* Meinert, as given by Hennig (1948), is erroneous.

**Miastor metraloas** Meinert

*Morphology.* The spatula is comparatively weakly sclerotized, only the teeth are clearly distinct. The remaining part of the head as well as the manubrium are sclerotized to a lesser degree, and their outline is quite indistinct. The spinule fields on the dorsal surface consist of 7-19 rows of spinules. The number of rows lowers from the middle segments towards the anterior and posterior end of the body. The spinule fields on the ventral surface comprise 12-23 rows.

*Biology.* The larvae are found below rotting bark of *Quercus*, *Fagus*, *Betula*, *Tilia*, and other broad-leaved trees, as well as in strongly decomposing wood of *Fagus* and *Tilia*.

*Material.* Vicinity of Tula, 29 May 1958, six larvae; Novo-Prokhladnoe, Krasnodarskiy kray, 8 May 1959, four larvae.

*Comment.* According to Pritchard, the genus consists of only one reliable species, *M. metraloas* Meinert (Pritchard 1960).

15. Genus **Brittenia** Edwards (fig. 33,1-4)

The genus comprises a single species, *B. fraxinicola* Edwards, whose larvae were found in Great Britain below the rotting bark of an oak.

*Morphology.* The larvae strongly resemble those of *Miastor* Meinert. They have a simple head capsule (fig. 33,1). On the ventral surface of the metathoracic and the first seven abdominal segments as well as on the dorsal surface of the metathoracic and the

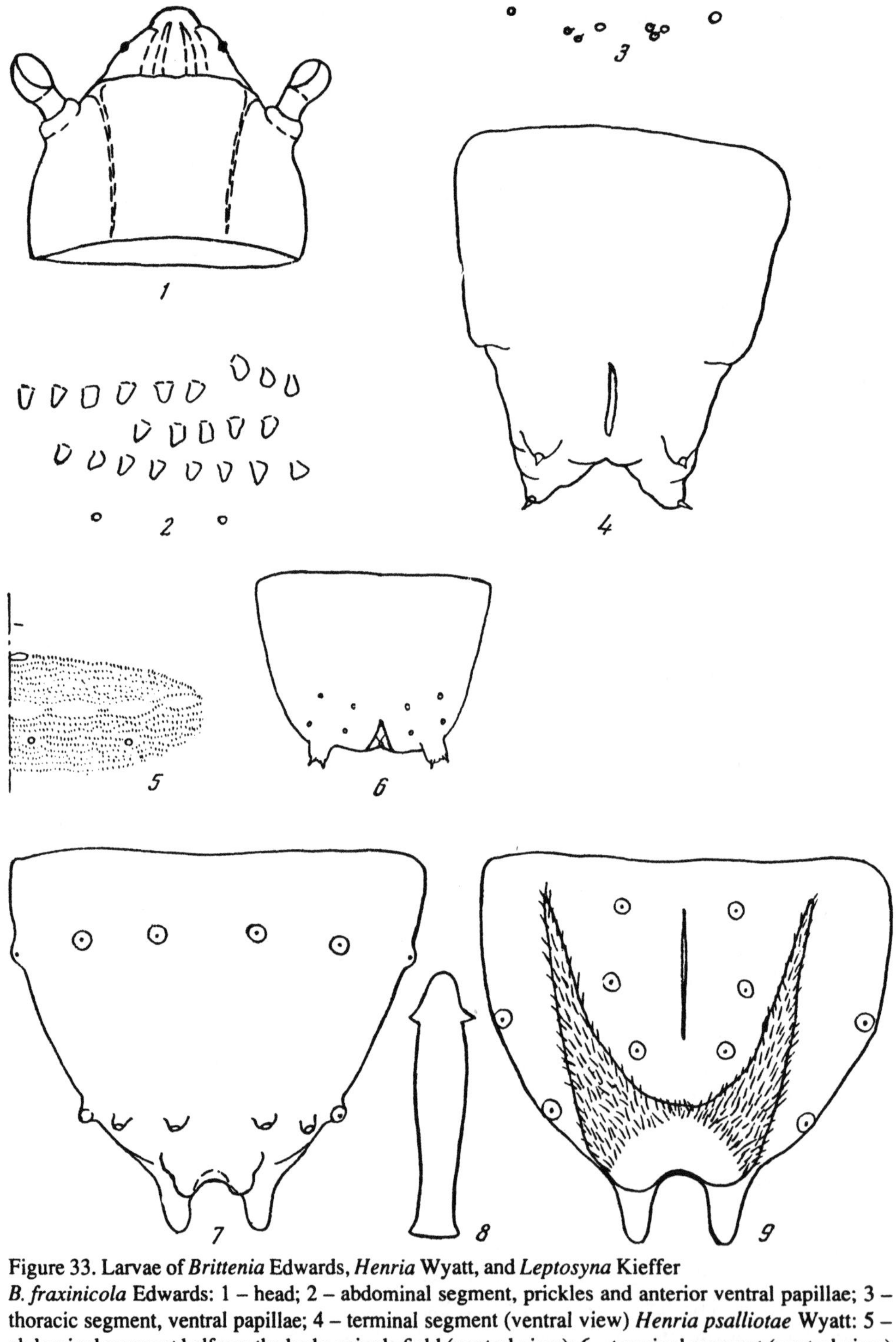


Figure 33. Larvae of *Brittenia* Edwards, *Henria* Wyatt, and *Leptosyna* Kieffer
*B. fraxinicola* Edwards: 1 – head; 2 – abdominal segment, prickles and anterior ventral papillae; 3 – thoracic segment, ventral papillae; 4 – terminal segment (ventral view) *Henria psalliotae* Wyatt: 5 – abdominal segment halfway the body, spinule field (ventral view); 6 – terminal segment (ventral view) *Leptosyna* sp.: 7 – terminal segment (dorsal view); 8 – spatula; 9 – terminal segment (ventral view) (1-4, 7-9 – orig.; 5-6 – after Wyatt 1961)

first four abdominal segments the naked body is provided with well-developed spinule fields (fig. 33,2). The terminal segment (fig. 33,4) bears four processes, each of which bears a bristle. Two of the processes are somewhat better developed than the other two. J. Wyatt (Institute for Plant Cultivation, Glasgow, Great Britain) sent us the data of an investigation of a series of larvae. He is presently preparing an exhaustive description of the larvae for publication and kindly permitted us to use his material.

## 16. Genus **Henria** Wyatt (fig. 14,4; 17,3; 33,5-6)

*Morphology*. The conical head is highly sclerotized. The antennae are more than three times as long as wide.

The spatula is one-toothed. The anterior half as well as the base of the manubrium are broadened.

The smooth body is covered with simple papillae. The thoracic and the first seven abdominal segments each bear six normal dorsal papillae. On the eighth abdominal segment only four dorsal papillae are found. Six simple papillae occur on the dorsal surface of the tip of the terminal segment. Three lateral papillae are located on each lateral surface of the segments (except on the terminal one). On the ventral surface of each thoracic segment are easily visible: two sternal papillae, four groups, each consisting of three intercalary papillae, and two simple sublateral papillae. On the first seven abdominal segments four anterior ventral papillae, located within the spinule fields, and four posterior ventral papillae are found. On the eighth abdominal segment occur only four simple ventral papillae. Eight simple papillae are located on the ventral surface of the tip of the terminal segment.

Spinule fields occur on the central surface of the meso- and metathoracic and the first eight abdominal segments. The short spinules have a widened base. The spinule field on the mesothorax comprises only two rows of spinules and is hardly visible. The field on the metathorax is compact. The fields on the first seven abdominal segments contain eight bare oval spots (fig. 33,5). The small field on the eighth abdominal segment consists of only a few rows. Two or three separate rows of spinules are found at some distance in front of the fields. The segments are moderately wide. The width of the segments only slightly surpasses their length.

The small and rounded terminal segment (fig. 33,6) gradually narrows to the posterior margin. Two small unsclerotized outgrowths are found on its tip. The apex of each outgrowth is divided into two, and bears two setiform papillae. The terminally positioned, rounded anal pore (fig. 33,6) is situated directly on the surface of the segment, not on a retractable tube.

*Biology*. The larvae develop in compost and feed on mycelium.

*Comment*. Wyatt published an accurate diagnosis of the genus (Wyatt 1961).

### **Henria psalliotae** Wyatt

*Morphology*. The morphology matches the diagnosis of the genus. On the abdominal segments each spinule field contains up to 22 rows of spinules, on the thoracic segments up to 13.

*Biology*. The larvae develop on the mycelium of edible mushrooms.

*Material.* We made use of the description of the larva published by Wyatt, and of a series of larvae which Wyatt (Institute for Plant Cultivation, Glasgow, Great Britain) sent us.

## 17. Genus **Leptosyna** Kieffer (fig. 33,7-9)

The larvae of *Leptosyna* sp., which J. Wyatt sent us, stand out by their extremely long, smooth body. The segments are 1.5 times as long as broad.

The simple, short head capsule has thin tentorial arms. The antennae have two segments. The spatula (fig. 33,8) is one-toothed. The cuticle of the first five body segments is transversely striated. The terminal segment (fig. 33,7,9) is provided with two well-developed, dorsally bent hooks with a rounded apex. Two short, blunt, sclerotized protuberances are located on the dorsal surface in front of the hooks.

J. Wyatt is presently preparing for publication an exhaustive description of the larvae of this genus. We gratefully acknowledge his permission to use the material he sent us.

## 18. Genus **Heteropeza** Winnertz

*Morphology.* The white, more or less transparent, larvae are very tiny and slender. The spatula is not clearly outlined, and has no well-developed head.

The small head capsule is comparatively weakly sclerotized and about as long as wide. The dorsal part of the capsule is slightly elongated compared to its ventral part. The tentorial arms are not pronounced. The antennae are relatively short. The second antennal segment is two to three times as long as the first, and two times as long as the width of its base. The apex of the second segment is rounded.

The shape of the short spatula is irregular; the base is slightly widened, the head and the tooth are weakly pronounced (in some cases only the tooth is manifest).

The smooth body bears simple papillae. On the dorsal surface of the thoracic and the first seven abdominal segments the usual set of six simple dorsal papillae is found; on the eighth abdominal segment only four dorsal papillae occur. On the ventral and dorsal surface of the terminal segment fourteen terminal papillae are located. The ventral surface of the thoracic segments bears the usual sternal, sublateral, and intercalary papillae. On the first seven abdominal segments the four anterior ventral papillae, which are located behind the spinule fields, as well as the four posterior ventral papillae are clearly visible. On the eighth abdominal segment only four simple ventral papillae are found.

On the ventral surface of the metathoracic and the first seven abdominal segments occur compact spinule fields.

The terminal segment gradually tapers off to the broadly rounded tip. The terminal pair of stigmata is located on the lateral surfaces of the segment and has the appearance of small sclerotized verrucae. It is not shifted to the posterior margin.

*Biology.* The larvae develop underneath rotting bark and in wood, more rarely they are found in other decomposing vegetable substrates. They are able to reproduce paedogenically.

*Comment.* Möhn gave an accurate diagnosis of this genus (Möhn 1960).

**Heteropeza pygmaea** Winnertz

*Morphology*. See the diagnosis of the genus. The spinule fields on the metathorax consist of 10-11 rows, those on the abdominal segments of 14-18 rows, the first three of which are separated from the others.

*Biology*. A large colony of larvae has been found below the bark of an oak.

*Material*. Vicinity of Tula, 29 May 1958, five larvae.

19. Genus **Heteropezina** Pritchard

*Morphology*. The small, white larvae are more or less transparent.

The head capsule is slightly elongated, and longer than the width of its base. The latero-posterior margin of the head capsule is heavily sclerotized and nearly black. The tentorial arms are not pronounced. The antennae are well-developed. Their second segment is as wide, and two or three times as long, as the first. The morphology of the spatula is the same as in the genus *Heteropeza* Winnertz.

The smooth body bears simple papillae. The compact spinule fields are located on the ventral surface of the metathoracic and the first seven abdominal segments. The spinules of the last three rows are smaller than those of the preceding rows.

The oval terminal segment does not bear any protuberances, outgrowths, etc. The stigmata have the appearance of small, sclerotized verrucae. The posterior pair of stigmata is situated on the central part of the penultimate segment and is not shifted to the dorsal surface.

*Biology*. The larvae develop below the bark of and within strongly decaying wood.

*Comment*. The diagnosis of the genus is given for the first time.

**Heteropezina cathistes** Pritchard

*Morphology*. See the diagnosis of the genus. The spinule field on the metathorax consists of six, on the first abdominal segment of eight, and on the other segments of thirteen rows of spinules. The first three rows are found at some distance from the other rows.

*Biology*. The larvae have been observed in strongly decayed, damp, grey wood of alder.

*Material*. Volkovo, Yaroslavskiy oblast, 20 June 1961, five larvae.

*Comment*. The diagnosis of the larvae is given for the first time.

## SUBFAMILY CECIDOMYIINAE

Adult gall midges belonging to the subfamily Cecidomyiinae (in which the first tarsal segment is four to ten times shorter than the second) can be reliably distinguished from those belonging to the Lestremiinae (in which the length ratio of the first tarsal segments is reversed). The larvae of the subfamily Cecidomyiinae are, however, very heterogeneous as Möhn in particular pointed out (Möhn 1955b). He proposed to split off the tribe Porricondyliini as a subfamily of its own. Clearly, this situation forces us to reconsider the system of the gall midges.

We, however, are of the opinion that still insufficient material has been collected to warrant such a far-reaching revision of the family. Therefore, we refrain from erecting new subfamilies and retain five tribes in the subfamily Cecidomyiinae. In the following text we will provide accurate descriptions of two of these tribes.

Key to the tribes of Cecidomyiinae

1(2) Eighth abdominal segment bearing four dorsal papillae, situated between the stigmata. The first seven abdominal segments provided with four posterior ventral papillae, apart from the anterior ones. Exclusively mycetophages. **Porricondyliini** (p. 126)

2(1) Eighth abdominal segment bearing two dorsal papillae between the stigmata. The first seven abdominal segments provided with two posterior papillae, apart from the anterior ones. Mostly phytophages, less frequently mycetophages or predators.

3(4) Posterior ventral papillae on the first seven abdominal segments simple. If the posterior ventral papillae are setiform, then the exocephalic arms are long and highly developed (*Monobremia* Kieffer, *Endaphis* Kieffer, *Aphidoletes* Kieffer, *Lestodiplosis* Kieffer, *Arthrocnodax* Rübsaamen). Body, generally, bare, without cuticular structures. . . . . . . . . . . . . . . . . . . . . . . . **Cecidomyiini** (p. 181)

4(3) Posterior ventral papillae on the first seven abdominal segments setiform. If the posterior ventral papillae are simple, then the exocephalic arms are reduced (genera group of *Rhopalomyia* Rübsaamen) or the mandibles are needle-shaped (*Rhizomyia* Kieffer). Body always bearing conspicuous cuticular structures.

5(6) Ventral papillae on the eighth abdominal segment setiform; if simple then the spatula is two-toothed or absent and the terminal segment bears no sclerotized hooks or teeth, positioned on robust outgrowths. Eighth abdominal segment always bearing two dorsal papillae. . . . . . . . . . . . . . . **Oligotrophini, Lasiopterini**

6(5) Ventral papillae on the eighth abdominal segment simple; if setiform (*Asphondylia* H. Loew) then the eighth abdominal segment bears four dorsal papillae and all ventral papillae on the thoracic segments have bristles. Spatula one-toothed, if two-toothed then the terminal segment bears a pair of sclerotized hooks or teeth, positioned on robust outgrowths. . . . . . . . . . . . . . . . . . . . . **Asphondyliini**

## TRIBE PORRICONDYLIINI

Morphologically the larvae of the tribe Porricondyliini constitute one of the most multiform and interesting groups of gall midge larvae. In some species the head may be comparatively small, though well-developed, simple, and provided with weakly pronounced tentorial arms; whereas in other species the internal skeleton may be highly pronounced, showing massive exocephalic arms.

The body may be bare or covered with various cuticular structures. The papillae may just as well be simple as complex. The location of the stigmata strongly varies. The spatula may be three-, two-, or one-toothed, or may be entirely lacking.

The rounded terminal segment is smooth or bears various outgrowths and sclerotized processes. The only character that is uniform in the majority of larvae is the position of the anal pore. This has the appearance of a clearly outlined, narrow, long slit on the ventral side of the terminal segment.

A comparison of the larvae of the various groups within the tribe shows a definite correlation of characters. This allows one to fundamentally subdivide the larvae of the tribe into two morphological types.

The first, rather large, group comprises the larvae of: *Ulmomyia* Mamaev, *Dicerura* Kieffer, *Winnertzia* Rondani, *Camptomyia* Kieffer, *Rhipidoxylomyia* Mamaev, *Trichoxylomyia* Mamaev, *Asynapta* H. Loew, *Synepidosis* Mamaev, and *Colomyia* Kieffer. They are united by structural aspects of the head.

The simple, short head capsule of these larvae is not provided with any sclerotized formation protruding beyond the capsule. The head is relatively weakly sclerotized, light yellow, small, and somewhat triangular. It is as long as or slightly longer than the width of the capsule at its base. The posterior edge of the capsule is not clearly outlined. It is smooth, and does not show any processes or arms. In many representatives of this group (*Trichoxylomyia* Mamaev, *Asynapta* H. Loew, *Dicerura* Kieffer) the internal skeleton of the head is inconspicuous. This is clearly due to the very weak sclerotization of the head capsule itself as well as of all parts belonging to it. Small rods running transversely along the posterior edge of the head capsule are developed in the larvae of *Camptomyia* Kieffer and particularly in *Winnertzia* Rondani and *Rhipidoxylomyia* Mamaev. Massive anterior tentorial arms are only well-developed in the larvae of *Colomyia* Kieffer. However, as is the case in the other representatives of this group, these arms are located within, and not extending beyond, the head capsule. Of those with a simple head capsule, nearly all papillae are simple in the following genera: *Winnertzia* Rondani, *Camptomyia* Kieffer, *Asynapta* H. Loew, *Rhipidoxylomyia* Mamaev, *Trichoxylomyia* Mamaev, and *Synepidosis* Mamaev. The only complex papillae are found on the thoracic segments where a part of the intercalary papillae bears short bristles. The larvae of *Dicerura* Kieffer, *Ulmomyia* Mamaev, and *Solntsevia* Mamaev, which bear papillae with well-developed, though short, bristles, form a rather distinct group. Thus, on the larvae of *Ulmomyia* Mamaev, for example, apart from the intercalary papillae, also the central pair of the four anterior ventral papillae bears bristles. On the larvae of *Solntsevia* Mamaev all dorsal, lateral, and ventral papillae are provided with a short bristle.

The body of the larvae of this group is naked, without any protuberances or verrucae on the dorsal and lateral surfaces of the segments. In some cases the body is longitudinally striated as, for instance, in *Winnertzia* Rondani and, to a lesser degree, in *Rhipidoxylomyia* Mamaev. The spinule fields, which are located on the anterior part of the segments, constitute the sole cuticular formations.

The largest subgroup among the larvae with a simple head capsule is formed by the closely related genera: *Winnertzia* Rondani, *Camptomyia* Kieffer, *Trichoxylomyia* Mamaev, *Rhipidoxylomyia* Mamaev, *Ulmomyia* Mamaev, and the affiliated genera *Asynapta* H. Loew and *Dicerura* Kieffer. All of these have spinule fields on the dorsal as well as on the ventral surface of the meso- and metathoracic and the first seven abdominal segments. The fields on the thoracic segments are, as a rule, compact. The abdominal segments of the larvae of *Asynapta* H. Loew, *Winnertzia* Rondani, and *Rhipidoxylomyia* Mamaev have spinule fields with bare oval spots on the dorsal surface, but compact fields on the ventral surface. The first seven abdominal segments of the larvae of *Asynapta* H. Loew and *Winnertzia* Rondani, and the abdominal segments of *Rhipidoxylomyia* Mamaev from the second onwards bear on the dorsal surface spinule fields with four bare oval spots.

The larvae of *Dicerura* Kieffer, *Camptomyia* Kieffer, and *Ulmomyia* Mamaev have dorsally as well as ventrally spinule fields with some bare oval spots. On all investigated larvae, except those of the genus *Camptomyia* Kieffer, the spinule fields on the dorsal surface each contain two bare spots. On *Camptomyia* Kieffer only the field on the first abdominal segment contains two bare spots, those on the other segments – four. The fields on the ventral surface of *Dicerura* Kieffer and *Ulmomyia* Mamaev each contain eight, those on *Camptomyia* Kieffer seven bare spots.

The larvae of *Colomyia* Kieffer, on which the spinule fields are distributed on the meso- and metathoracic, and the first seven abdominal segments stand apart. The fields on the thoracic segments are compact, those on the dorsal surface of the abdominal segments each contain four bare spots, those on the ventral surface – six. This distribution is specific for the representatives of this genus.

The distribution of the spinule fields nicely distinguishes the genera *Synepidosis* Mamaev and *Solntsevia* Mamaev (which bear spinule fields only on the ventral surface of the segments) from all larvae that were discussed above.

The terminal segment of most species belonging to the group under discussion (apart from the representatives of the genera *Camptomyia* Kieffer, *Dicerura* Kieffer, and *Ulmomia* Mamaev) is rounded and bears two or four pointed processes or lobes at its tip. The terminal segment is not shifted to the ventral side of the penultimate one, from which it is separated by a nearly straight or slightly bow-shaped suture. The narrow, slit-like anal pore is located on the anterior half of the ventral side of the segment. The anal slit is bare and not surrounded by longitudinal rows of tiny spinules. The posterior pair of stigmata is found at the same level as, or considerably in front of, the anterior edge of the anal slit.

The tip of the terminal segment of the larvae of *Colomyia* Kieffer, *Trichoxylomyia* Mamaev, *Winnertzia* Rondani, *Rhipidoxylomyia* Mamaev, *Solntsevia* Mamaev, and *Camptomyia* Kieffer bears small sclerotized hooks. At the tip of the body of the larvae of *Asynapta* H. Loew two, and at that of *Synepidosis* Mamaev four soft loboform processes are found.

The terminal segment of the larvae of *Camptomyia* Kieffer, *Dicerura* Kieffer, and *Ulmomyia* Mamaev is shifted to the ventral side of the body; dorsally it is partly covered by outgrowths of the penultimate segment. In the larvae of *Camptomyia* Kieffer and *Ulmomyia* Mamaev the outgrowths are comparatively broad and short, i.e., no longer than the terminal segment. In the larvae of *Dicerura* Kieffer the outgrowths of the penultimate segment are extremely long and considerably surpass the terminal segment. The posterior pair of stigmata, which in the larvae of *Camptomyia* Kieffer, *Ulmomyia* Mamaev, and *Dicerura* Kieffer is found at the apices of the outgrowths of the penultimate segment, is situated at the level of the posterior half of the anal slit or noticeably further caudally. In the larvae of *Camptomyia* Kieffer and *Ulmomyia* Mamaev the tip of the terminal segment bears small sclerotized outgrowths. In the larvae of *Dicerura* Kieffer the terminal segment is provided with unsclerotized outgrowths of a similar shape as those on the penultimate segment.

The second distinct group comprises the larvae of *Holoneurus* Kieffer, *Porricondyla* Rondani, *Parepidosis* Kieffer, and *Dicroneurus* Kieffer. The larvae of this group are characterized by the specific shape of the head, the presence of various cuticular formations, complex papillae, and some other structural peculiarities.

The highly developed head capsule is heavily sclerotized, dark brown, in most cases rather robust, and comparatively short and broad. It is, generally, as long as or slightly longer than the width of its base. Both, dorsal and ventral side are well-developed; the latter is somewhat shorter than the former. A clearly defined unsclerotized spot is found near the base of the antennae posterior edge of the capsule originate thick, straight exocephalic arms. The arms either are equally thick throughout their entire length, or noticeably widened at the tip. Their length is equal to or surpassing that of the head capsule, whereas they are as thick as or slightly less than the diameter of the second antennal segment. The arms run parallel.

The larvae of this group characteristically have complex papillae. These may be setiform, clavate, conical, or dendroid. As a rule, all or a large part of the dorsal, one of the three lateral (namely the one that is closest to the stigma), and the sublateral, as well as the lateral two posterior papillae are complex. All papillae, except the sternal and intercalary ones, in the larvae of *Porricondyla* Rondani and *Parepidosis* Kieffer are provided with well-developed bristles. In *Porricondyla* Rondani the ventral papillae on the eighth abdominal segment have well-developed bristles, but in *Parepidosis* Kieffer these papillae bear either short bristles or none at all. In the larvae of *Dicerura* Kieffer the four dorsal papillae on the eighth abdominal segment stand in a single transverse row, but in *Parepidosis* Kieffer and *Porricondyla* Rondani they are distributed over two rows.

The body of the larvae belonging to the group under discussion is, contrary to that of all other porricyndyline larvae, not bare, but on the dorsal and lateral surfaces more or less densely covered with cuticular processes, platelets, verrucae, etc. The ventral surface is, however, to a large extent smooth. Thus, the entire dorsal surface of the larvae of *Holoneurus* Kieffer, *Parepidosis* Kieffer, and *Porricondyla* Rondani is densely covered with oval or polygonal platelets bearing small protuberances. Closer to the lateral surfaces the platelets pass into tooth-like protuberances. These protuberances are largest along the lateral surfaces and become gradually smaller near the transition to the ventral surface. On the larvae of *Dicroneurus* Kieffer the dorsal surface is provided with weakly pronounced, variously shaped platelets bearing protuberances. However, on the ventral part of the lateral surfaces the protuberances pass into highly conspicuous, long and sturdy, setiform formations. The terminal segment of the larvae of this group also bears hairs, bristles, and verrucae of various shapes. The segments that are dorsally densely covered with various cuticular formations are devoid of dorsal spinule fields. Spinule fields are located on the ventral surface of the meso-, and metathoracic, as well as on all or a part of the abdominal segments. All fields are well-developed and compact, without bare spots. Spinule fields are lacking on the larvae of *Holoneurus* Kieffer. But in this case the ventral surface of the meso- and metathoracic, the first five abdominal segments, and also of the terminal segment is provided with well-developed, large, sclerotized hooks.

The terminal segment of the larvae of this group has a broadly rounded tip (some species of *Holoneurus* Kieffer), or bears two broad, conical, unsclerotized outgrowths. The terminal segment is on its dorsal surface densely covered with various cuticular protuberances, as is the entire body; the ventral surface of the terminal segment is, at least on the anterior half, smooth. In most cases the outgrowths are densely covered with bristles, hairs, spinules, etc. The anal slit is situated on the posterior half of the ventral side of the segment. The slit is surrounded by longitudinal rows of spinules or slender hairs, making a dense tuft.

The stigmata, generally, appear as well-pronounced sclerotized verrucae, wich bear an opening at the apex. All stigmata, apart from the terminal ones, are situated on the anterior half of the lateral sides of the segments. The terminal pair is shifted to the dorsal side of the segment.

Analysing the morphological characteristics we may say that both groups are fairly consistent. The traits that doubtlessly have a phylogenetic significance are correlated with a whole series of secondary features. Thus, the larvae with a simple head capsule characteristically have a naked body devoid of the various cuticular formations; the majority of species belonging to this group bears spinule fields on the dorsal as well as on the ventral surface. The papillae are predominantly simple and only as an exception setiform, in which case the bristles are extremely short and hardly visible. The posterior pair of stigmata is situated on the lateral surfaces of the segment, and is not shifted to the dorsal side.

The body of the larvae with a highly developed, heavily sclerotized head capsule and massive exocephalic arms is dorsally and laterally densely covered with variously shaped cuticular formations, whereas the spinule fields are confined to the ventral surface. In all species a part of the papillae on the dorsal and lateral surfaces is complex (setiform, clavate, conical, etc.). The posterior pair of stigmata is shifted to the dorsal side of the penultimate segment. In all these features the larvae are clearly distinct from those of the other groups and, doubtlessly, more complexly organized, compared to the larvae with a simple head capsule.

Key to the genera of Porricondyliini

1(20) Head capsule (fig. 35,4) simple, comparatively short, as long as or only slightly longer than its width at the base. Tentorial arms weakly pronounced, small, short, not protruding beyond the head capsule. All segments, apart from the terminal, bare, smooth, lacking complex papillae (only in a few cases papillae with short bristles present) as well as various cuticular formations. If the thoracic segments bear small, rounded, sclerotized platelets, then the stigmata on the eighth abdominal segment (fig. 35,2) are situated on very long (longer than the terminal segment) outgrowths. In most cases spinule fields present on the dorsal as well as on the ventral surface (sometimes only on the ventral surface) of the metathoracic and all or a part of the abdominal segments.

2(3) Head capsule very short, weakly sclerotized. The two short anterior tentorial arms well-developed (fig. 34,4), not protruding beyond the head capsule; tips of the arms interconnected. Spatula (fig. 34,5) two-toothed; its manubrium short, widened towards the base. Terminal segment (fig. 34,7) bearing two small processes, distance between them at least twice the width of a process at its base. . . . . . . . . . . . . . . . . . . . . . . . . . . . . *Colomyia* Kieffer (p. 133)

3(2) Head capsule (fig. 35,6) without clear tentorial arms. Spatula (fig. 35,5) one, or three-toothed, sometimes weakly pronounced, but never two-toothed.

4(9) Terminal segment (fig. 35,7) separated from the penultimate one by a bow-shaped line. Highly developed processes of the penultimate segment embracing the terminal segment laterally. Posterior pair of stigmata situated on these processes and thus found rearwards of the level of the anterior edge of the anal slit. Anal field bearing longitudinal spinule rows along the margin of the anal slit.

5(6) Stigmata of the eighth segment (fig. 35,2) situated on the apex of strongly elongated outgrowths that sometimes surpass the length of the ninth abdominal segment, so that the stigmata are found far behind the anal slit and even the anal segment itself. Thoracic segments provided with small, rounded, sclerotized platelets. Dorsal and ventral papillae partly bearing a short bristle. Spatula one-toothed. Terminal segment (fig. 16,3) provided with two broad outgrowths, each of them bearing four setiform terminal papillae. . *Dicerura* Kieffer (p. 136)

6(5) Stigmata of the eighth abdominal segment situated on the apex of relatively short outgrowths that do not protrude beyond the edge of the ninth segment. Thoracic segments without rounded sclerotized platelets. Dorsal and ventral papillae without bristles. Spatula variously shaped. Terminal segment (fig. 35,7) provided with two small processes which taper off to the tip, each of them bearing one to several sclerotized spines (fig. 36,2; 35,7).

7(8) Stigmata of the eighth abdominal segment (fig. 35,7) robust, appearing as long, sclerotized tubes, at the base of each of which two setiform papillae are located. Terminal segment (fig. 35,7) provided with two short, pointed processes each bearing three sclerotized spines and a single setiform papilla. Spinule fields occurring on the dorsal and ventral surface of the meso- and metathoracic, and the first six abdominal segments, as well as on the ventral surface of the seventh abdominal segment. Fields on the dorsal surface of the middle abdominal segments enclosing two, on the ventral surface eight, bare oval spots. Spatula (fig. 35,8) provided with three, nearly identical teeth, the central one of which is slightly longer than the lateral ones; all teeth separated from the remainder of the spatula by a clearly demarcated suture below which two more weakly sclerotized protuberances are located. Manubrium well-developed, its base widened. *Ulmomyia* Mamaev (p. 139)

8(7) Stigmata on the eighth abdominal segment (fig. 36,2) small, appearing as rounded verrucae. Terminal segment (fig. 36,6) provided with two small, pointed, apically sclerotized processes. Above these processes four nipple-shaped dorsal papillae are located at considerable interstices. Spinule fields occurring on the dorsal as well as on the ventral surface of the meso- and metathoracic, and the first seven abdominal segments. Fields on the dorsal surface of the middle abdominal segments enclosing four elongated bare spots, those on the ventral surface seven oval bare spots. Spatula (fig. 36,3) one-toothed; its manubrium weakly pronounced, without clearly outlined base. *Camptomyia* Kieffer (p. 140)

9(4) Terminal segment (fig. 37,1) separated from the penultimate one by a weakly bent line. Position of the segments normal, i.e. penultimate segment without processes embracing the terminal one. Terminal pair of stigmata situated laterally on the penultimate segment, not shifted beyond the level of the anterior edge of the anal slit. Anal field bare, without longitudinal spinule rows along the margins of the anal pore, though sometimes transverse rows of spinules may be found behind the anal slit. Spatula (fig. 37,3) one-toothed.

10(13) Tip of the terminal segment (fig. 38,9) not provided with sclerotized processes but with two small, rounded lobes, which are not distinctly demarcated from the segment.

11(12) Tip of the terminal segment (fig. 37,1) provided with four broad, unsclerotized lobes. Posterior third part of the anal field bearing transverse rows of sclerotized spinules. Spinule fields occurring only on the ventral surface of the meso- and metathoracic, and all abdominal segments. Second antennal segment (fig. 37,2) triangular, short (no longer than twice its width). . *Synepidosis* Mamaev (p. 142)

12(11) Tip of the terminal segment (fig. 38,9) provided with two broad, unsclerotized lobes. Anal field bare, without spinule fields. Spinule fields occurring on the dorsal as well as on the ventral surface of the meso- and metathoracic, and the first seven abdominal segments. Second antennal segment (fig. 38,8) elongated-oval, more than twice as long as wide. . . . . . . . *Asynapta* H. Loew (p. 143)

13(10) Tip of the terminal segment (fig. 39,4) provided with various sclerotized processes looking like teeth, hooks, etc.

14(15) Spinule fields located only on the ventral surface of the meso- and metathoracic, and the first seven abdominal segments, but absent from the dorsal surface. Dorsal, lateral, and ventral papillae on the eighth abdominal segment bearing very short bristles. Terminal segment (fig. 39,2) provided with two small, broad processes with a small tooth on the apex. Processes not separated, but their bases touching one another. . . . . . . . . . . . . . . *Solntsevia* Mamaev (p. 147)

15(14) Spinule fields located on the dorsal as well as on the ventral surface. All dorsal and lateral papillae simple. Processes of the terminal segment (fig. 39,4) separated at the bases, if close together, then pointed and conical.

16(19) Cuticula, especially on the last segments, longitudinally striated. Spinule fields located on the dorsal as well as on the ventral surface of the meso- and metathoracic, and the first seven abdominal segments. Processes of the terminal segment (fig. 40,4) comparatively robust, well-pronounced.

17(18) Terminal segment (fig. 39,4) broad, nearly rectangular. Processes of the terminal segment with a broad base, strongly narrowed at the upper third part, bearing a small sclerotized tooth on the tip; the bases not touching one another.
*Rhipidoxylomyia* Mamaev (p. 148)

18(17) Terminal segment (fig. 40,4) with a broad anterior half, strongly narrowed in the direction of the processes. Processes of the terminal segment conical, pointed, heavily sclerotized at the apex; bases touching one another.
*Winnertzia* Rondani (p. 150)

19(16) Cuticle not longitudinally striated. Spinule fields on the meso- and metathoracic, and the first five abdominal segments located on the dorsal as well as on the ventral surface; on the next two segments only on the ventral surface. Processes of the terminal segment (fig. 40,6) very small, tooth-shaped; distance between the processes equal to four times the diameter of the base of a process.
*Trichoxylomyia* Mamaev (p. 150)

20(1) Head capsule (fig. 41,1; 43,1) heavily sclerotized, provided with massive, well-developed, long and massive, exocephalic or tentorial arms originating on the latero-posterior edges of the capsule and located outside the head capsule. Arms straight, with a free-ending, widened tip, or bow-shaped and interconnected. Arms, as a rule, longer than the head capsule and as broad as, or slightly less than, the width of the base of the second antennal segment. Body, generally, densely covered with various cuticular formations, appearing as platelets, protuberances, bristles, etc. The larvae of some species have sickle-shaped, ventral hooks (fig. 15,1). The dorsal and ventral papillae are complex: clavate, brush-shaped, setiform, etc. Spinule fields only developed on the ventral surface of the meso- and metathoracic, and the first eight or nine abdominal segments, or lacking altogether. Apex of the terminal segment quite often split by a deep emargination.

21(22) Ventral side of the head capsule (fig. 41,1) broadly emarginated and provided with bow-shaped, interconnected tentorial arms. Dorsal surface of the body densely covered with acicular or dendroidly branched protuberances of various shapes (fig. 42,2). Stigmata situated on conical outgrowths of the lateral parts of the segments (fig. 42,2). Size of the outgrowths increasing towards the tip of the body; outgrowths on the eighth abdominal segment very large, nearly as long as the segment itself. . . . . . . . . . . . . . . . . . *Gynapteromyia* Mamaev (p. 153)

22(21) Ventral side of the head capsule not emarginate. Exocephalic arms (fig. 43,1) straight, free-ending. No acicular or dendroid protuberances among the structures covering the dorsal surface of the body. Stigmata situated directly on the surface of the segments.

23(24) Dorsal papillae (fig. 43,3) brush-shaped or conical, narrowed at the base and widened at the apex. A transverse row of sickle-shaped hooks (fig. 15,1) occurring on the meso- and metathoracic, and some abdominal segments. Terminal segment (fig. 43,5) rounded or provided with small lobe-shaped outgrowths. Anal slit surrounded by longitudinal rows of spinules. *Holoneurus* Kieffer (p. 156)

24(23) Dorsal papillae (fig. 46,5) setiform or clavate, never conical or brush-shaped. Spinule fields on the ventral surface well-developed; sickle-shaped hooks lacking. Terminal segment (fig. 46,7) by a deep emargination divided into two broad, conical outgrowths, which are covered with various cuticular formations. Anal slit bordered by long hairs.

25(26) Dorsal papillae on the eighth abdominal segment (fig. 46,7) arranged into a single transverse row between the stigmata. Dorsal and lateral papillae on the thoracic and the first seven abdominal segments clavate (fig. 46,5). *Dicroneurus* Kieffer (p. 162)

26(25) Dorsal papillae on the eighth abdominal segment arranged into two rows (fig. 50,4), generally the anterior pair more weakly pronounced. Dorsal and lateral papillae on the thoracic and the first seven abdominal segments (fig. 48,4) setiform or rod-like.

27(28) A part of the dorsal and lateral papillae (fig. 4,2) provided with extremely long bristles, surpassing the width of the body. . . . . . . *Parepidosis* Kieffer (p. 166)

28(27) All dorsal and lateral papillae (fig. 54,1) setiform or rod-like, small, considerably shorter than the width of the segments. . . . . . *Porricondyla* Rondani (p. 175)

1. Genus **Colomyia** Kieffer (fig. 34)

*Morphology*. The orange, flattened larvae are comparatively wide (length-width ratio 3:1). The body is smooth and bears hardly any outgrowths; only on the terminal segment two small, tooth-like protuberances are found.

Figure 34. Larvae of the genus *Colomyia* Kieffer
*C. clavata* Kieffer: 1 – spatula; 2 – terminal segment, dorsal view; 3 – terminal segment, ventral view; *Colomyia* sp.: 4 – head; 5 – spatula; 6 – spinule rows on the ventral surface of the abdominal segments; 7 – last two segments, dorsal view (1-3 – after Möhn 1955b; 4-7 – orig.)

The simple head capsule (fig. 34,4) is relatively wide and short. The length slightly surpasses the width at the base. Within the capsule occur two anterior tentorial arms, which do not protrude beyond the capsule, and whose ends are connected to the margin of the capsule. The antennae are cylindrical, two-segmented, rounded at the tip, and comparatively short. The second segment is no more than twice as long as it is wide at the base.

The spatula (fig. 34,1,5) bears two teeth. The triangular teeth are apically pointed and separated by a shallow emargination.

The body is smooth, without outgrowths. The papillae are simple, very weakly pronounced, and barely visible. The usual set of papillae is found, comprising six dorsal papillae on the thoracic and the first seven abdominal segments, and four dorsal papillae on the eighth abdominal segment. Near each stigma still another papilla, which other authors regard as a lateral one (Möhn 1955b), occurs. On each lateral surface of the segments three lateral papillae are found. The sternal and sublateral papillae are simple, and bear no bristles. On the first seven abdominal segments only two anterior, as well as two posterior, ventral papillae occur. The former are located within, the latter behind, the spinule fields.

Large spinule fields can easily be seen on the dorsal as well as on the ventral surface of the meso- and metathoracic, and the first eight abdominal segments.

The terminal segment (fig. 34,2,3) bears apically two short, conical, sclerotized processes. The distance separating these processes is at least equal to twice the width of a process at its base. The bare anal slit is found on the ventral side of the segment.

The stigmata of the last pair are small and slightly shifted to the posterior margin of the segment.

*Biology.* The larvae are acknowledged inhabitants of various decaying organic substrates. They are encountered in moss, litter, decaying wood, and in dead leaves of grasses.

Key to the species of *Colomyia* Kieffer

1(2) Spatula (fig. 34,1) comparatively short. Manubrium small and narrow; along its entire length as wide as the head of the spatula; not widened at the base. Teeth of the spatula as well as the emargination between them, rounded. Anal pore surrounded by 8-10 longitudinal rows of spinules. . . . . . . . . . . . . . . . *C. clavata* Kieffer

2(1) Spatula (fig. 34,5) comparatively long. Manubrium robust, noticeably widened at its base; its largest width considerably surpassing the width of the head of the spatula. Teeth of the spatula pointed. Anal pore bare, without longitudinal rows of spinules. . . . . . . . . . . . . . . . . . . . . . . . . . . . . . . . . . *Colomyia* sp.

**Colomyia clavata** Kieffer

*Morphology.* The elliptic body is dorso-ventrally flattened.

The spatula (fig. 34,1) is comparatively short. The manubrium is small and narrow. Along its entire length it is as wide as the head of the spatula; it is not widened at the base. The teeth of the spatula as well as the emargination between them are rounded.

The smooth body bears simple papillae. The anterior ventral papillae are located between the fifth and the sixth row (counted from the rear edge) of the spinule fields. The spinule fields on the dorsal surface consist of 22-24 (on the eighth abdominal segment only of 7-9) rows. The spinule fields on the ventral surface comprise 28-32 (on the eighth abdominal segment only 10-12) rows.

The anal pore is surrounded by 8-10 longitudinal rows of spinules (fig. 34,3).

*Biology.* The larvae live in litter on twigs that are covered with mould.

*Material.* We used the description of the larva given by Möhn (1955b).

**Colomyia** sp.

*Morphology.* The antennae are comparatively short. The antennal segments are small and of about equal length. The segments are no more than twice as long as wide. The robust manubrium (fig. 34,5) is conical and noticeably widened at the base. The teeth of the spatula and the emargination separating them are sharp.

The short and broad spinules (fig. 34,6) form large fields. On the dorsal and ventral surface of the meso- and metathoracic segments the fields comprise 11-14 rows, on the first seven abdominal segments 14-27. The number of rows is largest on the fifth abdominal segment. On the eighth abdominal segment only 5-7 rows are found. The spinule fields on the dorsal as well as on the ventral surface of the meso- and metathoracic as well as on the eighth abdominal segment are compact, without clear bare spots. On the first seven abdominal segments the fields on the dorsal surface each enclose four elongated bare spots, while those on the ventral surface contain six. On the seventh abdominal segment the bare spots, usually, are very weakly pronounced.

The bare anal pore (fig. 34,7) is situated on the ventral side of the terminal segment. It is not surrounded by rows of spinules.

*Biology*. The larvae develop in the mycelium in clefts in dying stubs of oaks that have been felled with an axe.

*Material*. Khust, Zakarpatskaya oblast, Ukr. S.S.R., 13 July 1963, three larvae.

*Comment*. The larva is described for the first time.

## 2. Genus **Dicerura** Kieffer (fig. 14,3; 16,3; 35,1-5)

*Morphology*. The cream coloured larva is flattened and comparatively long. The head is well-developed. The eighth abdominal segment is provided with two long outgrowths which bear the stigmata. The length-width ratio of the body is 4:1.

The head capsule (fig. 35,4) is short, its length is slightly less than its width at the base. There are no clearly pronounced tentorial arms. The antennae are well-developed. The first antennal segment is very short, it is no less than one and a half times as wide as it is long. The second segment is elongated-oval. It is nearly twice as long as its width in the middle.

The spatula (fig. 35,5) is one-toothed. The manubrium is weakly pronounced and parallel-sided. Sometimes its base is slightly widened.

The smooth body bears simple or setiform papillae. The dorsal and lateral papillae are simple or bear short bristles. The sternal papillae are simple (except on the metathorax where they are provided with short bristles). The intercalary papillae are of the usual shape. The sublateral papillae on the prothorax lack bristles, those on the meso- and metathorax are simple or bear short bristles. The ventral papillae on the abdominal segments (including the terminal one) bear short bristles. The anterior ventral papillae are located within the spinule fields.

Spinule fields occur on the dorsal as well as on the ventral surface of the meso- and metathoracic, as well as of the first six abdominal segments, but on the ventral surface only of the seventh abdominal segment. The spinules (fig. 14,3) are wide and predominantly blunt in the anterior, but slightly tapering off and pointed in the posterior part of the fields.

The last two segments (fig. 16,3) have a peculiar shape. The penultimate, eighth, abdominal segment is robust, as large as the seventh. It is provided with two long, soft outgrowths, which bear the terminal pair of stigmata. The latter appear as small, sclerotized verrucae. A single verruca-like papilla is found on the middle part of the outgrowths rather close to the outward facing surface. The rounded terminal segment is very small, nearly half as large as the preceding one. It is shifted to the ventral side of the penultimate segment, by which it is dorsally partly covered. The posterior edge of the terminal segment is provided with two broad outgrowths, which are half as long as those of the penultimate segment. On the tip of the outgrowths, four verruca-like papillae bearing short bristles are found. The anal slit is located on the ventral side of the terminal abdominal segment.

*Biology.* The larvae are semi-aquatic. They occur in leaf sheaths of irises, in damp, decaying wood, etc.

Key to the species of *Dicerura* Kieffer

1(2) Thoracic segments covered with rounded, sclerotized platelets. Spatula (fig. 35,1) provided with a small triangular tooth with a rounded tip. Base of the manubrium only slightly widened. Spinule fields on the dorsal surface consisting of 6-8, on the ventral surface of 10-12 rows of spinules. Anal slit surrounded by rows of spinules. In front of the anal slit 2-3 transverse rows of spinules are found. *D.kaltenbachi* (Rübsaamen)

2(1) Thoracic segments smooth. Tooth of the spatula (fig. 35,5) broad and rounded. Base of the manubrium strongly widened, at least twice as wide as the manubrium. Spinule fields on the dorsal surface of the abdominal segments consisting of 12-15, on the ventral surface of 16-25 rows of spinules. Spinule fields on the thoracic segments comprising 10-11 rows. Anal slit smooth. Terminal segment in front of the anal slit also smooth, without spinules. . . . . . . . . . . . . . . *Dicerura* sp.

**Dicerura kaltenbachi** (Rübsaamen)

*Morphology.* The spatula (fig. 35,1) is one-toothed. The nearly triangular tooth gradually narrows from the base to the apex. The base of the manubrium is not widened.

The dorsal surface of the body is smooth, only the thoracic segments are covered with rounded platelets. The dorsal and ventral papillae on the abdominal segments bear short bristles. The anterior ventral papillae are located within the spinule fields. The sternal papillae on the pro- and mesothorax are simple, those on the metathorax bear short bristles. The sublateral papillae on the prothorax are simple, whereas those on the meso- and metathorax are provided with bristles. The spinule fields consist on the dorsal surface of 6-8, on the ventral surface of 10-12 rows of spinules. The anal slit is surrounded by longitudinal rows of spinules. Two or three transverse rows of spinules occur on the terminal segment in front of the anal slit (fig. 35,3).

*Biology.* The larvae inhabit the leave sheaths of irises.

*Comment.* We made use of the description by Möhn (1955b).

**Dicerura** sp.

*Morphology.* The spatula (fig.35,5) is one-toothed. The very wide tooth has a rounded tip. The length of the tooth is slightly less than its width at the base. The base of the well-developed manubrium is widened; its width is no less than twice that of the manubrium.

The cuticle of the dorsal surface of the otherwise bare body is longitudinally striated. The dorsal as well as the ventral papillae on the abdominal segments bear short bristles. The anterior ventral papillae on the abdominal segments are located within the spinule fields. The sternal, sublateral, and a part of the intercalary papillae are simple. The remaining intercalary papillae bear short bristles. The spinule fields on the thoracic segments are compact and consist of 10-11 rows of spinules. The spinule fields on the

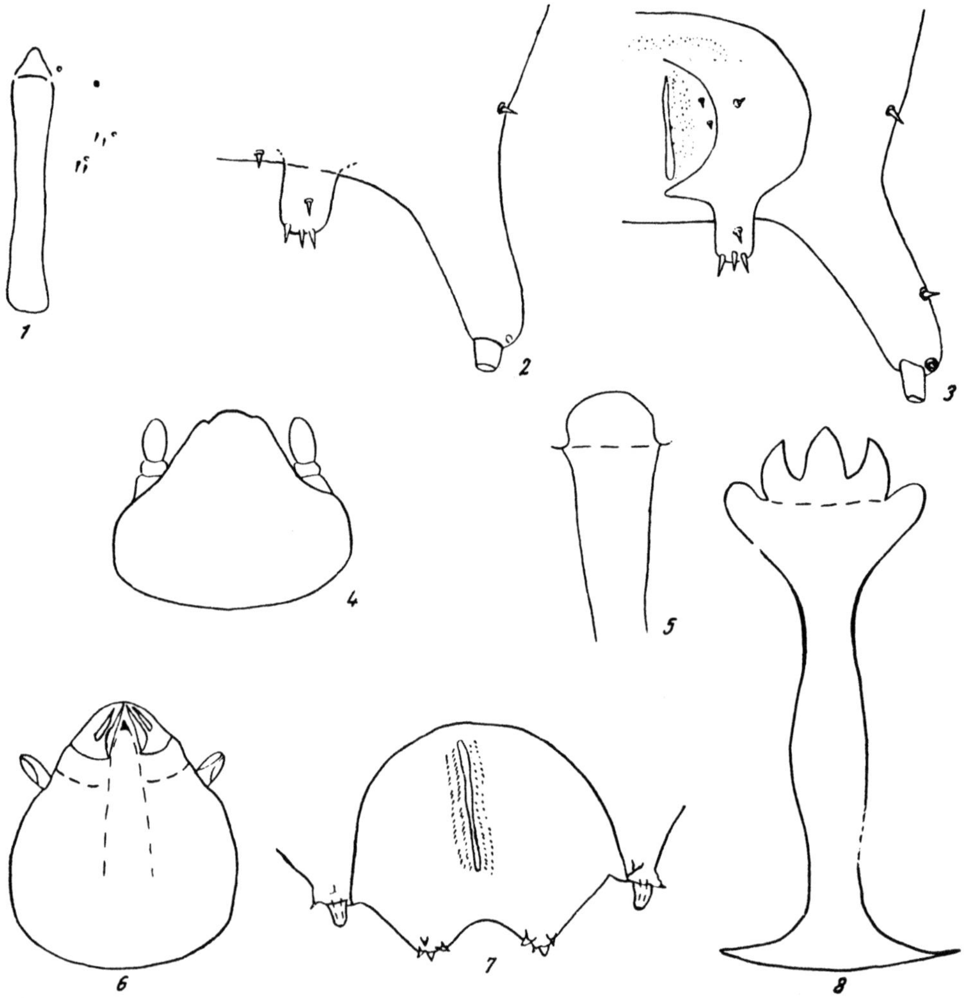


Figure 35. Larvae of *Dicerura* Kieffer and *Ulmomyia* Mamaev
*D. kaltenbachi* (Rübsaamen): 1 – spatula; 2 – last two segments, dorsal view; 3 – last two segments, ventral view; *Dicerura* sp.: 4 – head; 5 – spatula; *U. fungicola* Mamaev: 6 – head; 7 – last two segments, ventral view; 8 – spatula (1-3 – after Möhn 1955b; 4-8 – orig.)

dorsal surface of the first five abdominal segments contain two rounded, bare spots and consist of 12-15 rows. On the sixth abdominal segment only two or three rows of spinules are found. On the ventral surface of the first seven abdominal segments, the spinule fields each enclose eight rounded, bare spots. Each field is made up of 16-25 rows. The highest number of rows is found on the segments in the middle of the body. The anal slit (fig. 16,3) is smooth and not surrounded by longitudinal and transverse rows of spinules.

*Biology*. The larvae have been found in damp, strongly decomposed beech wood.

*Material*. Kvasy, Zakarpatskaya oblast, Ukr. S.S.R., 19 June 1963, two larvae.

*Comment*. The larva is described for the first time.

3. Genus **Ulmomyia** Mamaev (fig. 35,6-8)

*Morphology*. The elongated-oval larvae have a length-width ratio of 6:1. The end of the body is provided with four processes, which bear stigmata and sclerotized verrucae.

The simple head capsule (fig. 35,6) is short and oval, without emargination at the ventral side. It is at most only slightly longer than wide. The tentorial arms are not pronounced. The antennae are short. The first antennal segment is vestigial. The oval, second segment slightly tapers off at the apex and is one and a half times as long as its width at the base.

The spatula (fig. 35,8) bears three distinct teeth of the same shape, the central one slightly surpasses the lateral ones. The teeth are separated from the rest of the spatula by a clearly pronounced suture. The manubrium is well-developed.

The body is smooth, without processes. The papillae are predominantly simple. As a rule, the dorsal surface of the segments bears six simple papillae. Only four small, simple papillae are found on the dorsal surface of the eighth abdominal segment. Three lateral papillae are located on each of the lateral surfaces of the thoracic and the first seven abdominal segments. On the ventral surface of the thoracic segments occurs the usual set of papillae: two simple sternal, twelve intercalary (arranged into groups of three, two bearing a short bristle and a simple one), and two simple sublateral papillae. On the first seven abdominal segments four conspicuous anterior ventral papillae are found. The central ones of these bear short bristles and are located within the spinule fields; the two simple outer ones are situated behind the spinule fields. The four posterior ventral papillae are simple and form a single transverse row. Four simple ventral papillae are found on the eighth abdominal segment.

Rows of tiny spinules are distributed on the dorsal and ventral surface of the meso- and metathoracic, and the first six abdominal segments. On the seventh abdominal segment only the ventral surface bears such rows of spinules. These spinules form fields analogous to those found in *Winnertzia* Rondani and some related genera. The fields on the thoracic segments are compact. On the dorsal surface of the first six abdominal segments each field contains two, on the ventral surface eight, oval bare spots, whereas the field on the ventral surface of the seventh abdominal segment contains six oval bare spots.

The terminal segment (fig. 35,7) is separated from the penultimate one by a curved fold. Two wide processes occur on the tip of the terminal segment. Each process bears four small, sclerotized verrucae. The stigmata (which appear as long, sclerotized tubes) are located on the tips of small processes of the latero-posterior edges of the penultimate segment. Two small, bristle-bearing verrucae are found on each of the processes, in addition to the stigmata. The posterior pair of stigmata is located at the level of the anal slit, or more rearwards. All other stigmata appear as small, sclerotized processes which are situated halfway on the lateral surfaces of the segments. The anal slit is found on the ventral side of the terminal segment and is surrounded by 6-7 longitudinal rows of spinules. Six simple anal papillae are distributed around the anal slit. Two simple papillae are found on the processes of the ventral side of the segment.

*Biology*. The larvae live beneath rotting bark and deep within decaying wood.

*Comment*. The diagnosis of the genus is given for the first time.

**Ulmomyia fungicola** Mamaev

*Morphology*. Two rather weakly sclerotized processes occur on the spatula (fig. 35,8), behind the three large teeth. The base of the conspicuous manubrium is broadened.

The spinule fields are large and wide. The fields on the ventral surface of the thoracic segments comprise 17-20, those on the dorsal surface 13 rows of spinules. The fields consist of 28-29 rows on the ventral surface of the first and seventh abdominal segments, and of 33-36 on that of the other segments. On the dorsal surface of the first six abdominal segments the fields are made up of 20-24 rows. The spinules within the fields are sturdy and broad, in particular those on the ventral surface.

*Biology*. The larvae have been found on patches of white mould beneath the rotting bark of a spruce tree.

*Material*. Lake Glubokoe, Moskovskaya oblast, 23 August 1960, three larvae.

*Comment*. The larvae are described for the first time.

4. Genus **Camptomyia** Kieffer (fig. 16,1; 36)

*Morphology*. The body of the cream-coloured or orange larvae gradually tapers off to the posterior end. The smooth body is not longitudinally striated.

The base of the simple head capsule (fig. 36,1) is strongly widened. The head capsule is as long as, or slightly longer than, it is wide. The second segment of the two-segmented antennae is two or more times as long as its width at the base.

The spatula (fig. 36,3) is one-toothed. The tooth is of the same type as that of *Winnertzia* Rondani. The base of the weakly pronounced manubrium is not distinct.

With the exception of the penultimate and the terminal segments, the smooth body does not bear outgrowths. The simple papillae are weakly pronounced. The segments each bear only ten dorsal papillae. Three lateral papillae occur on each lateral surface. On the ventral surface of the segments the usual set of weakly developed simple papillae is found. A row of four ventral papillae is located on the eighth abdominal segment.

Spinule fields (fig. 36,4-5) occur on the anterior part of the dorsal as well as the ventral surface of the meso- and metathoracic, and the first seven abdominal segments. The fields on the thoracic segments are compact, the one on the dorsal surface of the first abdominal segment contains two indistinct oval bare spots, and those on the other segments (up to and including the seventh abdominal) contain four elongated bare spots. The field on the ventral surface of the first abdominal segment contains four weakly visible bare spots, whereas those on the other segments (up to and including the seventh abdominal) contain seven oval bare spots.

The small terminal segment (fig. 36,2) gradually narrows into the apex, which bears two wide, sclerotized processes with a pointed tip. The bases of the processes are, in most cases, rather widely apart and separated by a wide, rounded emargination. The latero-posterior margins of the penultimate segment form small outgrowths, which bear the terminal pair of stigmata. As a result of this situation the terminal pair of stigmata is usually located at the level of the central part of the anal pore. It is, in any case, not found in front of the anal pore, as is the case in the larvae of *Winnertzia* Rondani and some other genera. The anal slit is surrounded by a few rows of weakly visible spinules.

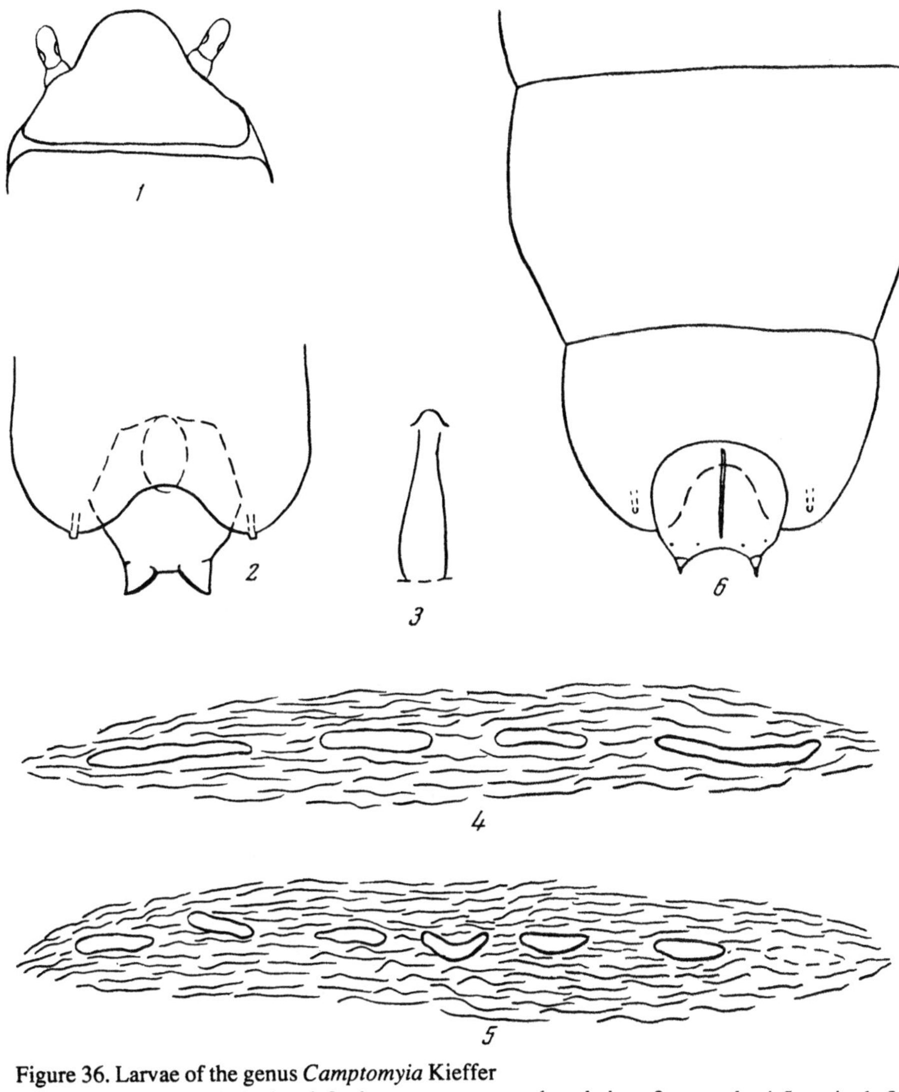


Figure 36. Larvae of the genus *Camptomyia* Kieffer
*C. corticalis* (H. Loew): 1 – head; 2 – last two segments, dorsal view; 3 – spatula; 4-5 – spinule fields on abdominal segment (4 – dorsal view, 5 – ventral view); *C. maxima* Mamaev: 6 – last three segments, ventral view (orig.)

*Biology*. The larvae occur in the forest zone and are exclusively inhabitants of decaying bark, wood, and fruiting bodies of polyporaceous fungi.

*Comment*. Up to this time the genus encompasses more than 50 European species, including new material. Characters that allow one to reliably distinguish the larvae have, however, not yet been found. Therefore no key for the identification of the species is presented.

5. Genus **Synepidosis** Mamaev (fig. 37)

*Morphology*. The cream-coloured body is smooth and elongated.

The small head capsule (fig. 37,2) is simple, weakly sclerotized and very short. Its length is less than its width at the base. Though the capsule is weakly sclerotized around the bases of the antennae, the membranous spots are not clearly demarcated. Tentorial arms are lacking. The antennae are comparatively short. The first antennal segment is small. The second one is nearly triangular and tapers off from the base to the tip; it is no more than twice as long as its width at the base.

The spatula (fig.37,3) is one-toothed. The broad tooth has a rounded tip. At the borderline with the manubrium the edges of the tooth spread outwards.

The smooth body bears no processes, except at the tip of the terminal segment. The simple papillae are weakly pronounced. Six simple dorsal papillae are located on the dorsal surface of each of the thoracic and the first seven abdominal segments. Only four simple papillae occur on the eighth abdominal segment. Three simple lateral papillae are found on each lateral surface of the segments. The usual set of papillae occurs on the ventral surface of the thoracic segments. The first seven abdominal segments bear two rows of ventral papillae. The four anterior ventral papillae are simple and situated within the spinule fields; the four posterior ones are also simple but located behind the spinule fields. On the eighth abdominal segment only four simple ventral papillae occur. The dorsal surface of the body lacks spinule fields.

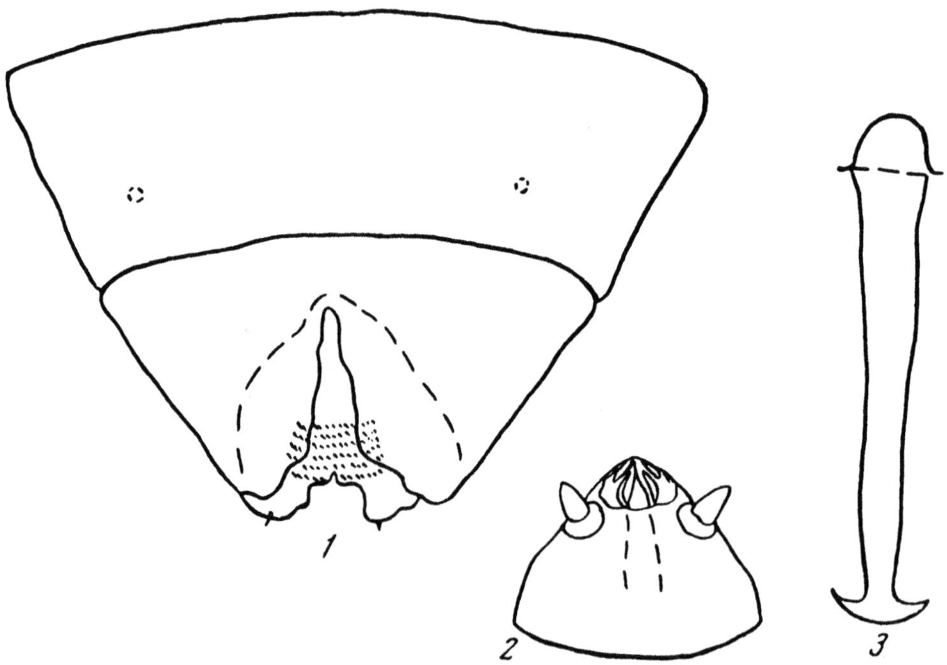


Figure 37. Larva of *Synepidosis longiventris* Mamaev
1 – last two segments, ventral view; 2 – head; 3 – spatula (orig.)

Spinule fields, consisting of a few rows of spinules, are found on the anterior part of the ventral surface of the meso- and metathoracic, as well as all abdominal segments.

The tip of the terminal segment (fig. 37,1) is provided with four wide, soft, unsclerotized, lobe-shaped processes, each apically bearing a single simple papilla. The posterior edge of the anal slit, which is situated on the ventral side of the segment, is not closed. A few transverse rows of spinules are found at the level of the lower half of the anal slit. The posterior pair of stigmata is slightly shifted to the dorsal side.

*Biology.* The larvae have been recorded from rotting bark.

*Comment.* The diagnosis of the genus is given for the first time.

**Synepidosis longiventris** Mamaev

*Morphology.* The diagnosis is the same as that of the genus. The spinule fields on the meso- and metathoracic, and the last two segments are compact and consist of ten to twelve rows. The fields on the first seven abdominal segments contain eight oval bare spots and comprise twenty to thirty rows. The largest fields are found on the middle abdominal segments.

*Biology.* The larvae have been observed on the mould beneath the very decayed bark of a maple log.

*Material.* Tellermanov forestry, Voronezhskaya oblast, 27 September 1960, two larvae.

*Comment.* The larva is described for the first time.

## 6. Genus **Asynapta** H. Loew (fig. 14,2; 15,2; 38)

*Morphology.* The long-oval larvae taper off at both ends. The body is smooth. The papillae are inconspicuous. The rounded terminal segment bears two small, wide outgrowths.

The simple head capsule (fig. 38,1,8) is small, weakly sclerotized, and short. Its length measures less than the width at the base; neither the ventral nor the dorsal side are emarginate. The small anterior tentorial arms are situated within the head capsule. The antennae are well-pronounced. The first antennal segment is very short and wide; the second is three and a half times as long as its width at the base. The spatula is lacking.

The papillae on the bare body are simple. Each of the thoracic and the first seven abdominal segments bears the usual set of six simple dorsal papillae, whereas only four dorsal papillae occur on the eighth segment. Three simple lateral papillae are found on each of the lateral surfaces of the segments. On the ventral surface of the thoracic segments occur: two simple sternal, two sublateral, and four groups of intercalary papillae (each group consisting of two setiform and one simple papilla). The first seven abdominal segments each bear four simple anterior ventral papillae (fig. 38,2). These may be either paired and situated on two spherical processes, or simple. In any case, they are situated within the spinule fields. The four simple posterior ventral papillae are located behind the spinule fields. Only eight papillae are found on the eighth abdominal segment. Eight conspicuous simple terminal papillae occur on the tip of the terminal segment.

Spinule fields (fig. 38,6) are located on the dorsal as well as the ventral surfaces of the meso- and metathoracic, and the first seven abdominal segments. In a few cases spinule fields are also found on the dorsal surface of the eighth abdominal segment.

The small, apically rounded terminal segment (fig. 38,9) is provided with two small, wide outgrowths. The anal slit is situated on the ventral side of the segment.

All stigmata, including the terminal pair, are found on the lateral sides of the segments; they are not shifted to the dorsal side.

*Biology.* The larvae live beneath the loose bark of decaying logs. They have also been recorded from spruce cones.

*Comment.* Since there are no characters that allow one to satisfactorily identify the adult phases of these two genera, we consider the genus *Rübsaamenia* Kieffer as synonymous with *Asynapta* H. Loew.

Key to the species of *Asynapta* H. Loew

1(2) Anterior ventral papillae of the first seven abdominal segments (fig. 38,2) single, located on oval bare spots within the spinule fields, and forming a single transverse row. Spinule fields on the dorsal surface of the first seven abdominal segments provided with two oval bare spots. Spinule fields on the ventral surface with eight oval bare spots (not counting the bare spots bearing papillae). *A. turaui* Rübsaamen

2(1) Anterior ventral papillae on the first seven abdominal segments (fig. 38,7) paired, located within the spinule fields on two large spherical welts. Fields on the dorsal surface of the first seven abdominal segments containing four oval bare spots; fields on the ventral surface compact (not counting the locomotive welts, which lack spinules).

3(4) Locomotive welts, bearing the anterior ventral papillae (fig. 38,7), on the first two abdominal segments large, close together, without rows of spinules between them. The pointed, conical spinules on the ventral surface of the mesothorax (fig. 38,5) are close together. Interstices between them noticeably smaller than the width of the spinules themselves. Spinule fields on the thoracic segments consisting of 5-11, on the abdominal segments of 15-23 (on the middle segments usually of 21-22) rows. *A. strobi* (Kieffer)

4(3) Locomotive welts bearing the anterior ventral papillae, on the first two abdominal segments smaller. Interstices between them equal to the diameter of the welts and bearing five or six rows of clearly discernable spinules. Spinules on the ventral surface of the mesothorax (as on fig. 38,3), wide and blunt. Interstices between them about equal to the width of the spinules themselves. Spinule fields on the thoracic segments consisting of 9-14 rows, on the abdominal segments of 20-30 rows. . . . . . . . . . . . . . . . . . . . . . . . . . . . . . . . . . . . *Asynapta* sp.

**Asynapta turaui** Rübsaamen

*Morphology.* The body is smooth. The anterior ventral papillae on the abdominal segments (fig. 38,4) stand alone on oval bare spots in a single transverse row within the posterior half of the spinule fields.

The spinule fields are distributed on the dorsal and ventral surface of the meso- and metathoracic, and the first seven abdominal segments, as well as on the ventral surface of

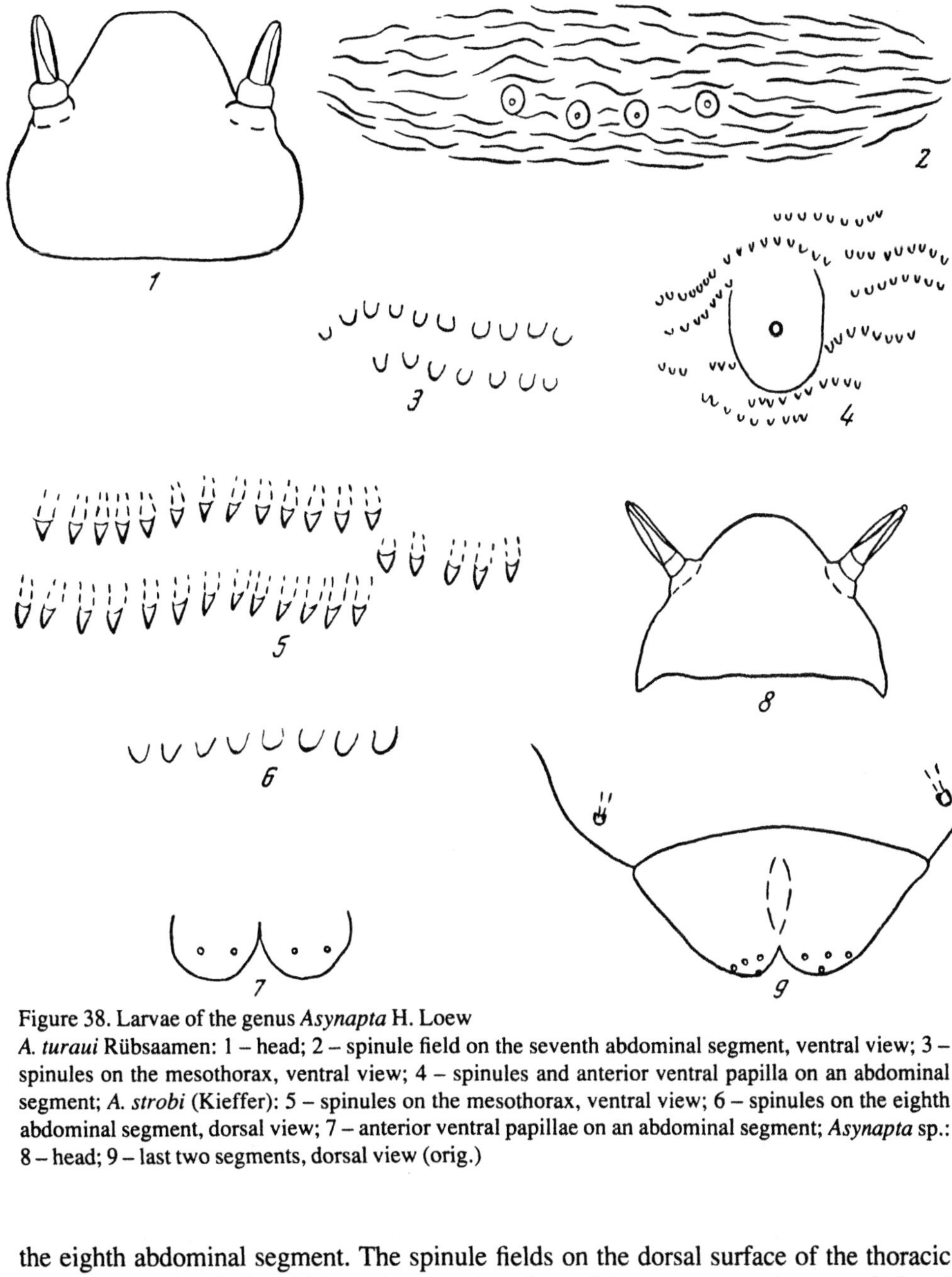


Figure 38. Larvae of the genus *Asynapta* H. Loew
*A. turaui* Rübsaamen: 1 – head; 2 – spinule field on the seventh abdominal segment, ventral view; 3 – spinules on the mesothorax, ventral view; 4 – spinules and anterior ventral papilla on an abdominal segment; *A. strobi* (Kieffer): 5 – spinules on the mesothorax, ventral view; 6 – spinules on the eighth abdominal segment, dorsal view; 7 – anterior ventral papillae on an abdominal segment; *Asynapta* sp.: 8 – head; 9 – last two segments, dorsal view (orig.)

the eighth abdominal segment. The spinule fields on the dorsal surface of the thoracic segments consist of 10 and 17, on the ventral surface of the respective segments of 19 and 24 rows of spinules. The fields on the dorsal surface of the first six abdominal segments comprise 28-38, on the ventral surface 37-48 rows. Just 18 rows occur on the dorsal surface of the seventh abdominal segment. On the ventral surface of the seventh and eighth abdominal segment 37 and 15 rows are found respectively. The fields on the thoracic and the last abdominal segments are compact. The fields on the dorsal surface of the first six segments contain two clearly demarcated oval bare spots; the fields on the ventral surface – eight (not counting the bare spots bearing simple papillae). The spinules on the ventral surface of the mesothorax are wide and blunt. The interstices between them are as wide as the spinules themselves (fig. 38,3).

*Biology.* The larvae develop in hollow stems of *Calamagrostis.*

*Material.* Vicinity of Prague (Czechoslovakia), 24 October 1963, six larvae (M. Skuhravá and B. Mamaev).

**Asynapta strobi** (Kieffer)

*Morphology.* The smooth body bears clearly discernable, simple papillae. The anterior ventral papillae on the first seven abdominal segments are paired (fig. 38,7). They are found in the central part of the spinule fields on two large spherical welts. The welts themselves, however, are not covered with spinules. The welts on the first two abdominal segments are close together, without rows of spinules between them.

The spinule fields are distributed on the dorsal as well as the ventral surface of the meso- and metathoracic, and the first seven abdominal segments. The fields comprise: 5-11 rows on the thoracic segments, 15-23 rows on the dorsal surface of the abdominal segments, 17-23 rows on the ventral surface of the abdominal segments. The fields on the thoracic segments are compact; those on the dorsal surface of the first seven abdominal segments contain four oval bare spots. Barring the rounded welts (which lack spinules), the fields on the ventral surface of the abdominal segment are compact.

The pointed, conical spinules on the ventral surface of the mesothorax are close together. The interstices between them are noticeably less wide than the spinules themselves. The spinules on the last segments are blunt (fig. 38,6).

*Biology.* The larvae develop in spruce cones and are apparently mycetophages.

*Material.* Pavlova Sloboda, Moskovskaya oblast, 27 May 1963, one larva; two larvae were received from M. Skuhravá and V. Skuhravý (Entomological Institute of the Czechoslovakian Academy of Sciences, Prague).

**Asynapta** sp.

*Morphology.* The smooth body is provided with weakly demarcated platelets. The cuticle is obliquely striated. The rounded locomotive welts which bear the anterior ventral papillae (fig. 15,2) are on the first two abdominal segments separated by a distance equal to the diameter of the welts. Five or six rows of clearly discernable spinules are found on the interstice between the welts.

Large spinule fields occur on the dorsal as well as on the ventral surface of the meso- and metathoracic, and the first seven abdominal segments. The fields comprise on the thoracic segments 9-14, on the abdominal segments 20-30 rows of spinules. The fields on the thoracic segments are compact; those on the dorsal surface of the abdominal segments contain four elongated bare spots, the inner ones of which are smaller than the outer. The fields on the ventral surface of the abdominal segments are compact, barring the rounded welts which lack spinules. The interstices between the wide, blunt spinules on the ventral surface of the mesothorax are no smaller than the width of the spinules themselves.

*Biology.* The larvae live below the dead bark of beech logs which is, however, only slightly decomposed.

*Material.* Khust, Zakarpatskaya oblast, Ukr. S.S.R., 11 July 1963, one larva.

*Comment.* The description of the larva is given for the first time.

7. Genus **Solntsevia** Mamaev (fig. 39,1-3)

*Morphology*. The cream-coloured, elongated larvae taper off at both ends. The terminal segment is provided with two small processes each bearing a small sclerotized tooth. The length-width ratio of the body equals 5:1.

The simple, small head capsule (fig. 39,1) is weakly sclerotized and shows no clearly discernable tentorial arms. The first antennal segment is very short; the second segment is elongated-oval (no more than twice as long as wide). The spatula (fig. 39,3) is one-toothed.

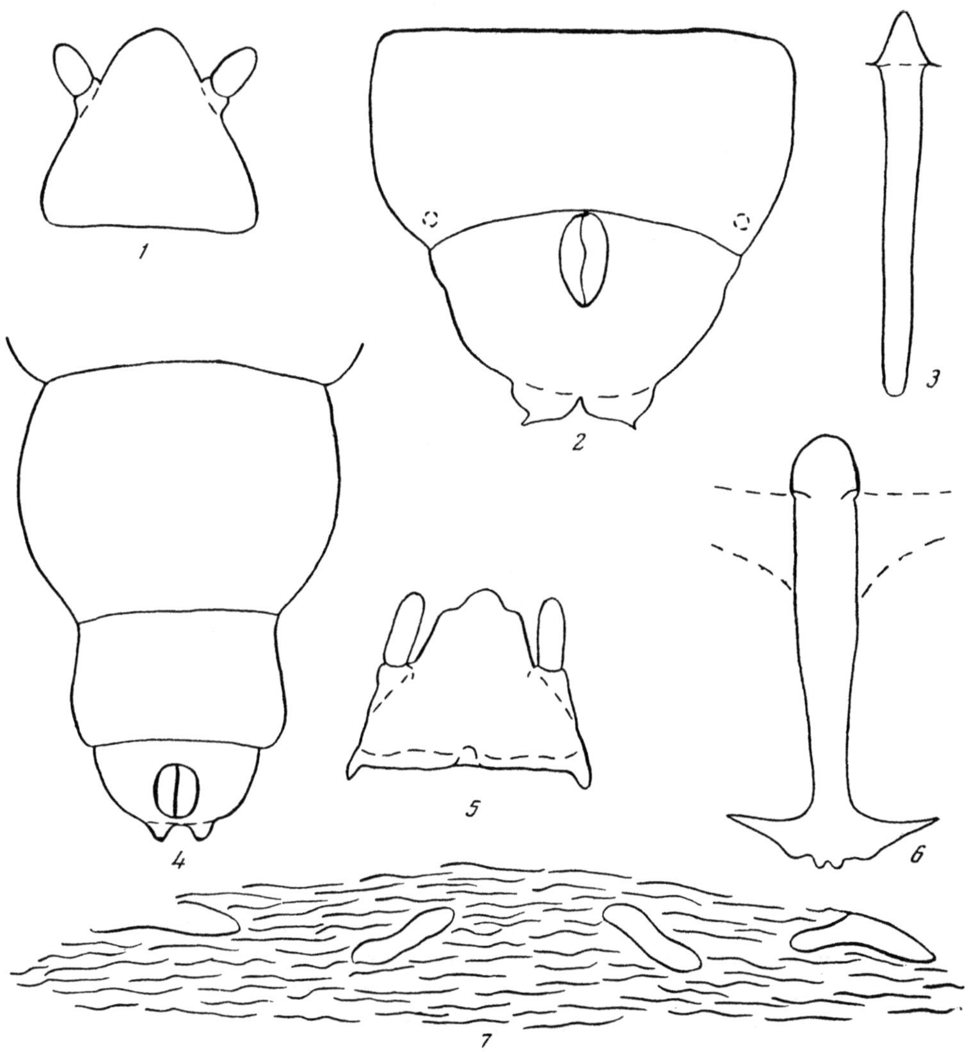


Figure 39. Larvae of the genera *Solntsevia* Mamaev and *Rhipidoxylomyia* Mamaev
*S. nigripes* Mamaev: 1 – head; 2 – last two segments, ventral view; 3 – spatula; *Rh. rubella* Mamaev: 4 – last three segments, ventral view; 5 – head; 6 – spatula; 7 – spinule field on an abdominal segment, dorsal view (orig.)

The papillae on the smooth body are weakly developed. The thoracic and the first seven abdominal segments are, on the dorsal surface, each provided with six dorsal papillae bearing very short bristles. Only four papillae are found on the eighth abdominal segment. The following papillae are easily visible on the ventral surface of each of the thoracic segments: two setiform sternal, two setiform sublateral, four groups of three normal, shortly setiform intercalary papillae. The first seven abdominal segments bear four easily visible setiform anterior ventral papillae, as well as four setiform posterior ventral papillae. The former stand in a single transverse row and are located within the spinule fields (between the fourth and sixth row, counting from the posterior edge). Only four, weakly developed, setiform papillae occur on the eighth abdominal segment.

Spinule fields are only found on the ventral surface of the meso- and metathoracic, and the first seven abdominal segments.

The rounded terminal segment (fig. 39,2) is shorter than the penultimate one, and apically bears two small, wide, toothed processes. The processes do not stand apart, but touch one another at the base. They are only apically sclerotized. The bare anal slit (fig. 39,2) is not surrounded by longitudinal rows of spinules. It is situated on the anterior half of the ventral side of the terminal segment. The very small and weakly developed stigmata are located along the lateral sides of the body on the middle part of the segments. The terminal pair is the only one that is shifted to the posterior margin of the segment.

*Biology.* The larvae inhabit decaying wood.

*Comment.* The diagnosis of the genus is given for the first time.

**Solntsevia nigripes** Mamaev

*Morphology.* The only tooth of the spatula (fig. 39,3) is pointed and triangular. The manubrium is very long and slightly narrower than the tooth. The base of the spatula is not widened.

The smooth body bears no processes. Spinule fields occur only on the ventral surface of the meso- and metathoracic, and the first seven abdominal segments. The fields on the thoracic segments are compact, comprising respectively 13 and 23 rows of spinules. The fields on the abdominal segments contain eight elongated-oval bare spots and consist of 40-47 rows on the first six abdominal segments and of 25 rows on the seventh.

*Biology.* The larvae have been found in a piece of a strongly decayed spruce stub.

*Material.* Pavlova Sloboda, Moskovskaya oblast, 27 May 1963, four larvae.

*Comment.* The larva is described for the first time.

8. Genus **Rhipidoxylomyia** Mamaev (fig. 39,4-7)

*Morphology.* The white larvae turn orange before the pupation. Behind the widened anterior part the body gradually narrows into the posterior end. The last two segments taper off more strongly as compared with the preceding ones.

The length of the simple, small head capsule (fig. 39,5) is equal to its width at the base. The tentorial arms are not pronounced.

The one-toothed spatula (fig. 39,6) stands out clearly. At the transition to the manubrium the oval tooth has rounded edges.

The naked body is provided with simple, weakly pronounced papillae. The usual set of six papillae is found on the dorsal surface of the segments. Just four papillae occur on the eighth abdominal segment. The ventral surface of each thoracic segment bears two simple sternal, twelve intercalary papillae arranged into groups of three (two of which are provided with a short bristle), and two sublateral papillae. The first seven abdominal segments each bear two well-developed anterior ventral papillae, which are located, widely separated from one another, behind the spinule fields. The four posterior ventral papillae are well-developed. The two central ones are found slightly in front of the lateral ones. The eighth abdominal segment bears only four ventral papillae.

Spinule fields occur on the dorsal surface of the metathoracic and the first seven abdominal segments as well as on the ventral surface of the meso- and metathoracic, and the first seven abdominal segments. The spinule fields on the dorsal surface of the metathoracic and the first abdominal segment contain two oval bare spots. The fields on the next six abdominal segments contain four oval bare spots, the two central ones of which are situated in the centre of the field, the lateral ones in the anterior half (fig. 39,7). The fields on the ventral surface are compact.

The length of the wide terminal segment (fig. 39,4) is less than its width. Its lateral sides are nearly parallel. Two broad processes occur at the tip of the segment. The interstice between them is equal to half the width of the base of a process. The processes are apically lightly sclerotized. Two simple terminal papillae are found on the dorsal surface of each process; on the ventral surface only a single papilla occurs. The smooth anal slit is located on the ventral side of the body. It is not surrounded by rows of spinules. There are six anal papillae.

The stigmata are located on the lateral surfaces of the segments. The terminal pair is found on the posterior half of the segment, the other ones on the anterior half.

*Biology.* The larvae of the known species occur in the forest zone, where they exclusively inhabit decaying wood.

*Comment.* The diagnosis of the genus is given for the first time.

**Rhipidoxylomyia rubella** Mamaev

*Morphology.* The first segment of the short antennae (fig. 39,5) is weakly developed and considerably shorter than wide. The second segment is elongated-oval and more than twice as long as wide.

The spatula (fig. 39,6) is provided with a broadly rounded tooth. The manubrium is fairly well-pronounced. It has a clearly demarcated base and lightly traced wing-like appendices at the upper part (near the transition to the tooth).

The spinule fields on the dorsal surface of the abdominal segments comprise 25-29 rows of spinules, the one on the metathorax comprises 23 rows. Though the field on the mesothorax is rather large, the spinules are scattered and do not form distinct rows. The ventral field on the mesothorax comprises 15 rows, on the metathorax 23, and on the abdominal segments 27 to 38 (on the sixth segment).

*Biology.* The larvae have been observed in comparatively dry, strongly decomposed elm-wood.

*Material.* Vicinity of Tula , 6 June 1958, three larvae.

*Comment.* The larva is described for the first time.

9. Genus **Winnertzia** Rondani (fig. 3,3; 6,1; 12,1; 40,1-4)

*Morphology.* The white or yellowish, longitudinally striated body is somewhat widened in the middle and tapers off at the ends. The terminal segment (fig. 3,3) bears two sclerotized processes.

The simple, small head capsule (fig. 6,1) is nearly triangular, and has a pointed anterior end. It is as long as the width of its base. The antennae are well-developed. The second antennal segment has a rounded tip and is two or more times as long as wide.

The spatula (fig. 40,1) bears a single oval tooth. Though the lower part of the tooth is not clearly demarcated, its margins spread laterally at the transition to the manubrium. The base of the manubrium is well-pronounced.

The smooth body is not provided with processes. It is longitudinally striated over a considerable distance. The simple papillae do not bear bristles. The usual set of six dorsal papillae occurs on the thoracic and the first seven abdominal segments. Besides the four usual papillae, two small additional papillae are found in the middle of the eighth segment. The sternal, intercalary, and sublateral papillae on the ventral surface of the thoracic segments have the usual shape. The abdominal segments each bear four anterior ventral papillae, situated behind the spinule fields, as well as four posterior ventral papillae. Eight simple anal papillae are found on the terminal segment.

Spinule fields are distributed on the dorsal and ventral surface of the meso- and metathoracic, and on the first seven abdominal segments. The spinules stand densely in transverse rows (fig. 40,3). Each field comprises 10-35 rows. The fields on the ventral surface of all segments, as well as those on the dorsal surface of the thoracic segments, are compact. The fields on the dorsal surface of the first seven abdominal segments contain four oval bare spots.

The anterior half of the terminal segment (fig. 40,4) is widened, whereas the tip strongly tapers off. The tip bears two conspicuous, pointed processes, whose bases, though separated by a narrow, deep emargination, are close together.

The terminal pair of stigmata is located on the lateral sides of the penultimate segment, not shifted to the dorsal side. The anal slit is shifted to the anterior margin of the segment. It is more than half as long as the segment. There are no spinules around the anal slit.

*Biology.* The known species live in the forest zone beneath the bark of, and within, decaying wood.

*Comment.* The genus is very rich of species, which cannot be distinguished during the larval phase. Therefore, no key to the species is presented.

10. Genus **Trichoxylomyia** Mamaev (fig. 11,3; 18,4; 40,5-9)

*Morphology.* The white larvae with their blue, translucent intestine are highly characteristic. The length-width ratio of the body equals 5:1.

The well-developed, triangular head capsule (fig. 40,7) is weakly sclerotized. A small emargination is found at the ventral side. The tentorial arms are not pronounced. The first segment of the well developed antennae is short. The cylindrical second antennal segment is apically rounded and nearly three times as long as wide.

The single tooth of the spatula (fig. 40,8) is triangular and rounded at the tip. The relatively weakly sclerotized manubrium has no pronounced base.

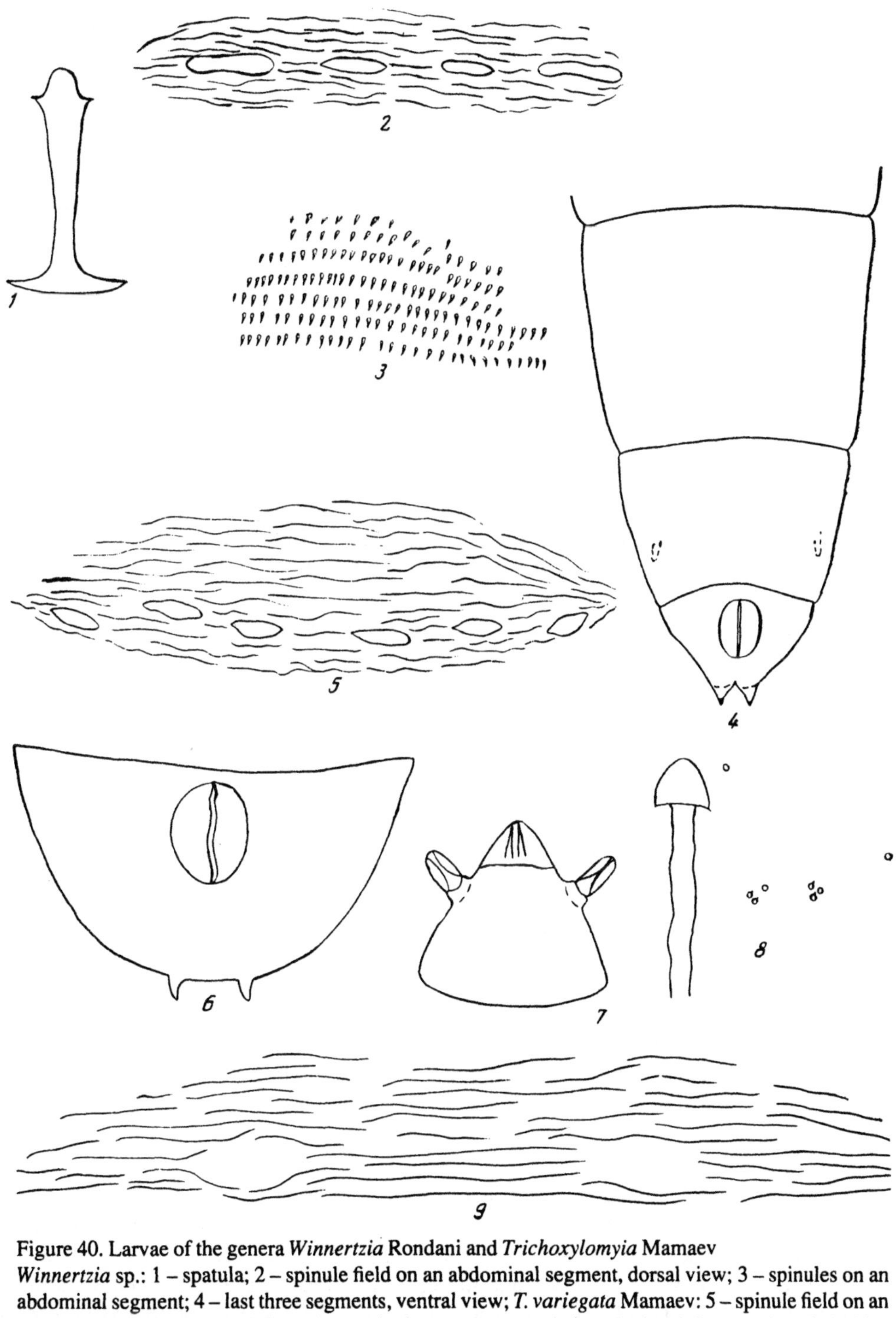


Figure 40. Larvae of the genera *Winnertzia* Rondani and *Trichoxylomyia* Mamaev
*Winnertzia* sp.: 1 – spatula; 2 – spinule field on an abdominal segment, dorsal view; 3 – spinules on an abdominal segment; 4 – last three segments, ventral view; *T. variegata* Mamaev: 5 – spinule field on an abdominal segment, ventral view; 6 – terminal segment, ventral view; 7 – head; 8 – spatula and papillae on the prothorax; *T. pilosa* Mamaev: 9 – spinule field on an abdominal segment, dorsal view (orig.)

The smooth body has no processes and bears simple, weakly developed papillae. Six simple dorsal papillae occur on each of the thoracic and the first seven abdominal segments. Just four simple dorsal papillae are found between the stigmata on the eighth abdominal segment. The terminal segment bears six dorsal papillae. The ventral surface of each of the thoracic segments is provided with two simple sternal and four groups of intercalary papillae (each group comprising two setiform and one simple papilla). Apart from these papillae, a few simple sublateral papillae are located nearer to the lateral sides of the segment. Four anterior ventral papillae, situated behind the spinule fields, as well as four posterior ventral papillae occur on the first seven abdominal segments. Just four ventral papillae are located on the eighth abdominal segment. No less than three lateral papillae are found on all segments of the body.

Spinule fields (fig. 40,5,9) occur on the dorsal as well as on the ventral surface of the meso- and metathoracic, and the first four or five abdominal segments. On the ventral surface spinule fields are found only on the fifth or sixth to seventh abdominal segment. The spinule fields on the thoracic segments are compact; those on the abdominal segments may contain bare spots.

The rounded terminal segment (fig. 40,6) bears two apical teeth. The distance between the teeth is at least four times as large as the width of a tooth. There is no emargination between the teeth. The small, oval anal slit is located on the anterior half of the ventral side of the segment. The penultimate segment is separated from the terminal one by a slightly bent line. The stigmata on the penultimate segment are situated near the posterior edge, where they are found at the level of the proximal end of the anal slit. These stigmata are small and provided with four respiratory openings (fig. 18,4).

*Biology.* The larvae develop deep within damp, highly decomposed wood of broad-leaved trees.

*Comment.* The diagnosis of the genus is given for the first time.

Key to the species of *Trichoxylomyia* Mamaev

1(2) Spinule fields on the dorsal surface of the first four abdominal segments consisting of 10-12 rows; each field containing two bare spots (fig. 40,9). Spinule fields on the ventral surface of the abdominal segments compact, without bare spots; fields consisting of 20-24 rows. . . . . . . . . . . . . . . . . . . . . . . . *T. pilosa* Mamaev

2(1) Spinule fields on the dorsal surface of the first five abdominal segments consisting of 18-21 rows; each field containing two bare spots. Spinule fields on the ventral surface of the abdominal segments (fig. 40,5) containing six elongated bare spots; fields on the middle segments consisting of 30-34 rows. . . . *T. variegata* Mamaev

**Trichoxylomyia pilosa** Mamaev

*Morphology.* The triangular head capsule is somewhat longer than its width at the base. The spatula (fig. 11,3) bears a well-pronounced tooth with a slightly spreading base.

Not counting the fields on the thoracic segments, the spinule fields are distributed on the dorsal surface of the first four, and the ventral surface of the first seven, abdominal segments. The dorsal fields on the abdominal segments (fig. 40,9) contain two bare spots. The dorsal as well as the ventral fields on the thoracic segments, and the ventral fields on the abdominal segments are compact. The fields on the dorsal surface of the thoracic segments comprise 6-10 rows, those on the ventral surface 14-17. The fields on the dorsal and ventral surface of the abdominal segments comprise respectively 10-12 and 20-26 rows.

*Biology.* The larvae have been found beneath the bark and in the outer parts of the strongly rotting sap-wood of a maple-log.

*Material.* Tellermanov forestry, Voronezhskaya oblast, 25 September 1960, one larva.

*Comment.* The larva is described for the first time.

**Trichoxylomyia variegata** Mamaev

*Morphology.* The triangular head capsule (fig. 40,7) is as long as its width at the base. The spatula (fig. 40,8) bears a well-pronounced oval tooth and is somewhat more pointed than in the preceding species.

Spinule fields, barring those on the thoracic segments, occur on the dorsal surface of the first five abdominal segments and on the ventral surface of the first seven. The fields on the dorsal surface of the abdominal segments contain two bare spots; while the fields on the ventral surface (fig. 40,5) contain six elongated bare spots. The fields on the dorsal surface of the meso- and metathoracic segments consist of 9 and 10 rows, the respective fields on the ventral surface of 11 and 24. The dorsal spinule fields on the first four abdominal segments comprise 18-21 rows, the one on the fifth only 11. The ventral spinule fields on the abdominal segments consist of 24-34 rows.

*Biology.* The larvae occur deep within quite wet rotting beech-wood.

*Material.* Khust, Zakarpatskaya oblast, Ukr. S.S.R., one larva.

*Comment.* The larva is described for the first time.

## 11. Genus **Gynapteromyia** Mamaev (fig. 41, 42)

*Morphology.* The body of the broad, orange larvae is strongly flattened. The ratio of the length to the largest width equals 2:1. The stigmata are located on conical processes.

The ventral side of the small, though heavily sclerotized, head capsule (fig. 41,1) is deeply emarginate. The length of the capsule only slightly surpasses its width at the base. The wide tentorial arms protrude like a rounded bow beyond the head capsule. The two-segmented antennae are well-developed. The first segment is very short and wide; the apically rounded second segment is cylindrical and long (3.5-4 times as long as the width at the base). The spatula is lacking.

The setiform dorsal papillae (fig. 42,6) stand on a long, wide verruca. Their number and distribution do not deviate from the normal pattern. The ventral surface of each thoracic segment bears two setiform sternal (fig. 42,1), four groups of intercalary (each group consisting of two extremely long and a single short papilla), as well as two sublateral papillae with comparatively short bristles. The ventral surface of each abdominal segment, including the eighth, is provided with a single transverse row of four clearly discernable papillae bearing a fairly long bristle (fig. 42,8). The lateral papillae are of the same type as the dorsal ones. Since the body is densely covered with pointed needle- and bristle-shaped protuberances, the papillae are not always conspicuous.

The dorsal surface of the body (fig. 42,2) is densely covered with various, pointed cuticular protuberances. The lateral surfaces predominantly bear long, acicular cuticular protuberances. The ventral surface of the segments is densely covered with thin setiform and tooth-like protuberances, the dimensions of which decrease from the middle to the anterior and posterior edges of the segments. Spinule fields are lacking.

The terminal segment (fig. 41,3) is markedly smaller than the penultimate one and bears at its tip small, wide, rounded lobes. The rounded anal pore (fig. 42,5) is located on top of a wide swelling on the ventral side of the segment.

The stigmata (fig. 42,7) appear as rather sturdy, sclerotized verrucae with three or four respiratory openings at the tip. All stigmata are located on large, conical, dorso-lateral outgrowths the dimensions of which increase towards the end of the body. The

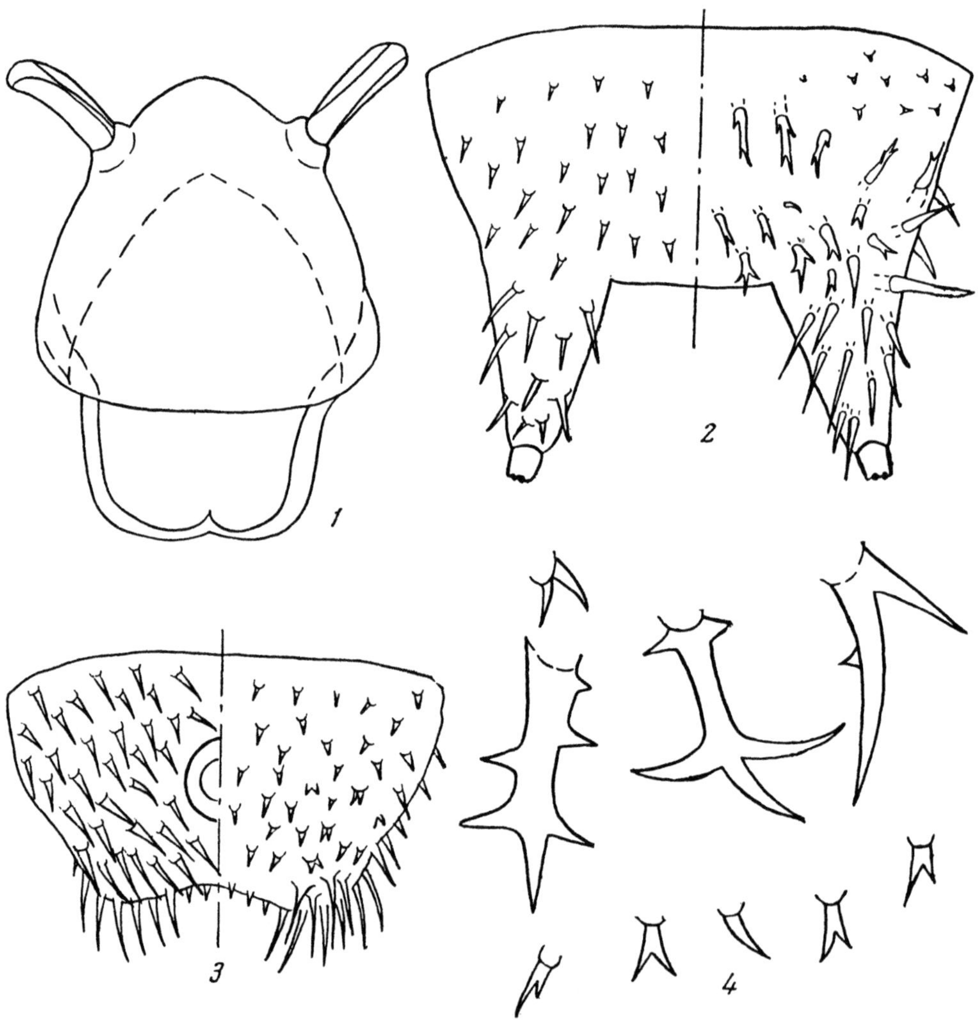


Figure 41. Larva of *Gynapteromyia carpatica* Mamaev
1 – head; 2 – eighth abdominal segment, dorsal (right) view and ventral (left) view; 3 – terminal segment, dorsal (right) and ventral (left) view; 4 – cuticular structures on the dorsal surface of the eighth abdominal segment (orig.)

outgrowths on the eighth segment are very large, only slightly shorter than the segment proper. The outgrowths bear sturdy, acicular, sometimes branched, protuberances (fig. 41,2).

*Biology*. The larvae live in decomposing vegetable remains.

*Comment*. The diagnosis of the genus is given for the first time.

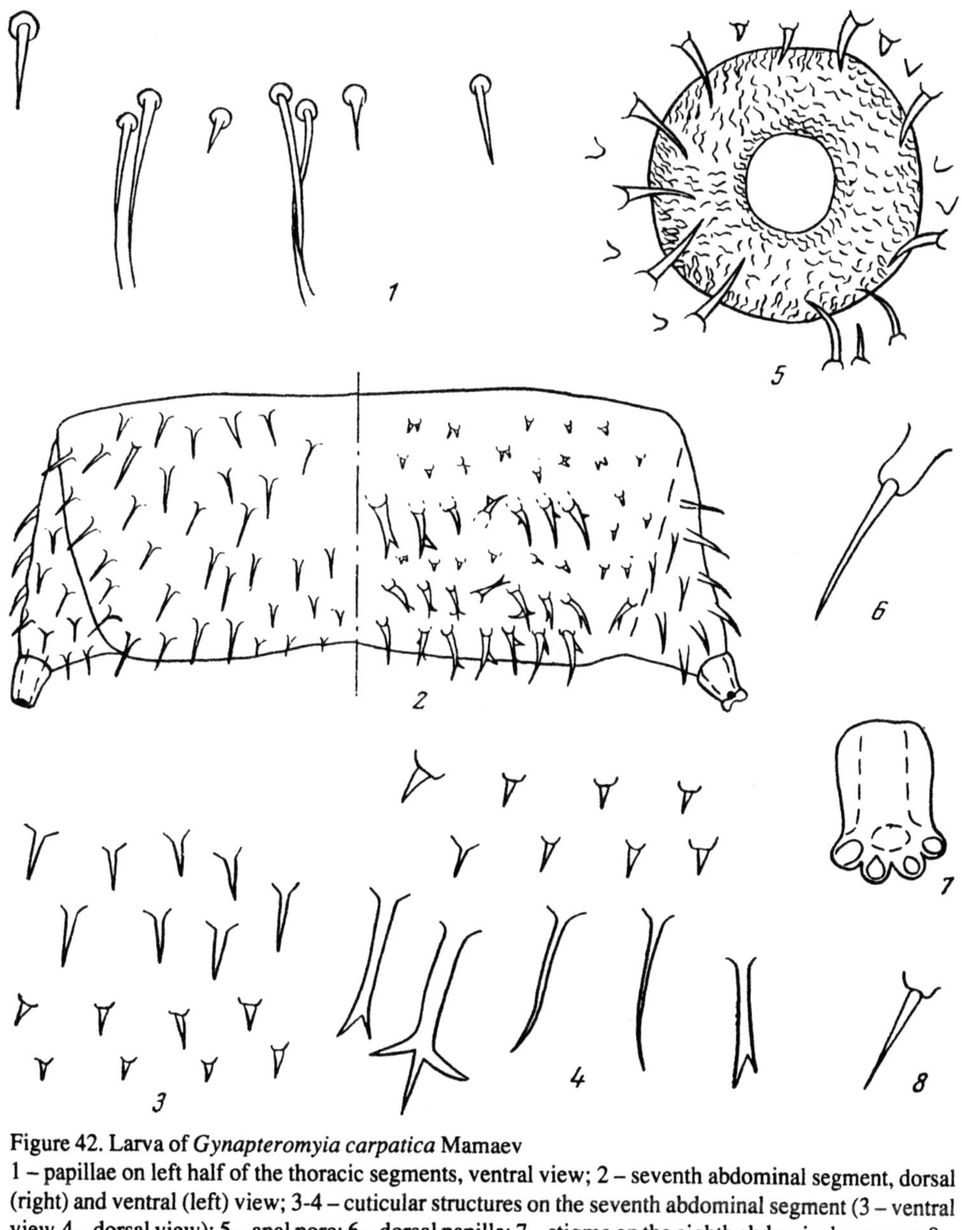


Figure 42. Larva of *Gynapteromyia carpatica* Mamaev
1 – papillae on left half of the thoracic segments, ventral view; 2 – seventh abdominal segment, dorsal (right) and ventral (left) view; 3-4 – cuticular structures on the seventh abdominal segment (3 – ventral view, 4 – dorsal view); 5 – anal pore; 6 – dorsal papilla; 7 – stigma on the eighth abdominal segment; 8 – ventral papilla on an abdominal segment (orig.)

**Gynapteromyia carpatica** Mamaev
*Morphology.* The body is densely covered with various cuticular formations. The protuberances on the dorsal surface of the segments (fig. 41,4; 42,4) are pointed and of varying sizes. The small protuberances are acicular, simple or two-branched, while the big ones, as a rule, are tree-like. Long, thin, setiform protuberances of uniform dimensions are evenly distributed over the whole ventral surface of the segments (fig. 42,3).

The dorsal surface of the terminal segment (fig. 41,3) is entirely covered with acicular protuberances the dimensions of which gradually increase towards the end of the segment.

*Biology.* The larvae have been found on the surface of a strongly rotting beech log which was covered with mycelium.

*Material.* Kvasy, Zakarpatskaya oblast, Ukr. S.S.R., 28 June 1963, two larvae.

*Comment.* The larva is described for the first time.

12. Genus **Holoneurus** Kieffer (fig. 12,5-6; 15,1; 43; 44; 45)

*Morphology.* The yellow-orange, elliptic larvae are flattened.

The small, short, nearly triangular head capsule (fig. 43, 1) is as long as its width at the base. Well-pronounced exocephalic rods, which are as long as the width of the base of the head capsule, are attached to the posterior edge of the head capsule. The antennae are well-pronounced. The long, narrow second antennal segment gradually tapers off to the tip and is two or three times as long as wide.

The spatula (fig. 43,2) is three-toothed. The large central tooth tapers off at the tip. The small lateral teeth are markedly shorter than the central one. Their inner margin (turned to the central tooth) is more clearly sclerotized than the outer one. The base of the well-pronounced, straight manubrium is not widened.

The entire dorsal surface of the body (fig. 43,3) is densely covered with oval or polygonal platelets, which are largest on the first and last segments. The platelets bear small, sclerotized, cuticular protuberances, which, near the lateral parts of the body, may be pointed. A few of the large platelets (fig. 43,5) with long, sharp teeth are distributed over the dorsal surface of the terminal segment.

The ventral surface is over a considerable distance bare and bears only small spine-like protuberances the dimensions of which increase in the direction of the lateral parts of the body. A transverse row of sickle-shaped hooks (fig. 15,1) is found on the anterior third part of the meso- and metathoracic, and the first five as well as the ninth abdominal segments. The number of hooks per row varies between 8 and 22. The rows on the meso- and metathoracic, and the terminal segment are compact, those on the other segments are split into two halves.

The dorsal and lateral papillae (fig. 43,4) are conspicuous. The thoracic and the first seven abdominal segments are each provided with six dorsal papillae, the eighth and ninth abdominal segments with four. The thoracic segments each bear on both lateral surfaces four lateral papillae, the first eight abdominal segments two, and the ninth a single very small, hardly discernable, one. The central dorsal papillae on the prothorax are rod- or tree-like; the other ones are setiform. One of the lateral papillae which are found on the thoracic and on the first eight abdominal segments is setiform, whereas all others, including the dorsal papillae on the meso- and metathoracic and the abdominal segments, are squamiform (fig. 44,5). Though a single pair of simple sternal papillae is conspicuous, the papillae on the ventral surface of the body, in general, are weakly developed. Four simple anterior ventral papillae, located at the edge of the spinule fields, as well as four posterior ventral papillae (the inner ones simple, the outer ones setiform) are found on the first seven abdominal segments.

The end of the body (fig. 43,5) bears two small, hardly visible outgrowths. The cuticle of the outgrowths is smooth and not covered with sclerotized platelets. The anal slit is surrounded by longitudinal rows of spinules.

The terminal pair of stigmata is, as a rule, shifted to the dorsal surface of the penultimate segment, where it is located at nearly the same level as the outermost pair of dorsal papillae.

*Biology*. The larvae live in forest litter and on the surface of decaying wood.

*Comment*. Kieffer (1894, 1896a) has given a diagnosis of the genus.

Key to the species of *Holoneurus* Kieffer

1(4) Dorsal papillae (fig. 44,5) squamiform, with a row of teeth at the widened apex; lateral ones more or less rod-like. Each of the middle segments provided with 20-22 sickle-shaped hooks. Outermost papillae on the meso- and metathoracic, and the first seven abdominal segments (fig. 43,3) largest (twice as large as those of the central pair).

2(3) Dorsal papillae 'a' and 'b' (fig. 1) on the meso- and metathoracic, and the first abdominal segment (fig. 43,3) elongated (no less than twice as long as its width at the tip). Lateral papillae located near the stigmata, rod-like, but apically slightly widened and split into three to five teeth. Sclerotized platelets on the central part of the first seven abdominal segments behind the central dorsal papillae, elongated and columnar. Interstice between papillae 'a' and 'b' on the metathoracic and the first five abdominal segments larger than that between 'b' and 'c'. Sclerotized verrucae bearing the prothoracic stigmata short, as long as the basal width of a verruca. *H. fulvus* Kieffer

3(2) Dorsal papillae 'a' and 'b' on the meso- and metathoracic, and the first abdominal segment (fig. 44,5) widened, at most 1.5 times as long as, mostly just as long as, the width of a papilla at the tip. Lateral papillae located near the stigmata rod-like, the tip neither widened nor split (though on the middle segments sometimes bearing two or three teeth). Sclerotized platelets on the central part of the first seven abdominal segments behind the central dorsal papillae rounded-oval or irregularly polygonal. Interstices between papillae 'a' and 'b' and papillae 'b' and 'c' on the metathoracic and the first five abdominal segments of equal length. Sclerotized verrucae bearing the prothoracic stigmata elongated-conical, longer than the basal width of a verruca. . . . . . . . . . . . . . . . . . . . . . . . . . *H. pini* (Mamaev)

4(1) Dorsal and lateral papillae (fig. 45,1,2,4) brush-shaped. No more than 14 sickle-shaped hooks on each of the middle segments. Outermost dorsal papillae on the meso- and metathoracic, and the first seven abdominal segments smaller than the central ones (which are largest). . . . . . . . . . . . . . . *H. corticalis* Mamaev

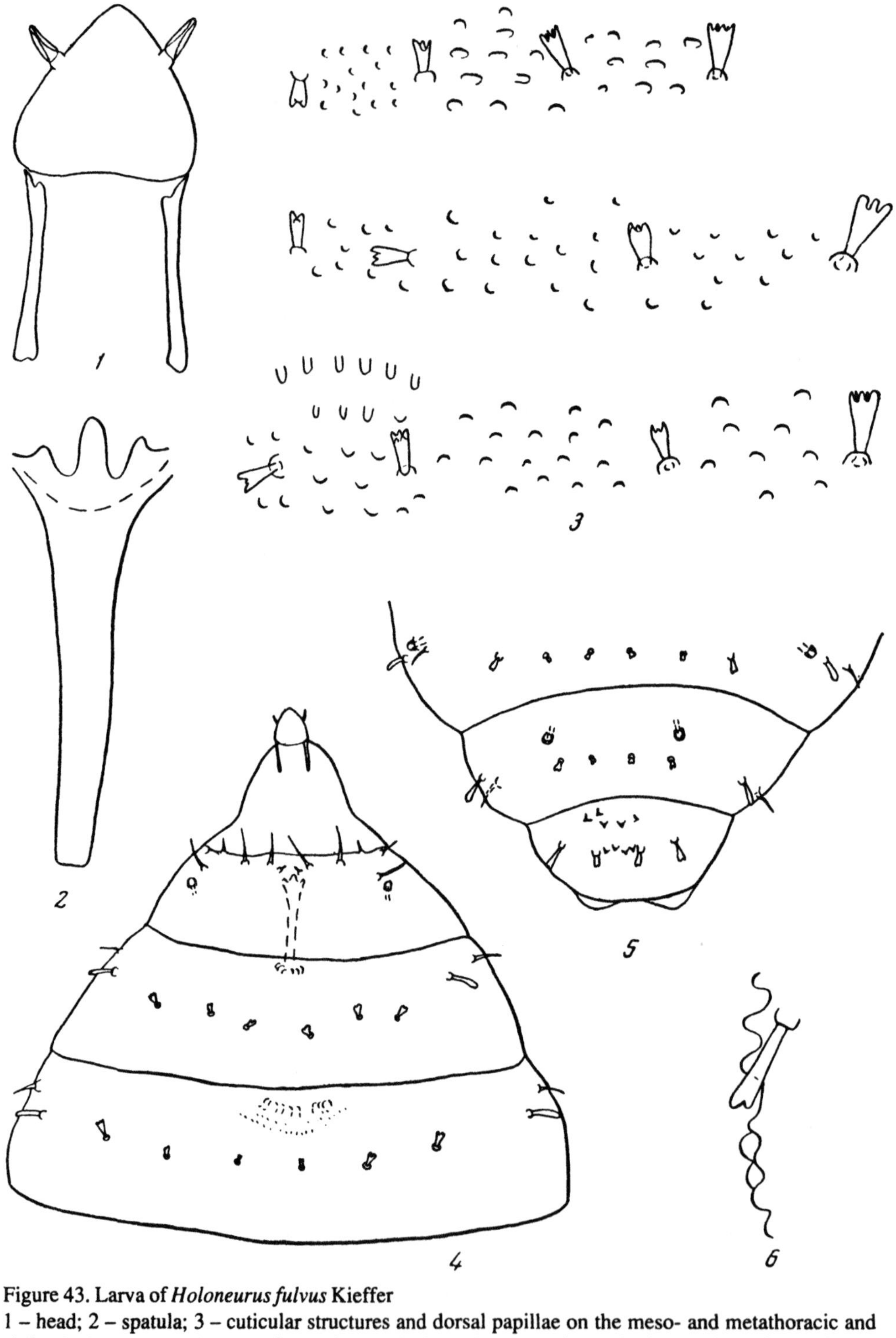


Figure 43. Larva of *Holoneurus fulvus* Kieffer
1 – head; 2 – spatula; 3 – cuticular structures and dorsal papillae on the meso- and metathoracic and abdominal segments; 4 – anterior end, dorsal view; 5 – posterior end, dorsal view; 6 – cuticular structures and lateral papilla on the second abdominal segment (orig.)

**Holoneurus fulvus** Kieffer

*Morphology.* The sclerotized platelets covering the dorsal surface of the body (fig. 44,1-2) are, as a rule, rounded-oval or irregularly polygonal. The platelets on the central part of the first seven abdominal segments (behind the central dorsal papillae) are elongated, columnar. The sclerotized platelets on the lateral parts of the body (fig. 43,6) appear as small cuticular protuberances, which are slightly pointed on the anterior part of the segment, but rounded on the posterior.

The dorsal papillae of the central pair are on the prothorax (fig. 43,4) rod-like, not apically widened, whereas those on the meso- and metathoracic, and the first eight abdominal segments are squamiform and provided with a row of teeth on the widened tip. The papillae 'c' (fig. 1) are largest, the papillae 'b' are considerably smaller. The papillae 'a' and 'b' on the meso- and metathoracic, and the first abdominal segment are elongated. They are no less than twice as long as their width below the teeth. The papillae 'a' have four or five teeth, the other ones bear, in general, a large number of teeth. The dimensions of the dorsal papillae 'a', as well as their number of teeth, increase from the metathoracic to the seventh abdominal segment. The size of the papillae 'b' decreases, and on the middle abdominal segments the papillae 'b' are markedly smaller than the papillae 'c'. The distance between 'a' and 'b' is on the metathoracic and the first two abdominal segments larger than that between 'b' and 'c'. The lateral papillae (fig. 43,6) that are located nearest to the stigmata are rod-like and slightly widened towards the tip, where they are split up into three to five teeth.

A row of 14 to 22 sickle-shaped hooks is found on the ventral surface of the meso- and metathoracic, the first five, and the ninth abdominal segments. The rows on the meso- and metathoracic, and the terminal segment are undivided, while the central part of the rows on the remaining segments is devoid of hooks.

The short, sclerotized verruca bearing the prothoracic stigma, is as long as its width at the base.

*Biology.* The larvae live among patches of mycelium and lichens, in particular on the surface of stubs and branches of beeches.

*Material.* Vicinity of Tula, 3 June 1958, four larvae.

*Comment.* The larva has been described by Kieffer (1896a).

**Holoneurus pini** (Mamaev)

*Morphology.* The larvae are structurally very similar to those of *H. fulvus* Kieffer. The sclerotized platelets (fig. 44,3-4) covering the dorsal surface of the body, though of various dimensions, are more or less of the same shape. They are predominantly oval, not columnar. The platelets on the lateral parts of the segments are similar to those of *H. fulvus* Kieffer.

The dorsal papillae of the central pair are, on the prothorax, rod-shaped, with a split tip. The dorsal papillae (fig. 44,5) on the meso- and metathoracic, and the first eight abdominal segments are squamiform and bear a row of teeth on the widened tip. The papillae 'c' (fig. 1) are largest, 'a' and 'b' are considerably smaller and on the meso- and metathorax of equal size. The papillae 'a' and 'b' on the meso- and metathoracic, and the first abdominal segment are widened. They are at most 1.5 times as long as they are wide below the apex, in most cases they are as long as wide. As a rule, the papillae 'a' bear as many teeth as the papillae 'b' and 'c', i.e., no less than six, mostly seven or eight. The

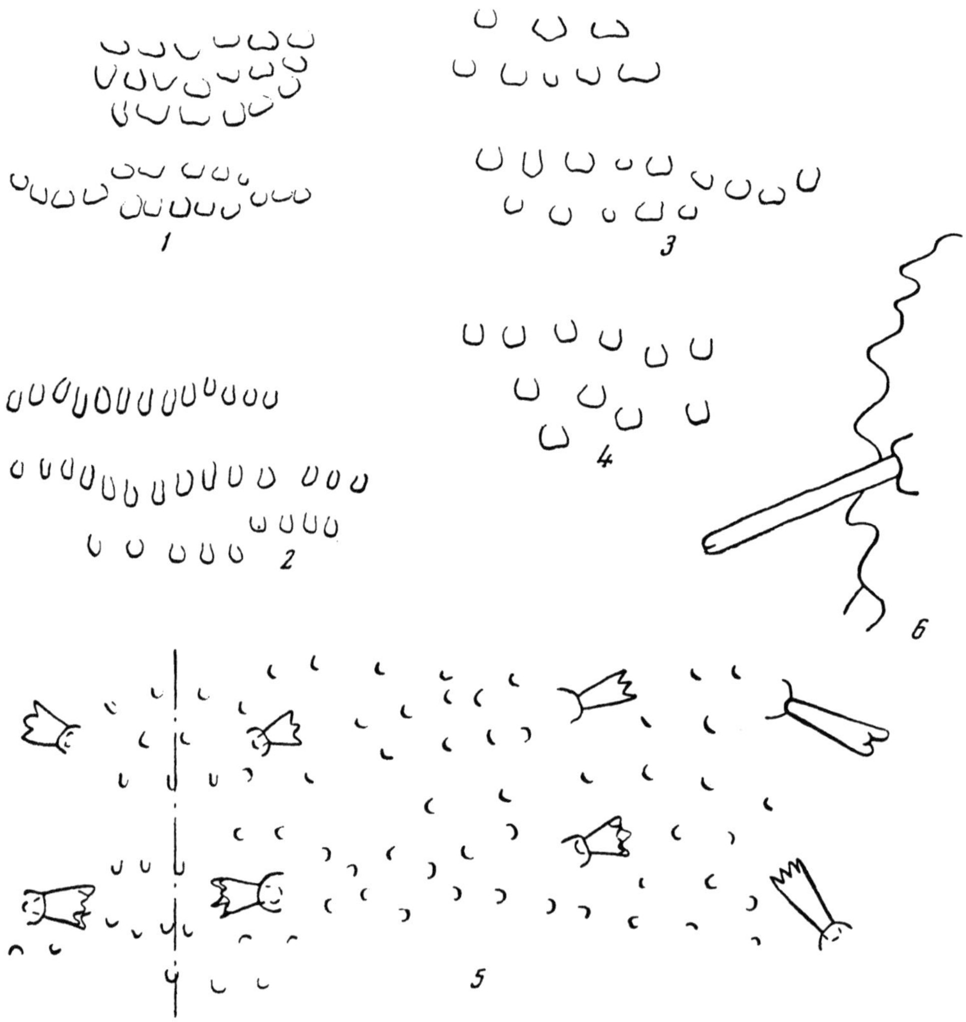


Figure 44. Larvae of the genus *Holoneurus* Kieffer
*H. fulvus* Kieffer: 1 – cuticular structures on the thoracic segments, dorsal view; 2 – cuticular structures on the abdominal segments, dorsal view; *H. pini* (Mamaev): 3 – cuticular structures on the thoracic segments, dorsal view; 4 – cuticular structures on the abdominal segments, dorsal view; 5 – cuticular structures and dorsal papillae on the meso- and metathorax; 6 – cuticular structures and lateral papilla on the first abdominal segment (orig.)

distance between the papillae 'a' and 'b' is on the metathoracic and the first seven abdominal segments equal to that between 'b' and 'c'; in some cases the distance 'a'-'b' may be less than 'b'-'c'. The lateral papillae (fig. 44,6) that are located closest to the stigmata are rod-like and do not have a split tip (they may, however, sometimes be widened only on the middle segments, in which case they bear two or three apical teeth). The lateral papillae are longest on the meso- and metathorax.

The number of sickle-shaped hooks which are distributed on the meso- and metathoracic, the first five, and the ninth abdominal segments varies between 12 and 22 a row. The sclerotized verrucae bearing the prothoracic stigmata are elongated-conical and longer than their width at the base.

*Biology*. The larvae live on the mycelium in clefts in pine and spruce stubs.

*Material*. Levinskaya, Moskovskaya oblast, 25 August 1958, two larvae.

*Comment*. The larva is described for the first time.

**Holoneurus corticalis** Mamaev

*Morphology*. The sclerotized platelets covering the dorsal surface of the body are irregularly polygonal or oval. Most of them are provided with small, pointed protuberances which are particularly conspicuous on inspection of the lateral parts of the segments. The sclerotized platelets on the anterior part of the segment, in front of the stigmata, are drawn out into pointed, hook-like protuberances (fig. 45,3); those on the posterior part are less pointed.

The central pair of dorsal papillae on the prothorax (fig. 45,2) is brush-shaped and looks like a dense tuft of basally fused hairs. This structure is shared by all dorsal papillae on the remaining segments, as well as by the terminal papillae on the dorsal surface of the terminal segment and the first pair of lateral papillae on the meso- and metathoracic, and the first eight abdominal segments. From the mesothorax onwards the papillae 'a' (fig. 1) are largest, and 'b' considerably smaller, though well-pronounced. The papillae 'b' and 'c' are of equal size on the meso- and metathorax; on the first seven abdominal segments the papillae 'b' are markedly smaller than the papillae 'c'. Though the central papillae on the eighth abdominal segment are well-pronounced, the lateral ones are hardly developed and vestigial. The terminal segment bears four large terminal papillae. The lateral

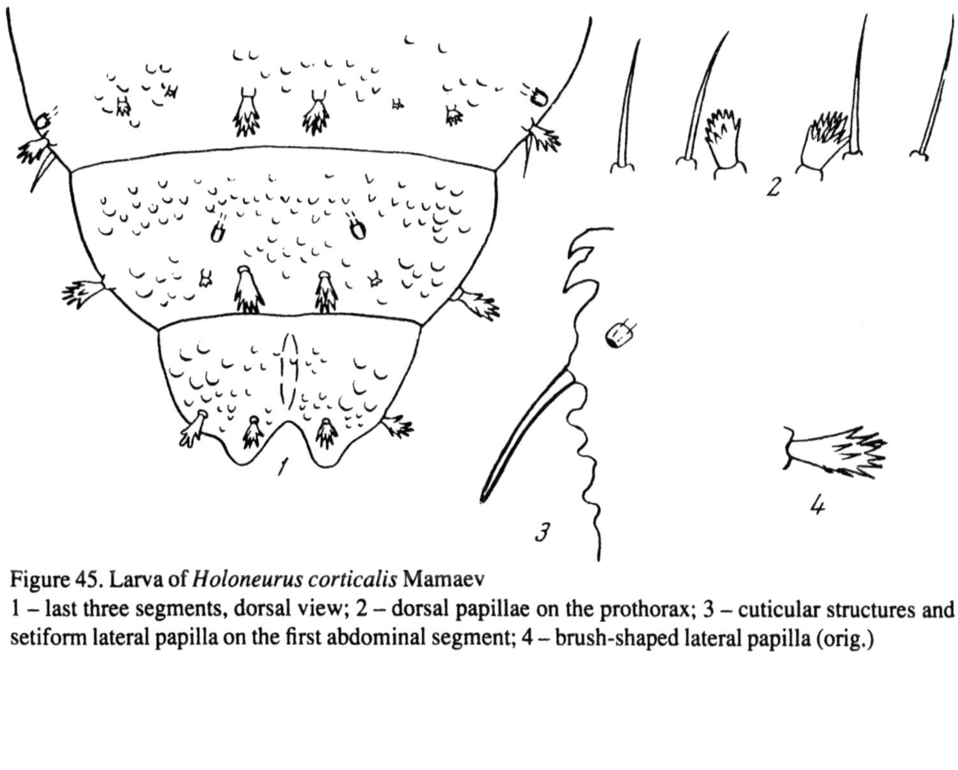


Figure 45. Larva of *Holoneurus corticalis* Mamaev
1 – last three segments, dorsal view; 2 – dorsal papillae on the prothorax; 3 – cuticular structures and setiform lateral papilla on the first abdominal segment; 4 – brush-shaped lateral papilla (orig.)

papillae, which are situated closest to the stigmata, are well-pronounced and of equal size on the meso- and metathoracic and the first eight abdominal segments. The distance between the papillae 'a' and 'b' is on the mesothorax equal to that between 'b' and 'c'. From the metathorax onwards the distance between 'a' and 'b' is markedly less than that between 'b' and 'c'; on the second and third abdominal segments distance 'a'-'b' is half as large as distance 'b'-'c'.

The rows of sickle-shaped hooks, which are found on the ventral surface, comprise on the pro- and mesothoracic segments eight, on the metathoracic and the first five abdominal segments fourteen, and on the ninth abdominal segment ten hooks.

The respiratory openings are situated on long sclerotized verrucae. The verrucae bearing the prothoracic stigmata are longer than they are wide at the base.

*Biology.* The larvae inhabit the dark mould below the bark of extremely decomposed oak logs.

*Material.* Vicinity of Tula, 27 May 1958, two larvae.

*Comment.* The larva is described for the first time.

## 13. Genus **Dicroneurus** Kieffer (fig. 12,4; 46; 47)

*Morphology.* The large, bright orange larvae are flattened. The length-width ratio of the body equals 3:1. The terminal segment ends in two fleshy, wide outgrowths.

The well-developed head capsule (fig. 46,3) is heavily sclerotized, broad and short. Its length is about equal to its width at the base. The dorsal shield of the head capsule is well-developed, and has a rounded posterior edge, while the ventral side of the capsule is reduced. Thick exocephalic rods with a rounded tip originate laterally on the posterior edge of the head. The rods are as long as the head capsule. The first segment of the well-developed antennae is very short and wide. The second, nearly cylindrical, and very long segment slightly narrows at the rounded tip. It is no less than three times as long as its diameter.

The spatula (fig. 46,6) is one-toothed. The tip of the tooth is rounded. The base of the long manubrium may be sclerotized.

The dorsal and lateral papillae (fig. 46,2,5) are setiform or clavate. The usual set of six dorsal papillae is found on the thoracic and the first seven abdominal segments. Just four dorsal papillae occur on the eighth abdominal segment. The last segment bears six terminal papillae. The lateral papillae are located on each of the lateral surfaces of the thoracic and the first eight abdominal segments. Two simple sternal, four groups of intercalary (each group comprising one simple and two setiform papillae), and two setiform or clavate sublateral papillae (fig. 46,4) are found on the ventral surface of each thoracic segment. Four simple anterior and four posterior ventral papillae (the two inner ones of which are simple, the two outer ones clavate or setiform) are located on each of the first seven abdominal segments (fig. 47,1). Just four simple ventral papillae occur on the eighth abdominal segment, whereas the terminal segment bears six or eight simple anal papillae.

The dorsal surface of the body (fig. 46,7) is covered with rounded or pointed protuberances. Compact spinule fields (fig. 46,7) are found on the anterior third part of the ventral surface of the meso- and metathoracic, and the first eight abdominal segments. The remaining part of the segments is smooth.

The terminal segment (fig. 46,7) apically bears two wide outgrowths. The anterior half of the segment is at the dorsal as well as the ventral surface predominantly smooth. The outgrowths and the neighbouring parts of the cuticle are covered with rounded or pointed protuberances, which are bluntest on the dorsal surface. The anal slit (fig. 46,7) is situated on the ventral side of the segment and surrounded by dense rows of hairs.

The stigmata are of the normal type; the terminal pair is shifted to the dorsal side.

*Biology*. The larvae develop in forest litter.

*Comment*. The identification of the majority of the species of *Dicroneurus* Kieffer is hindered by the heterogeneous descriptions of the species in this genus. The type species, *D. lineatus* Kieffer, obviously belongs to *Porricondyla* Rondani. Awaiting a revision of the genus, we include the species related to *D. tölgi* Kieffer, i.e. the species whose males have antennae with 15 segments and a clavate gonostyle, in *Dicroneurus* Kieffer.

Key to the species of *Dicroneurus* Kieffer

1(2) Dorsal, lateral, and sublateral papillae setiform, setae small. Dorsal surface of the segment (fig. 46,2) densely covered with slightly pointed protuberances. Spinule fields consisting of 26-28 rows. Tooth of the spatula (fig. 46,1) small, triangular, apically rounded. . . . . . . . . . . . . . . . . . . . . . . . . . . . . . *Dicroneurus* sp.

2(1) Dorsal, lateral, and sublateral papillae (fig. 46,5,9) clavate, comparatively large. Dorsal surface of the terminal segment only on the posterior half covered with pointed protuberances, the outgrowths, in particular, densely covered with protuberances. The anterior half of the segment smooth, without protuberances. Spinule fields consisting of 25-35 rows. Tooth of the spatula (fig. 46,6) extremely wide, rounded. . . . . . . . . . . . . . . . . . . . . . . . . . . . . . *D. montanus* Mamaev

**Dicroneurus** sp.

*Morphology*. The second antennal segment is nearly three times as long as wide. The spatula (fig. 46,1) bears a very small, triangular tooth. The base of the manubrium is widened.

The body is densely covered with platelets and verrucae. The dorsal and lateral papillae are setiform. The dorsal papillae on the eighth abdominal segment stand in a single transverse row. Two sublateral papillae bear short bristles, other papillae on the thoracic segments (the sternal and a part of the intercalary ones) are simple. The four anterior ventral papillae bear very short bristles (the two central ones are located immediately behind, the lateral ones within, the last three rows of the spinule fields). The central pair of posterior ventral papillae is simple, while the lateral pair is provided with short bristles. The ventral papillae on the eighth abdominal segment are simple.

Spinule fields, consisting of 26-28 rows, occur on the ventral surface of the segments. The terminal segment (fig. 46,2) bears two rounded lobe-shaped outgrowths. The dorsal surface of the outgrowths is densely covered with toothed platelets, the ventral surface with rounded ones. The anal slit is surrounded by 15-20 longitudinal rows of spinules, whereas 6-8 transverse rows are found in front of the slit.

*Biology*. The larvae develop among rotting leaves in forest litter.

*Material*. We made use of the description by Möhn (1955b).

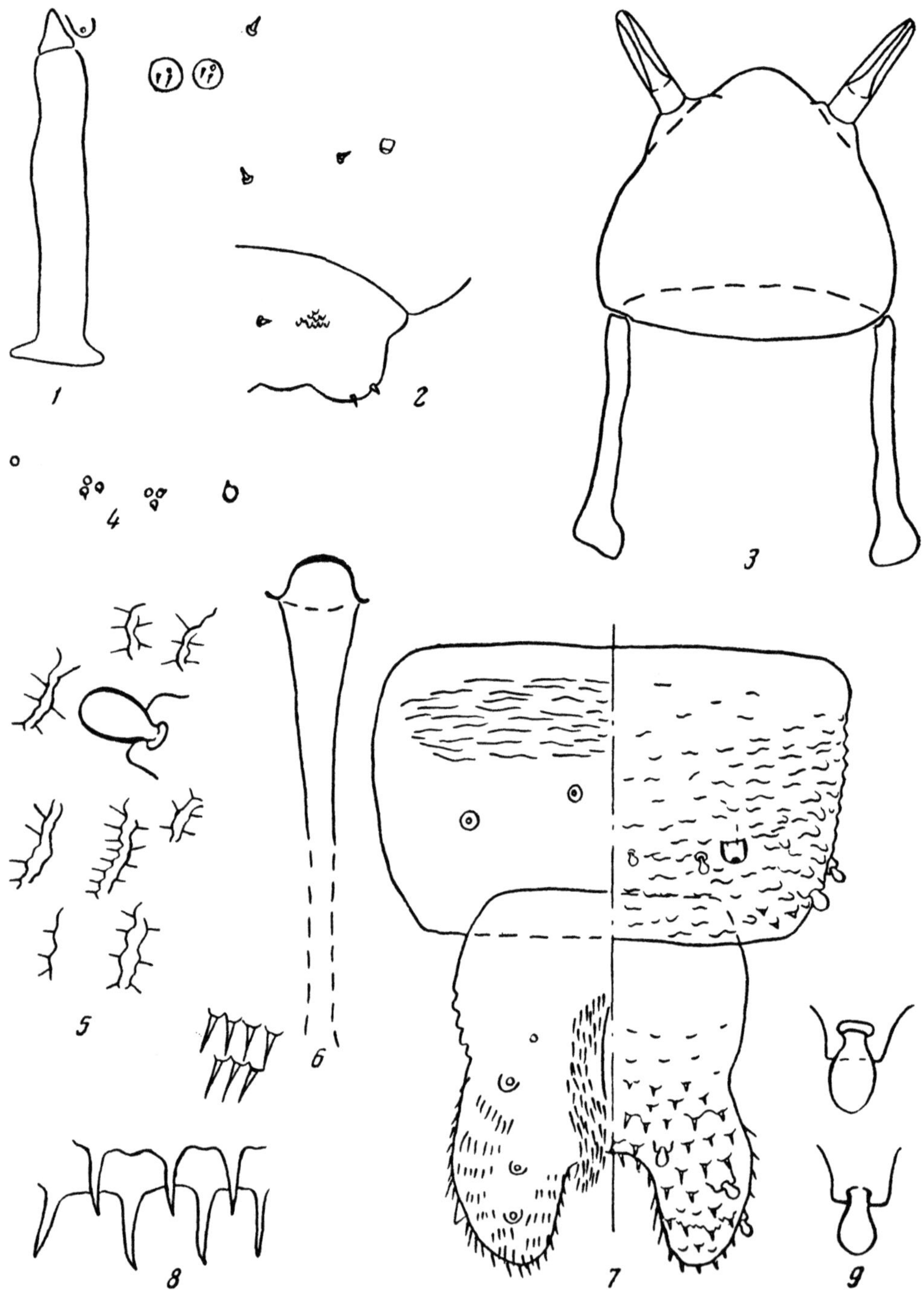


Figure 46. Larvae of the genus *Dicroneurus* Kieffer
*Dicroneurus* sp.: 1 – spatula; 2 – terminal segment, dorsal view; *D. montanus* Mamaev: 3 – head; 4 – papillae on the thoracic segments, ventral view; 5 – cuticular structures and dorsal papilla on the abdominal segments; 6 – spatula; 7 – last two segments, dorsal (right) and ventral (left) view; 8 – cuticular structures on the terminal segment; 9 – terminal papillae (1-2 – after Möhn 1955b; 3-9 – orig.)

**Dicroneurus montanus** Mamaev

*Morphology*. The head capsule is heavily sclerotized. The first antennal segment is short and wide; the nearly cylindrical second one slightly tapers off at the tip and is very long (almost four times as long as wide).

The single tooth of the spatula (fig. 46,6) is broad and rounded. The very long manubrium is, in particular basally, weakly sclerotized. Its width is somewhat less than that of the tooth.

The body is covered with cuticular formations of various shapes. The papillae (fig. 47,2) are clavate and located on broad verrucae, or are simple. Six clavate papillae, whose dimensions gradually decrease from the anterior segments towards the end of the body, are found on the dorsal surface of each of the thoracic and the first seven abdominal segments. The papillae on the seventh abdominal segment are very small. Just four clavate dorsal papillae, situated between the stigmata, occur on the eighth abdominal segment. Six clavate papillae are located centrally on the dorsal surface of the terminal segment. Three clavate lateral papillae are found on each of the lateral surfaces of the thoracic and the first eight abdominal segments. Two of these papillae are noticeably larger than the third one, which is shifted to the ventral side. The sternal papillae of the thoracic segments are simple, without bristles. The intercalary papillae form two groups

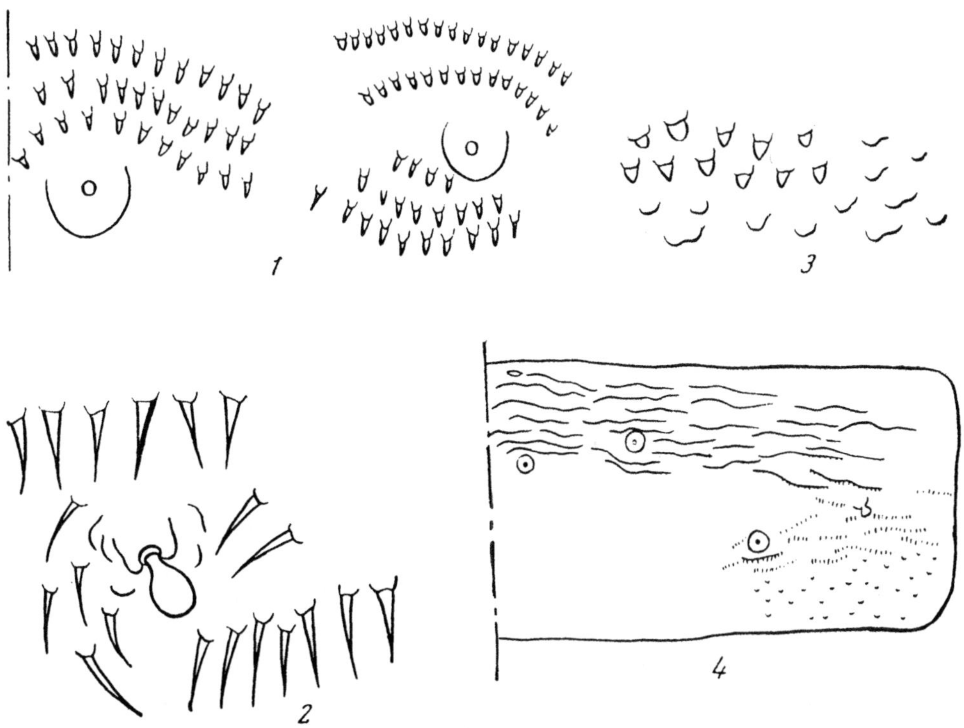


Figure 47. Larva of *Dicroneurus montanus* Mamaev
1 – spinules and anterior ventral papillae on the abdominal segments; 2 – cuticular structures and posterior ventral papilla; 3 – cuticular structures on the posterior part of the abdominal segments, ventral view; 4 – abdominal segment, ventral view (orig.)

of three papillae on either side of the sternal ones. In each group one papilla is simple, while the two others bear very short bristles. The sublateral papillae are clavate. A row of four simple anterior ventral papillae, running along the posterior edge of the spinule fields, occurs on the first seven abdominal segments (fig. 47,4). The two inner posterior ventral papillae are simple, the two outer ones clavate. Eight simple papillae are located on the ninth abdominal segment. Two of them are found on the base of the outgrowths, the other six on the outgrowths themselves.

The entire dorsal surface (fig. 46,7) is covered with small protuberances the sizes of which increase in the direction of the lateral surfaces, where they are extremely pointed (fig. 47,3). The compact spinule fields (fig. 47,1) on the ventral surface of the segments comprise 25-35 rows. The four or five anterior rows consist of wide, blunt, sparsely distributed spinules. The spinules of the markedly denser, following rows are narrower and longer. Nearer to the lateral surfaces the spinules gradually pass into the setiform protuberances which occupy a large part of the surface of the segments. The posterior ventral papillae are located within the spinule fields. The setiform protuberances pass into small spinules and verrucae along the posterior margin of the segment.

The last two segments are abruptly narrowed. The terminal segment is provided with two wide, lobe-shaped outgrowths (fig. 46,7) with a rounded tip. The dorsal surface of the terminal segment is covered with the same protuberances as the other segments. The anterior half of its ventral surface is smooth; whereas the posterior half (i.e. the outgrowths) and the lateral surfaces bear long hairs. The anal slit (fig. 46,7) is situated on the ventral side. It is covered with long hairs which are especially dense between the outgrowths. The stigmata on the eighth abdominal segment are shifted far to the dorsal side.

*Biology*. The larvae have been observed among rotting leaves in the litter of a beech forest.

*Material*. Kvasy, Zakarpatskaya oblast, Ukr. S.S.R., 28 June 1963, two larvae.

*Comment*. The larvae are described for the first time.

## 14. Genus **Parepidosis** Kieffer (fig. 4,2; 9,1; 12,2; 13,3; 16,4; 48; 49; 50; 51)

*Morphology*. The body of the larvae (fig. 4,2) is dorso-ventrally flattened. The ventral side is flat, the dorsal side vaulted. Rows of characteristic setiform papillae, which are considerably longer than the width of the body, are distributed on the dorsal surface. The terminal segment is forked; the anal slit is seamed with hairs.

The well-developed, heavily sclerotized, dark brown head capsule (fig. 48,1) is comparatively short (hardly longer than its width at the base) and not emarginate at the ventral side. Membranous, unsclerotized spots are found at the bases of the antennae. Two heavily sclerotized exocephalic rods with a widened posterior end, originate on the posterior edge of the head capsule. The rods are as long as, or longer than, the capsule. The second segment of the well-developed antennae is, in most cases, apically pointed. It is three to five times as long as its width at the base.

The spatula (fig. 48,2) bears a single tooth. This tooth is often its only conspicuous part.

The body is covered with platelets (fig. 50,1,3), rounded or tooth-like protuberances, unsegmented bristles, and setiform papillae of various degrees of development. All segments, not counting the last two, bear six setiform dorsal papillae. The penultimate segment bears four or six papillae. The papillae consist of a large, well-developed conical base and a bristle of varying length. The bristles on the dorsal papillae (fig. 50,4) are either extremely long (longer than the width of the body), or very short and hardly pronounced. The degree to which the dorsal papillae are developed is not equal among the various species. The most constant ratio among the papillae is found on the first five abdominal segments. There the papillae 'a' and 'c' (fig. 1) bear extremely long bristles, while the papillae 'b' are vestigial, bearing very short bristles. The papillae on the terminal segment (fig. 50,4) are provided with short bristles. Each of the thoracic segments bears on the ventral surface (fig. 48,2) two simple sternal papillae, four groups of intercalary, and two sublateral papillae. Each group of intercalary papillae consists of two papillae bearing short bristles and a single simple papilla. The setiform sublateral papillae stand on large verrucae. Four anterior ventral papillae are located within, or along the posterior edge of, the spinule fields on the ventral surface of the first seven abdominal segments. Four posterior ventral papillae are found behind the spinule fields. The ventral papillae of the anterior and central pair bear short bristles or are simple. The papillae of the marginal pair of posterior ventral papillae have the same appearance as the dorsal papillae with a short bristle. Just four simple or setiform ventral papillae occur on the eighth abdominal segment. Six setiform dorsal as well as six simple or setiform ventral papillae are located on the posterior part (some of them on the outgrowths) of the terminal segment.

Compact, wide spinule fields, comprising a considerable number of rows, are distributed on the ventral surface of the meso- and metathoracic, and the first eight abdominal segments. The field on the eighth abdominal segment is found on the posterior, instead of the anterior, half of the segment. A few species also have some rows of spinules on the anterior half of the ventral surface of the terminal segment. The spinules of the fields pass near the lateral surfaces into pointed protuberances or platelets and, in some species, into unsegmented bristles.

The last two segments (fig. 50,4) narrow abruptly as compared with the other ones. The eighth segment, though as long as the seventh, is about twice as narrow as the latter. The ninth segment is longer, but twice as narrow as the eighth one. It ends into two wide, lobe-shaped outgrowths the interstice between which is equal to the width of an outgrowth. The terminal segment is over a considerable distance smooth; only the outgrowths are covered with cuticular formations. The anal slit, which is situated on the ventral side of the segment, is densely covered with hairs.

The posterior pair of stigmata is shifted to the dorsal surface and found on the posterior half of the eighth segment. The stigmata on the other segments are found in the middle of the segments, rather close to their lateral surfaces.

*Biology.* The larvae are inhabitants of forest litter and are also found beneath loose, extremely decayed bark.

*Comment.* Kieffer (1900, 1913) has given a diagnosis of the genus.

Key to the species of *Parepidosis* Kieffer

1(2) On the prothorax (fig. 48,3) only the papillae 'b' (fig. 1) are highly developed and provided with long bristles, the other four papillae bear short bristles. On the meso- and metathoracic, and the sixth and seventh abdominal segment only the papillae 'a' are provided with long bristles, the other ones with short. The dorsal and lateral surfaces (fig. 48,4-6) are densely covered with highly developed, unsegmented bristles. The outgrowths of the terminal segment are covered with similar, though smaller, bristles. The spinule fields on the abdominal segments contain anterior ventral papillae with short, though well-developed, bristles. The spatula (fig. 48,2) is highly developed; the manubrium is long and basally widened.
*P. venusta* (Winnertz)

2(1) On the prothorax (fig. 50,1) the papillae 'b' and 'c' are highly developed and provided with long bristles, while the papillae 'a' bear short bristles. On the mesothorax the papillae 'a' and 'b' bear long, the papillae 'c' short bristles. On the metathorax only the papillae 'a' bear long bristles, the other papillae are short. The body may be either densely or sparsely covered with blunt or pointed protuberances (fig. 50,2). The outgrowths on the terminal segment are covered with hairs or protuberances. The anterior ventral papillae are simple. The central pair is located outside the spinule field, the two marginal papillae are found within the last two or three rows of spinules. The spatula (fig. 50,8) is weakly developed, only the small tooth is clearly pronounced, the manubrium is lacking, in some cases the spatula is not developed at all.

3(4) Not counting the outgrowths the terminal segment (fig. 50,4) is bare along its entire length. The outgrowths are densely covered with hairs. One of the lateral papillae on the metathorax and the seventh abdominal segment is half as long as the other one. On the mesothorax the interstice 'a'-'a' (fig. 1) is on the mesothorax twice as large as 'a'-'b'. The thoracic and abdominal segments are comparatively densely covered with blunt and pointed protuberances. . . . . . . . . *P. ulmicortis* Mamaev

4(3) The terminal segment (fig. 51,6) bears a few rows of tiny spinules on the middle part of the dorsal surface. The lateral papillae are on the metathoracic segment as long, or nearly as long, as those on the terminal segment. Interstice 'a'-'a' is equal to, or smaller than, 'a'-'b'. The thoracic and abdominal segments are sparsely covered with small, blunt or pointed protuberances.

5(6) A large field of tiny spinules is found on the anterior part of the ventral surface of the terminal segment (fig. 51,6). The spinule fields on the abdominal segments consist of 10-12 rows. The outgrowths of the terminal segment are covered with small, pointed protuberances. . . . . . . . . . . . . . . . . . . . . . . . . *P. arcuata* Mamaev

6(5) The ventral surface of the terminal segment is devoid of spinules. The spinule fields on the abdominal segments consist of 16-18 rows. The outgrowths of the terminal segment are bare. . . . . . . . . . . . . . . . . . . . . . . *P. argentifera* (De Meijere)

**Parepidosis venusta** (Winnertz)

*Morphology.* The length-width ratio of the large, yellow-orange larvae equals 3:1.

The exocephalic rods (fig. 48,1) are 1.5 times as long as the head capsule. The first segment of the highly developed antennae is broad and short (approximately twice as broad as long); the long, cylindrical second segment has a rounded tip and is three times as long as wide.

The spatula (fig. 48,2) bears a single, long, conical tooth. Wing-shaped processes spread from the base of the tooth. The very long manubrium may be weakly pronounced. It is as wide as the tooth. The base of the manubrium is widened.

The body is densely covered with various cuticular formations. Simple, as well as

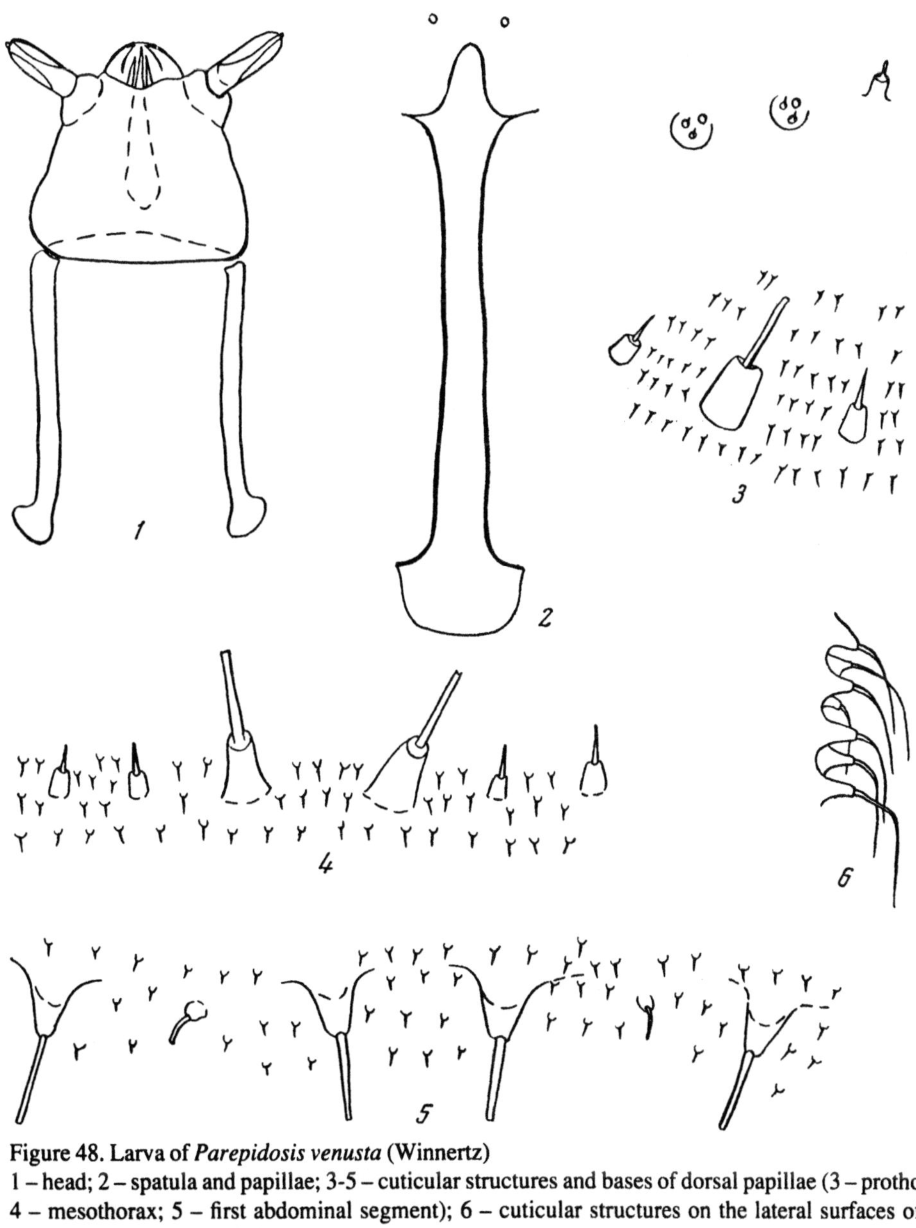


Figure 48. Larva of *Parepidosis venusta* (Winnertz)
1 – head; 2 – spatula and papillae; 3-5 – cuticular structures and bases of dorsal papillae (3 – prothorax; 4 – mesothorax; 5 – first abdominal segment); 6 – cuticular structures on the lateral surfaces of the abdominal segments (orig.)

setiform, papillae occur. Six setiform dorsal papillae are found on each of the thoracic and the first seven abdominal segments (fig. 48,3-5). The papillae 'b' (fig. 1) on the first thoracic segment bear long bristles; the bristles on the other papillae are noticeably shorter. The papillae 'a' bear long bristles (while the other papillae are short and setiform) on the meso- and metathoracic, and the sixth and seventh abdominal segments (fig. 48,4). On the first five abdominal segments (fig. 48,5) the papillae 'a' and 'c' are provided with long bristles, whereas the papillae 'b' bear very short ones. Moreover, six dorsal papillae are found in pairs, one behind the other, between the stigmata on the eighth abdominal segment (fig. 16,4). The anterior pair is setiform, the next pair simple, while the last, largest, pair bears extremely long bristles. The distance separating the papillae 'a'-'a' is, on the abdominal segments, 1.5-2 times less than the distance 'a'-'c'. Six small, dorsal, setiform papillae are located on the outgrowths of the terminal segment. Two lateral papillae of about equal length are found on the meso- and metathoracic, and the first seven abdominal segments. A single, extremely small, lateral papilla is found on the prothorax as well as on the eighth abdominal segment. The papillae on the ventral surface of the thoracic segments are typical for the genus, but the anterior and posterior ventral

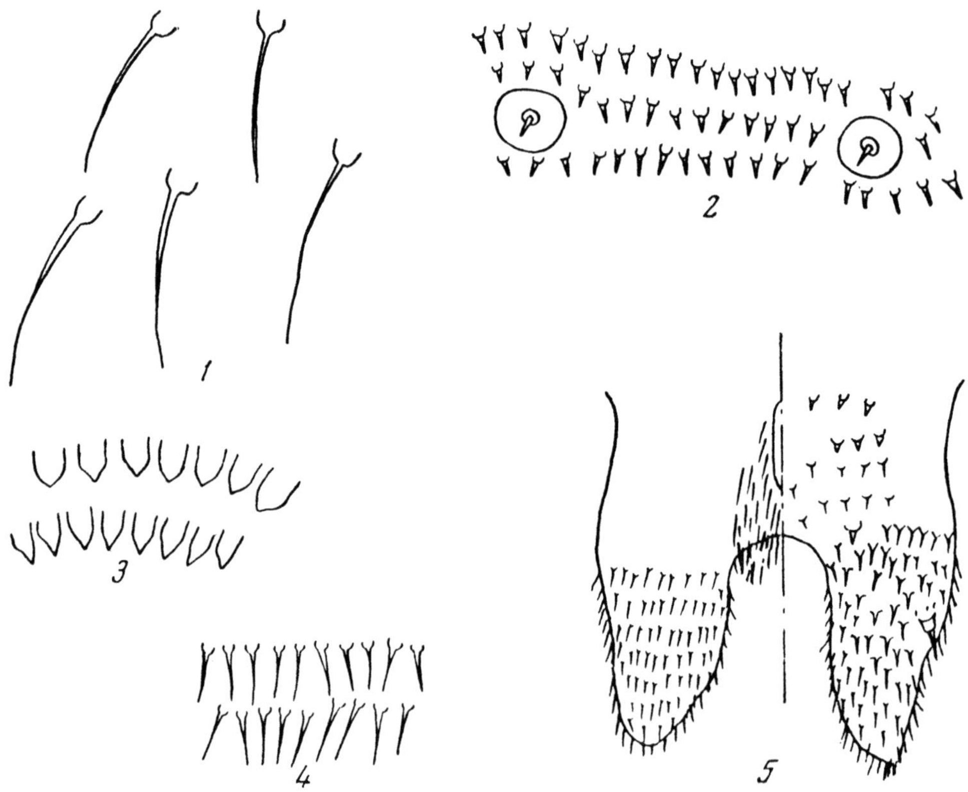


Figure 49. Larva of *Parepidosis venusta* (Winnertz)
1 – cuticular structures on the abdominal segments, dorsal view; 2-3 – spinules and anterior ventral papillae on the abdominal segments; 4 – cuticular structures on the abdominal segments, ventro-lateral view; 5 – terminal segment, dorsal (right) and ventral (left) view (orig.)

papillae on the abdominal segments bear short bristles instead of being simple. All anterior ventral papillae (fig. 49,2) are found within the boundaries of the spinule fields. The dorsal surface of the body is densely covered with setiform protuberances (fig. 49,1) whose dimensions increase towards the lateral surfaces of the segments (fig. 48,6; 49,4).

Compact spinule fields are distributed on the anterior half of the ventral surface of the meso- and metathoracic, and the first seven abdominal segments (fig. 49,3). On the eighth abdominal segment the spinules cover the posterior half of the segment, while the anterior half is bare.

A considerable portion of the dorsal as well as the ventral surface of the terminal segment is bare. The apical outgrowths alone are covered with cuticular structures (setiform on the dorsal and hair-like on the ventral surface).

*Biology.* The larvae have been observed on the surface of a stub of extremely decayed beech wood.

*Material.* Kvasy, Zakarpatskaya oblast, Ukr. S.S.R., 30 June 1963, three larvae.

**Parepidosis ulmicortis** Mamaev

*Morphology.* The exocephalic rods are as long as the head capsule. The second antennal segment is sharply pointed, its length-width ratio equals 4:1 (fig. 50,9). The tooth is the only conspicuous part of the spatula (fig. 50,8); the manubrium is not pronounced.

The bristles of the dorsal papillae (fig. 50,1-3) vary in length. The following papillae are vestigial and bear short bristles: on the prothorax – 'a' (fig. 1), on the mesothorax – 'c', on the metathorax – 'b' and 'c'. The papillae 'b' are undeveloped on the first six abdominal segments, 'b' and 'c' on the seventh. Just two well-developed dorsal papillae occur on the eighth abdominal segment (fig. 50,4). In front of these papillae two very small, vestigial papillae are found. The terminal segment bears the set of papillae that is usual in the genus. The distance 'a'-'a' is twice as large as 'a'-'b'. On the abdominal segments the distances 'a'-'a' and 'a'-'c' are about equal. Two lateral papillae are found on each of the lateral surfaces of the thoracic and the first eight abdominal segments. On the prothorax one of these papillae is very short, no more than one third as long as the second, whereas on the meso- and metathoracic, and the seventh abdominal segment one papilla is no less than half as long as the other. The lateral papillae on the first six abdominal segments are of approximately equal size. One of the lateral papillae on the eighth abdominal segment is extremely small and hardly pronounced. The papillae on the ventral surface of the body are typical for the genus. The anterior and posterior papillae are simple. The central pair of anterior ventral papillae is located behind the spinule field; the marginal pair within the field, along its posterior margin.

The entire body, barring the neck and the terminal segment, is densely covered with sclerotized platelets and tooth-shaped protuberances (fig. 50,6-7). Transverse rows of conical protuberances are densely distributed on the dorsal surface. The protuberances on the anterior part of the segments are hook-shaped and pointed, those on the posterior part are straighter and blunter. The protuberances tend to be larger nearer to the lateral surfaces.

Large spinule fields occur on the ventral surface of the meso- and metathoracic, and the first eight abdominal segments. Along the lateral and posterior surfaces the spinules of the fields pass into oval or slightly pointed protuberances, which, though somewhat blunter, resemble those on the dorsal surface (fig. 50,5). Along the posterior margin of the

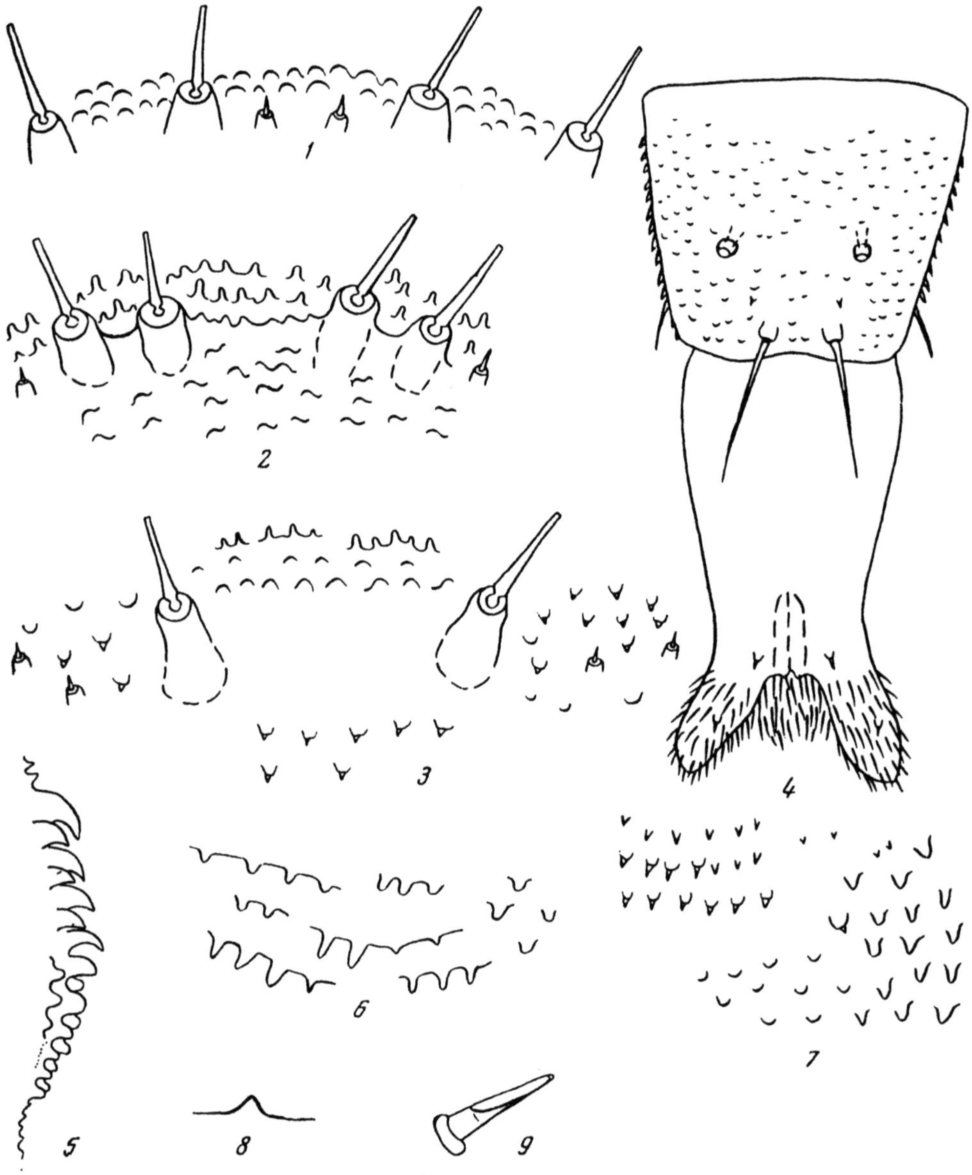


Figure 50. Larva of *Parepidosis ulmicortis* Mamaev
1-3 – cuticular structures and bases of dorsal papillae (1 – pro-; 2 – meso-; 3 – metathorax); 4 – last two segments, dorsal view; 5-7 – cuticular structures on the abdominal segments (5 – lateral; 6 – dorsal; 7 – ventro-lateral view); 8 – spatula; 9 – antennae (orig.)

segments small, smooth parts of the cuticle without protuberances are found. The terminal segment lacks spinules.

The strongly elongated terminal segment (fig. 50,4) is to a considerable extent smooth. The outgrowths of the terminal segment are covered with unsegmented hairs. The anal slit is densely haired. A dense tuft of hairs is located behind the slit, between the outgrowths.

*Biology*. The larvae were found beneath the extremely decomposed bark that was torn off from an elm log.

*Material*. Tellermanov forestry, Voronezhskaya oblast, 30 September 1960, two larvae.

*Comment*. The larva is described for the first time.

**Parepidosis arcuata** Mamaev

*Morphology*. The exocephalic rods (fig. 51,1-2) are as long as the head. The length-width ratio of the sharply pointed second antennal segment equals 5:1.

The conical tooth of the very weakly developed spatula (fig. 51,3) is rather broad.

The dorsal papillae are provided with bristles of varying lengths. The dorsal papillae on the prothorax are vestigial, only the papillae 'a' (fig. 1) bear short bristles. The papillae 'c' are incompletely developed on the mesothorax, while on the metathorax the papillae 'b' and 'c' bear short bristles. The papillae 'b' are incompletely developed on the first six abdominal segments, as are the papillae 'b' and 'c' on the seventh. The eighth abdominal segment (fig. 51,4) bears just six papillae, only two of which are well-developed, bearing long bristles; in front of these, two simple and two short, setiform papillae are situated. On the mesothorax the distance 'a'-'a' is equal to, or less than, the distance 'a'-'b'. On the abdominal segments the distances 'a'-'a' and 'a'-'c' are about equal. Two lateral papillae are found on each of the lateral surfaces of the thoracic and the first eight abdominal segments. One of the papillae on the prothorax is very short (no longer than one third as long as the other), while the corresponding papilla on the mesothorax is more than two thirds as long as the other. The lateral papillae on the mesothorax are (almost) equally long. The dimensions of the papillae on the first seven abdominal segments are also equal. One of the lateral papillae on the eighth abdominal segment is very small and vestigial. The papillae on the ventral surface are of the type that is normal for the genus (fig. 51,5). The anterior as well as the posterior ventral papillae are simple. The central pair of anterior ventral papillae is situated outside the spinule field, along its posterior edge, whereas the marginal pair is located within the posterior part of the field, between the last two or three rows.

The entire body, not counting the neck, is covered with sparse, weakly pronounced, blunt or pointed protuberances (fig. 51,7). A few weakly demarcated, blunt protuberances are often distributed on the anterior and posterior parts of the dorsal surface of the segments. Protuberances are lacking in the vicinity of the papillae. The protuberances tend to be more pointed, larger, and more densely distributed nearer to the lateral surfaces of the segments (fig. 51,7). They are more clearly pronounced on the anterior half of the segments than on the posterior.

Spinule fields (fig. 51,5) occur on the ventral surface of the meso- and metathoracic, and all abdominal segments. The spinules constituting the fields are small and rounded, and, nearer to the lateral surfaces, gradually pass into blunt, weakly pronounced protuberances. The spinule field on the eighth abdominal segment is shifted to the posterior margin. The fields on the abdominal segments comprise ten to twelve rows. They are not compact, but consist of three parts: a rounded central part and two elongated-elliptical lateral parts. Small, smooth spots without protuberances are found along the posterior margin of the segments. To a considerable extent the dorsal surface of the terminal

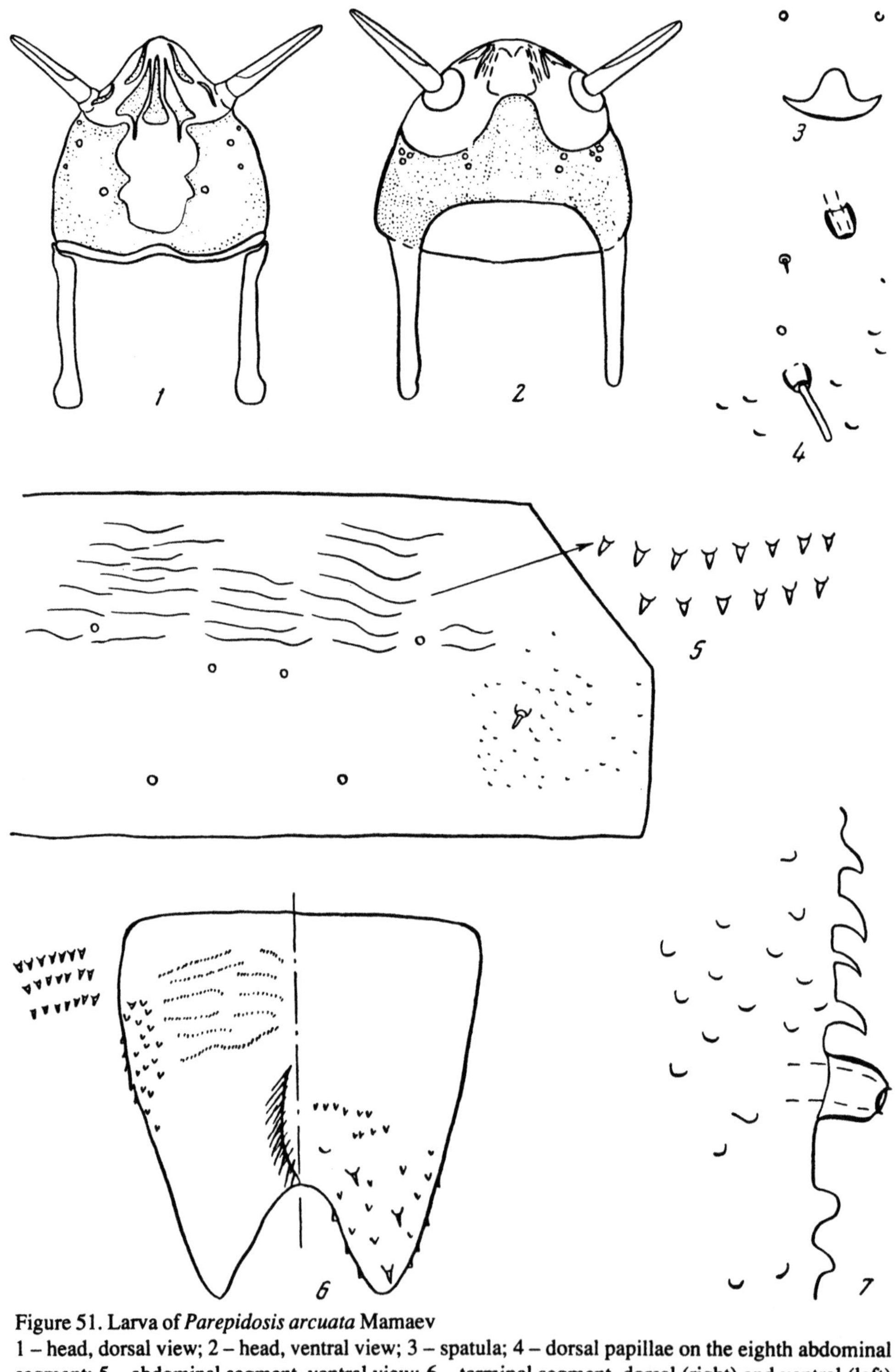


Figure 51. Larva of *Parepidosis arcuata* Mamaev
1 – head, dorsal view; 2 – head, ventral view; 3 – spatula; 4 – dorsal papillae on the eighth abdominal segment; 5 – abdominal segment, ventral view; 6 – terminal segment, dorsal (right) and ventral (left) view; 7 – cuticular structures on the lateral surfaces of the abdominal segments (orig.)

segment (fig. 51,6) is bare. A field of tiny spinules, which reach onto the lateral surfaces, is found on the anterior part of its ventral surface.

The terminal segment (fig. 51,6) bears, apart from the spinule fields on the ventral surface mentioned above, a few rows of tiny spinules on the central part of its dorsal surface. The outgrowths are sparsely covered with small, pointed protuberances. The anal slit (fig. 51,6) is provided with short hairs; there is no tuft of hairs between the outgrowths.

*Biology*. The larvae live among rotting leaves in the litter of beech forests.

*Material*. Kvasy, Zakarpatskaya oblast, Ukr. S.S.R., 17 June 1963, three larvae.

*Comment*. The larvae are described for the first time.

**Parepidosis argentifera** (De Meijere)

*Morphology*. The spatula is reduced. The body is provided with small, weakly pronounced, rounded platelets and blunt protuberances. Spinule fields consisting of 16-18 rows of spinules are easily visible on the ventral surface. The terminal segment bears on its dorsal surface a few rows of tiny spinules; its ventral surface is bare.

The dorsal papillae bear bristles of varying lengths. The papillae 'a' (fig. 1) on the prothorax are vestigial bearing short bristles, the papillae 'c' are undeveloped on the mesothorax, as are the papillae 'b' and 'c' on the metathorax. The papillae 'b' are undeveloped on the first six abdominal segments, and 'b' and 'c' on the seventh. The eighth abdominal segment bears only two well-developed papillae, which are provided with long bristles. In front of these papillae another pair with very short bristles is found. Two equally long papillae occur on each of the lateral surfaces of the segment. All anterior papillae are simple. The central ones are located behind the spinule fields, the marginal pair is situated between the second and third row of spinules. The two central ones of the four posterior ventral papillae are simple, as are all four ventral papillae on the eighth segment.

*Biology*. The larvae develop within rotting vegetable remains.

*Material*. The larvae of this species have been described by De Meijere (1907). We made use of the description by Möhn (1955b).

15. Genus **Porricondyla** Rondani (fig. 52, 53, 54)

*Morphology*. The slightly flattened, wide body of the larvae is abruptly narrowed at both ends. It is nearly three times as long as wide. The tip of the terminal segment is divided into two parts.

The well-developed, rather heavily sclerotized head capsule (fig. 52,1) is somewhat larger than broad. Large, weakly sclerotized spots occur on the dorsal shield around the bases of the antennae. A small, rounded emargination is found on the posterior edge of the dorsal side of the capsule; the ventral side has a nearly straight edge, and is more heavily sclerotized than the dorsal side. Two well-developed, exocephalic rods, which are as long as, or longer than, the head capsule, originate at the latero-posterior angles of the head capsule (closer to its dorsal side). The rods are nearly straight, with a strongly pronounced or slightly widened tip. The first segment of the well-developed, long antennae is very short; the second segment is strongly elongated and gradually tapers off from the base to the apex. The spatula is either not pronounced or one-toothed (fig. 52,2).

The dorsal and lateral surfaces of the body are covered with various cuticular structures (fig. 52,7). The papillae are predominantly setiform or rod-shaped (very rarely simple). Six setiform or rod-shaped dorsal papillae (fig. 53,4) occur on each of the thoracic and the first seven abdominal segments. Just two pairs, one behind the other, of dorsal papillae are found on the eighth abdominal segment. Three well-developed setiform lateral papillae are located on each of the lateral surfaces of the thoracic and abdominal segments. The following papillae occur on the ventral surface of each of the thoracic segments: two simple sternal papillae, four groups of intercalary papillae (each group consisting of two setiform and one simple papilla), two setiform sublateral papillae (fig. 53,5). Four well-developed setiform anterior ventral papillae (situated within or behind the spinule fields) and four setiform posterior ventral papillae are found on the ventral surface of the first seven abdominal segments. The bristles on the outer pair of posterior ventral papillae are more clearly pronounced than those on the inner pair. Just four ventral papillae with very short bristles are located on the eighth abdominal segment. Some simple anal papillae are found on the terminal segment around the anal pore. The entire dorsal and lateral surfaces of the body are densely covered with weakly pronounced, oval platelets or with cuticular protuberances of various shapes (fig. 54,5). These protuberances are tooth-shaped on the anterior half of the segments; on the posterior half they are blunter, resembling verrucae. Compact spinule fields are distributed on the meso- and metathoracic, and all abdominal segments. The spinules gradually pass into platelet- or verruca-shaped formations. The posterior half of the segments, rearwards of the spinule fields, is bare.

The terminal segment (fig. 52,5) is comparatively short. It is no longer than each of the two preceding segments. To a large extent the ventral surface of the segment is bare; just a few rows of spinules are found on its anterior half. The segment bears at its tip two broad, conical, apically rounded outgrowths, which are no longer than half the width of the segment. The slit-like anal pore is situated on the ventral side of the terminal segment. It is covered by a dense tuft or some rows of slender hairs.

The well-developed stigmata resemble sturdy, rounded verrucae. The posterior pair of stigmata is shifted to the dorsal surface of the eighth abdominal segment; the other ones are found on the lateral surfaces of the segments.

Key to the species of *Porricondyla* Rondani

1(2) The exocephalic rods (fig. 52,1) are apically widened. The latero-posterior angles of the penultimate segment (fig. 52,5) are provided with distinct, conical outgrowths. The large, tooth-shaped protuberances (fig. 52,3) fringing the lateral surfaces of the segments, are several times as large as the protuberances on the dorsal surface. The manubrium is well-pronounced. The outgrowths on the terminal segment (fig. 52,5) are wide apart; the distance separating them is larger than the width of an outgrowth. The anal slit (fig. 52,5) is sparsely fringed with hairs, which do not form a dense cover. . . . . . . . . . . . . . . . . . . . . . . . *Porricondyla* sp.

2(1) The exocephalic rods (fig. 53,1) are not apically widened. The latero-posterior angles of the penultimate segment are not provided with outgrowths. The protuberances (fig. 54,1), which densely cover the dorsal and lateral surfaces of the segments, are small and of equal size. The spatula is not pronounced or only represented by a weakly expressed tooth (fig. 54,2). The bases of the outgrowths on the terminal segment touch one another. The anal slit (fig. 54,1) is seamed by a cover of long, densely distributed hairs.

3(4) The exocephalic rods (fig. 53,1) are 1.5 times as long as the basal width of the head capsule. The dorsal surface, and to a large extent also the ventral surface, of the body segments are covered with tiny, pointed protuberances. The anterior ventral papillae (fig. 53,2) and the central posterior ones are setiform. From the fifth abdominal segment onwards the dorsal papillae (fig. 53,4) are stick-shaped with a slightly widened tip. . . . . . . . . . . . . . . . . . . . . . . . . . . *P. lata* Mamaev

4(3) The exocephalic rods are short, as long as, or shorter than, the width of the base of the capsule. The dorsal and lateral surfaces of the body-segments are comparatively densely covered with small, pointed protuberances. The ventral surface is to a considerable extent bare. The anterior and central posterior ventral papillae are simple. The dorsal papillae on the thoracic and the abdominal segments (fig. 54,5) are setiform. . . . . . . . . . . . . . . . . . . . . . . . . . . . . *P. neglecta* Mamaev

**Porricondyla** sp.

*Morphology*. The massive exocephalic rods (fig. 52,1) are apically widened. The rods are as long as, or slightly longer than, the width of the base of the head capsule. The cylindrical, apically rounded second antennal segment is comparatively short. It is no longer than half the width of the capsule at its base.

The one-toothed spatula (fig. 52,2) is well-pronounced. The rounded tooth is basally provided with long, pointed processes. The well-pronounced manubrium is as wide as the tooth. The base of the manubrium is slightly widened.

The dorsal papillae are setiform. Four anterior ventral papillae bearing short bristles are found within the spinule fields on the ventral surface of the first seven abdominal segments. The rest of body is smooth, but locally bears tooth-like spinules and small, oval platelets (fig. 52,6-7). The dorsal surface of the cuticle is sparsely covered with very weakly pronounced, oval platelets. A row of sturdy, tooth-shaped protuberances is located along the lateral margins of the segments (fig. 52,3). The dimensions of these protuberances decrease from the anterior to the posterior edge of the segment. Conical outgrowths (fig. 52,4), each of which bears a setiform lateral papilla, are found laterally on the posterior margin of the abdominal segment.

The spinule fields are distributed on the ventral surface of the meso- and metathoracic, and all abdominal segments. The small fields are of equal size and comprise eight to ten rows of spinules. Small, clearly outlined, oval platelets are distributed along the posterior and lateral margins of the segments.

The margin of the terminal segment (fig. 52,5) bears the same tooth-shaped protuberances as the other segments. The terminal segment ends into lobe-shaped outgrowths with a pointed tip. The interstice between the outgrowths is wider than the width of an outgrowth. The anal slit is located on the ventral side of the terminal segment; it is surrounded by rows of thin hairs.

*Biology*. The larvae develop on rotting leaves in litter.

*Material.* Pavlova Sloboda, Moskovskaya oblast, 25 May 1963, three larvae.

*Comment.* The larva is described for the first time. The peculiarities of the larvae open the possibility of their future splitting off as an independent genus, when the known species are revised.

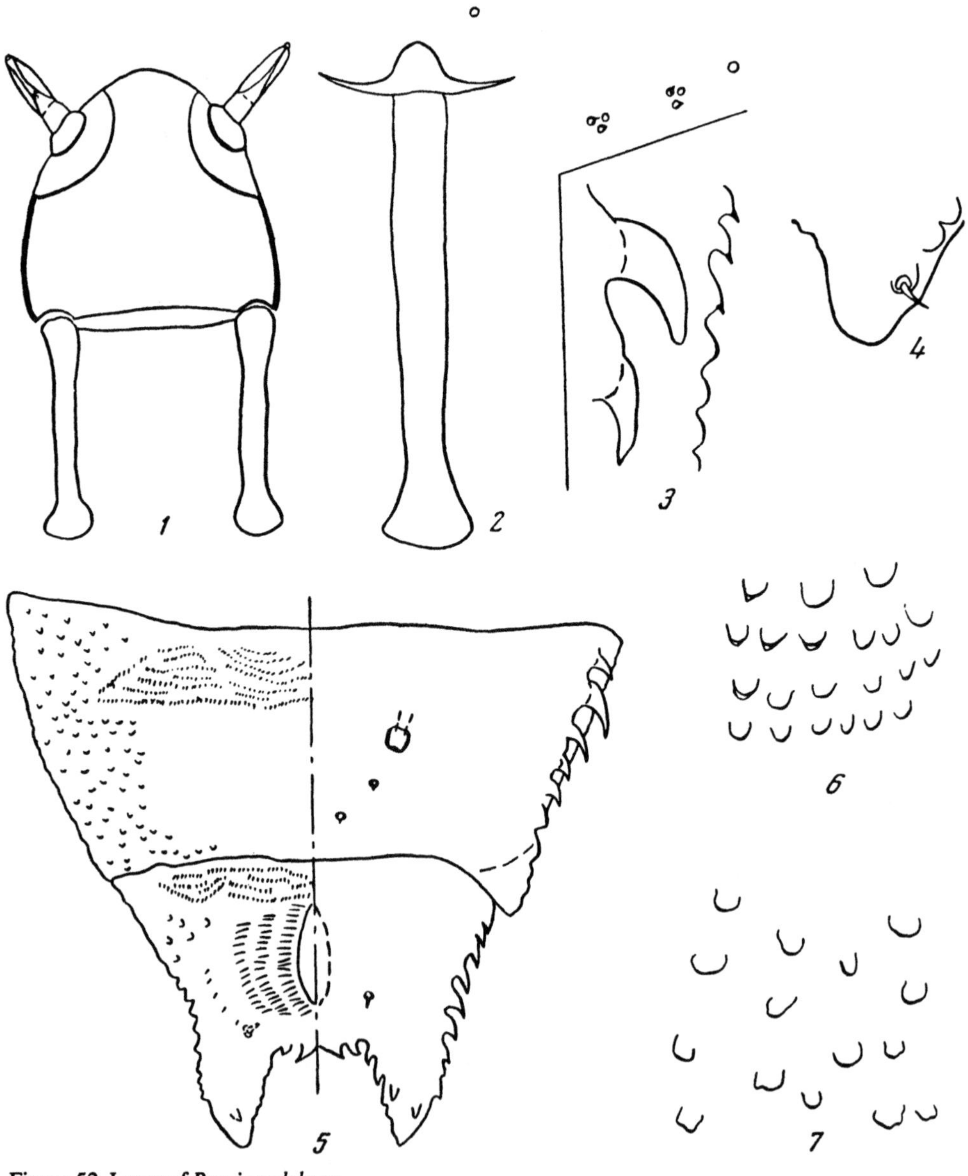


Figure 52. Larva of *Porricondyla* sp.
1 – head; 2 – spatula and papillae; 3 – cuticular structures on the lateral surfaces of the eighth abdominal segment; 4 – outgrowth of the eighth abdominal segment; 5 – last two segments, dorsal (right) and ventral (left) view; 6-7 – cuticular structures on the eighth abdominal segment (6 – ventral; 7 – dorsal view) (orig.)

**Porricondyla lata** Mamaev

*Morphology*. The straight exocephalic rods (fig. 53,1) are at most only very slightly widened at the tip. The rods are nearly 1.5 times as long as the width of the capsule at its base. The long second antennal segment gradually tapers off from the base to the apex. Its length surpasses half the width of the head capsule. The spatula is not pronounced.

The dorsal papillae (fig. 53,4), setiform on the thoracic segments, gradually become stick-like on the abdominal segments. They are particularly well-pronounced on the last two segments. The setiform anterior ventral papillae (fig. 53,2) form a single transverse

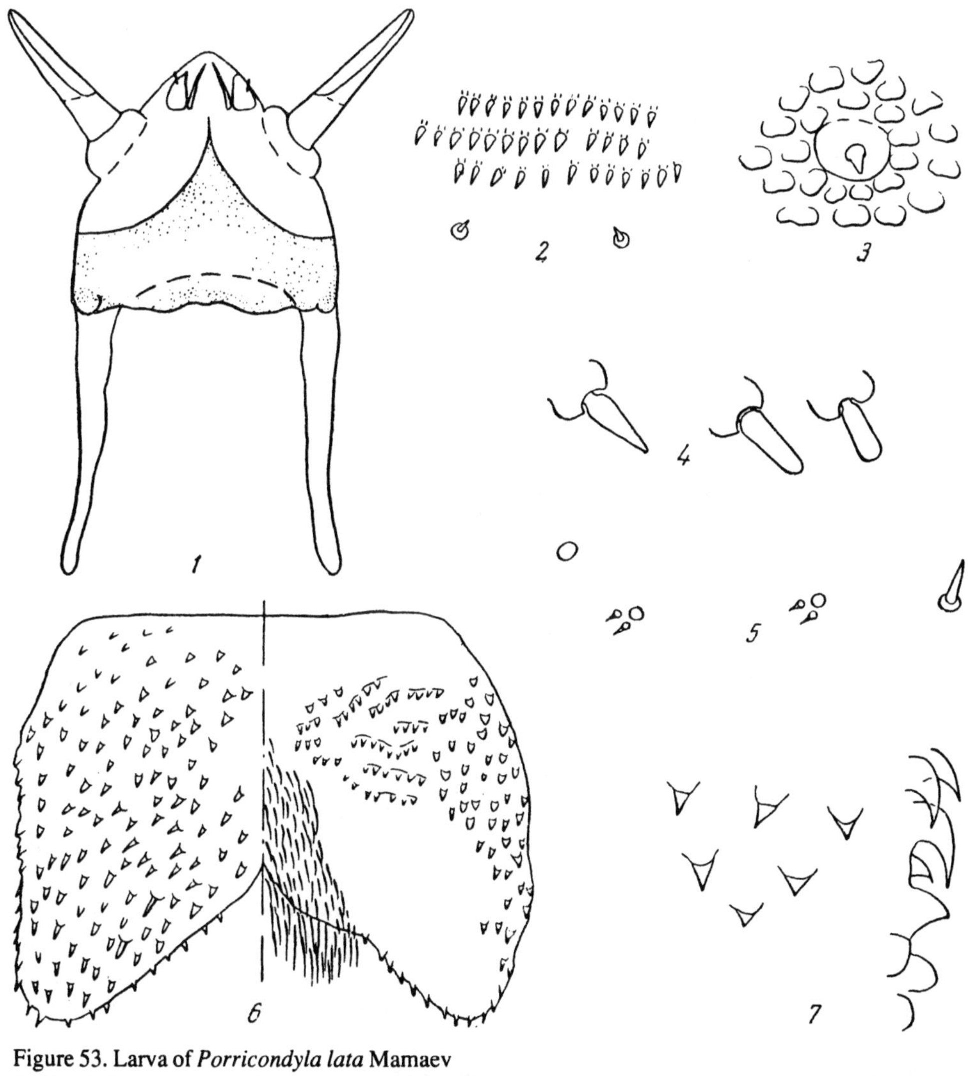

Figure 53. Larva of *Porricondyla lata* Mamaev

1 – head; 2 – spinules and anterior ventral papillae on the abdominal segments; 3 – posterior ventral papilla; 4 – shape of the dorsal papillae on the frontal (left) and rear (right) abdominal segments; 5 – papillae on the thoracic segments, dorsal view; 6 – terminal segment, dorsal (left) and ventral (right) view; 7 – cuticular structures on the lateral surfaces of the abdominal segments (orig.)

row behind the spinule fields. The posterior ventral papillae are setiform or stick-like (fig. 53,3). The whole dorsal and lateral surfaces of the body are densely covered with various cuticular protuberances (fig. 53,7). These are tooth-shaped on the anterior half of the segments and blunter, resembling verrucae, on the posterior.

Large spinule fields, comprising 18-19 rows on the thoracic and 20-26 on the first eight abdominal segments, are distributed on the ventral surface of the meso- and metathoracic, and all abdominal segments. The spinules, which are blunt, wide and comparably large on the anterior parts of the fields, are smaller and narrower on the posterior.

The terminal segment (fig. 53,6) is relatively short; it is no longer than any of the two preceding segments. The dorsal surface of the terminal segment is densely covered with pointed protuberances. The ventral surface is to a considerable extent bare; just a few rows of small spinules are found on its anterior half. The bases of the outgrowths on the terminal segment meet. The anal slit is densely covered with long hairs.

*Biology.* The larvae develop in the litter of coniferous forests.

*Material.* Pavlova Sloboda, Moskovskaya oblast, 15 August 1963, three larvae.

*Comment.* The larvae are described for the first time.

**Porricondyla neglecta** Mamaev

*Morphology.* The straight exocephalic rods are at most only slightly widened at the tip. The length of the rods is equal to, or less than, the width of the head capsule at its base. The small, triangular, apically rounded tooth is the only discernable part of the weakly pronounced spatula (fig. 54,2).

The dorsal papillae (fig. 54,5) are on all segments (including the eighth) setiform with a pointed tip. Though they are thin throughout on the thoracic and the first three abdominal segments (fig. 54,5), their middle part is slightly widened on the other segments (fig. 54,6). On the first seven abdominal segments (fig. 54,4) the four simple anterior ventral papillae form a single row behind the spinule fields. The posterior ventral papillae are distributed as usual. The central pair is simple; the shape of the setiform marginal papillae (fig. 54,4) resembles that of the dorsal papillae. The setiform lateral papillae (fig. 54,3) have the same shape as the dorsal ones.

The dorsal and lateral surfaces of the body (fig. 54,3) are entirely covered with small protuberances. Though these are blunt and rounded on the dorsal surface of the middle segments, they are pointed on the dorsal surface of the terminal segment as well as on the lateral surfaces of all segments. The protuberances on the anterior half of the segments are tooth-shaped, on the posterior half-straight. Spinule fields consisting of extremely small spinules, are located on the ventral surface of the metathoracic and all abdominal segments. The spinules are comparatively short and pointed.

The relatively short terminal segment (fig. 54,1) is no longer than any of the two preceding segments. The dorsal surface of the segment is densely covered with pointed protuberances; the ventral surface is bare, not counting a few spinules on its anterior half. The bases of the outgrowths touch one another. The anal slit is densely covered with long hairs.

*Biology.* The larvae develop in decomposing forest litter.

*Material.* Zavety Il'icha, Moskovskaya oblast, 25 June 1962, one larva.

*Comment.* The larvae are described for the first time.

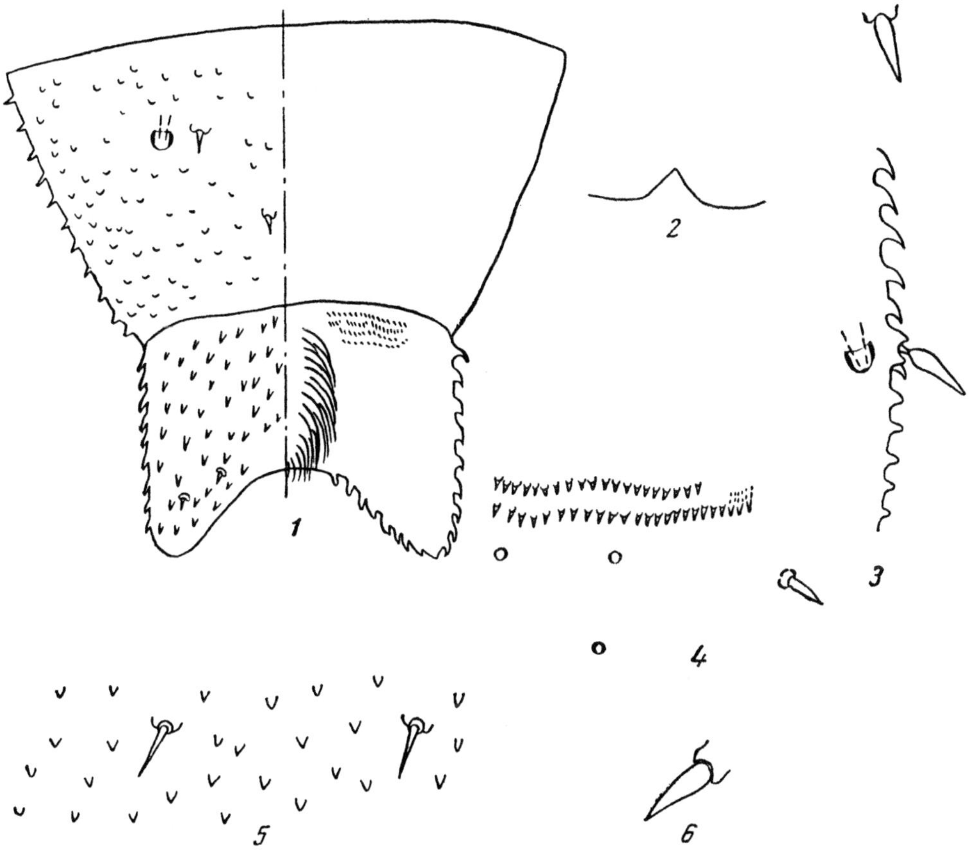


Figure 54. Larva of *Porricondyla neglecta* Mamaev
1 – last two segments, dorsal (left) and ventral (right) view; 2 – spatula; 3 – cuticular structures and lateral papillae on the abdominal segments; 4 – spinules and ventral papillae on the abdominal segments; 5 – cuticular structures and dorsal papillae on the mesothorax; 6 – dorsal papilla on the seventh abdominal segment (orig.)

## TRIBE CECIDOMYIINI

The larvae of the representatives of this tribe are elongated. Their anterior as well as their posterior end is rounded. The body is, as a rule, four to five times as long as wide. The head and the anterior body-segments of the predatory larvae (*Arthrocnodax* Rübsaamen, *Feltiella* Rübsaamen) are strongly elongated and narrowed. A spindle-shaped body with narrowed anterior and posterior ends is characteristic for the larvae of *Stelechodiplosis* Möhn.

An elongated-oval head capsule with clear, isolated exocephalic rods, which are equally wide along their full length or slightly widened at the posterior end, is characteristic of the majority of the species. The rods are as long as, or slightly longer than, the width of the capsule (*Clinodiplosis* Kieffer, *Mycodiplosis* Rübsaamen, *Contarinia* Rondani, etc.). The rods are comparatively short (no longer than half the width of the capsule) in a small group of larvae (*Thomasiniana* Strand, *Profeltiella* Kieffer). The exocephalic rods are strikingly elongated and nearly three times as long as the width of the head capsule in some species as, for example, *Itonida* Meigen, *Arthrocnodax* Rübsaamen, *Stelechodiplosis* Möhn. A relatively small group of larvae has a morphologically complex head capsule. The larvae of *Dichaetia* Mamaev have a large, elongated head capsule. The highly pronounced dorsal shield of the capsule is markedly longer than the width of the capsule at the base. The ventral side of the capsule is strongly reduced and half as long as the dorsal side. The comparatively large head capsule of the larvae of *Lestodiplosis* Kieffer is only weakly sclerotized, whereas its posterior half is reduced. The tentorial arms are powerfully developed and, in this genus, protrude beyond the head capsule.

The shape and structure of the antennae are highly variable within this tribe. The second antennal segment is in most species elongated-oval, slightly narrowed, and rounded at the apex. It is 1.5-3 times as long as the width of its base. The second antennal segment of some species gradually tapers off from the base to the tip and is nearly triangular (*Itonida* Meigen). The second antennal segment of most predatory larvae is thin, strongly elongated, in some cases nearly flagellar (*Feltiella* Rübsaamen, *Therodiplosis* Kieffer, *Lestodiplosis* Kieffer). The antennae of the larvae of *Buhriella* Stelter are bluntly truncated and lack the characteristic sensillae, which renders them quite peculiar.

The spatula is in the majority of species well-pronounced, and infrequently absent (*Feltiella* Rübsaamen, *Lestodiplosis* Kieffer, *Therodiplosis* Kieffer). The head of the spatula is predominantly provided with two conspicuous teeth; a one-toothed spatula is exceptional (*Arthrocnodax* Rübsaamen). Some species have a spatula with a widened head, without teeth (*Loewiola* Kieffer, *Itonida* Meigen). Though the shape and the size of the spatula are also rather strongly variable, the majority of larvae is characteristically provided with a comparatively small spatula with a well-pronounced, clearly outlined head and manubrium. On the one hand the two-toothed spatula may be very small in some larvae (*Acaroletes* Kieffer) and, on the other hand, extremely long and sturdy in some other species (*Putoniella* Kieffer, *Buhriella* Stelter). The larvae with a wide and massive, nearly square spatula, which lacks a separate head and manubrium (*Paurosphondylus* Rübsaamen, *Dishormomyia* Kieffer, *Planetella* Westwood, etc.), make a special group.

The body of most larvae is bare or provided with various, hardly developed, cuticular structures. Only a relatively small group of larvae is distinguished by well-developed, oval verrucae, platelets, and protuberances of irregular shape on the dorsal and lateral surfaces; but even then the ventral surface of the body is predominantly bare (*Clinodiplosis* Kieffer, *Ametrodiplosis* Rübsaamen, *Dichaetia* Mamaev, etc.). On the dorsal as well as on the ventral surface of the body the larvae of *Arthrocnodax* Rübsaamen are rather densely covered with verrucae.

The papillae that are distributed on the thoracic and the first eight abdominal segments are morphologically quite uniform. They are mostly simple or setiform, with a small verruca at their base. The bristles are, as a rule, several times shorter than the length of the segment; in a few larvae only, e.g. in *Echinella* Mamaev, they are very long. The dorsal papillae on the abdominal segments of *Dichaetia* Mamaev are spherical and superficially resemble the dorsal papillae of *Peromyia* Kieffer. The terminal papillae show the largest morphological diversity. Since their structure often determines the shape of the terminal segment, they deserve a closer inspection (see below).

The anterior ventral papillae are mostly located behind the spinule fields; they occur within the fields only in a few cases. Locomotive outgrowths – prolegs (*Lestodiplosis* Kieffer) or large rounded swellings bearing the anterior ventral papillae (*Arthrocnodax* Rübsaamen) on their tip – are found on the ventral surface of the segments in a small group of predatory larvae.

Spinule fields consisting of clearly discernable, transverse rows of spinules are found on most larvae that belong to this tribe. The fields occur on the ventral surface of the meso- and metathoracic, and all abdominal segments. Solely on the larvae of *Haplodiplosis* Rübsaamen fields are also found on the dorsal surface. Spinule fields are lacking only on a comparatively small group of larvae (*Loewiola* Kieffer, *Feltiella* Rübsaamen, *Lestodiplosis* Kieffer, *Therodiplosis* Kieffer, *Arthrocnodax* Rübsaamen).

The terminal segment is, as a rule, highly developed and only slightly smaller than the penultimate one. The terminal segment is noticeably shorter than the preceding one in just a few larvae (*Agevillea* Hubault). In the majority of species of this tribe the terminal segment bears two outgrowths, which are separated by an emargination of variable shape. Less often, the terminal segment is rounded or rectangular, without clear outgrowths at the apex (*Paurosphondylus* Rübsaamen, *Dishormomyia* Kieffer, *Plemeliella* Seitner). The terminal segment is shifted to the ventral side as a result of the lengthening of the penultimate segment in the larvae of *Itonida* Meigen and *Stelechodiplosis* Möhn.

The terminal papillae on the tip of the last segment may attain large dimensions. A part or all of the terminal papillae are in many species conical or verruca-shaped, and apically provided with bristle-, stick-like, or differently shaped appendages (*Clinodiplosis* Kieffer, *Ametrodiplosis* Rübsaamen, *Contarinia* Rondani, *Karschomyia* Felt). The length of the setiform terminal papillae strikingly varies. They are highly developed, bearing long bristles, in some genera (*Aphidoletes* Kieffer, *Echinella* Mamaev), whereas they are extremely small and hardly visible in others (*Dishormomyia* Kieffer and related genera).

The stigmata are, as a rule, provided with comparatively large, elongated-oval verrucae. In most larvae the stigmata on the prothoracic and the first seven abdominal segments are located on the lateral surfaces, while the terminal pair is shifted to the dorsal side of the eighth abdominal segment. The prothoracic pair of stigmata is found on the dorsal side of the larvae of *Dichaetia* Mamaev. The terminal pair of stigmata is well-developed in the larvae of *Triommatomyia* Mamaev, which inhabit very damp, rotting wood.

### Key to the genera of Cecidomyiini

1(20) Each of the first seven abdominal segments bears ten dorsal papillae.
2(9) The spatula has a well-pronounced, narrow manubrium (fig. 55,6).
3(4) The spatula bears a single tooth with a rounded tip (fig. 55,1).
*Dichrona* Rübsaamen (p.190)
4(3) The spatula bears two teeth.

5(6) The head of the spatula is provided with two large, lateral processes (fig. 55,4). *Pseudhormomyia* Kieffer (p. 191)

6(5) The head of the spatula lacks lateral processes.

7(8) The dorsal surface of the terminal segment bears pointed protuberances. The teeth of the spatula are small and pointed (fig. 55,6). *Dyodiplosis* Rübsaamen (p. 192)

8(7) The dorsal surface of the terminal segment is bare. The teeth of the spatula are large and rounded (fig. 55,7). . . . . . . . . *Proshormomyia* Kieffer (p. 192)

9(2) The spatula is wide and massive (fig. 56,1), and has no distinct manubrium.

10(13) The dorsal papillae bear long bristles, which are, on the eighth abdominal segment, as long as, or longer than, the stigmata.

11(12) The terminal papillae are simple (fig. 56,2). . . . *Planetella* Westwood (p. 193)

12(11) The terminal papillae are setiform. . . . . . . . *Dishormomyia* Kieffer (p. 195)

13(10) The dorsal papillae bear short bristles, which, when found on the eighth abdominal segment, are decidedly shorter than the stigmata.

14(15) The second antennal segment is nearly three times as long as wide. *Jaapiola* Rübsaamen (p. 195)

15(14) The second antennal segment is approximately as long as wide.

16(17) Between the teeth of the spatula (fig. 56,6) a small process of irregular shape is found. . . . . . . . . . . . . . . . . . . . . . . *Diplolaboncus* Rübsaamen (p. 196)

17(16) Between the teeth of the spatula (fig. 56,8) no process is found.

18(19) Two of the six terminal papillae bear a short bristle, the other four papillae are simple (fig. 56,9). . . . . . . . . . . . . . *Paurosphondylus* Rübsaamen (p. 197)

19(18) All six terminal papillae bear a well-developed bristle. Two of these bristles are widened and conical, the other four are not widened. *Taphodiplosis* Kieffer (p. 198)

20(1) Each of the first seven abdominal segments bears six dorsal papillae.

21(32) At least the sublateral papillae on the meso- and metathorax bear well-developed bristles; if the bristles are comparatively short or resemble verrucae, then the terminal segment bears two clear, pointed, tooth-like outgrowths.

22(25) The anterior ventral papillae on the first seven abdominal segments resemble verrucae or bristles. The terminal segment bears two pointed, tooth-like outgrowths (fig. 58,6).

23(24) The anal slit is shifted to the dorsal side. The body-segments are bare, without any cuticular formations except straight rows of extremely small spinules on the dorsal and ventral surface. The papillae on the ventral surface are setiform. *Endaphis* Kieffer (p. 198)

24(23) The anal slit is situated on the ventral side. The dorsal and ventral surface of the body-segments are uniformly covered with isolated oval verrucae. Spinule fields do not occur. The papillae on the ventral surface resemble small verrucae. *Monarthropalpus* Rübsaamen (p. 200)

25(22) The anterior ventral papillae on the first seven abdominal segments are simple. The terminal segment is oval or bears small, verruca-like protuberances.

26(29) The spinule fields are well-pronounced. There are no prolegs. The second antennal segment is four to five times as long as wide (fig. 59,1).

27(28) The four ventral papillae on the eighth abdominal segment are simple, without bristles. The sublateral papillae on the prothorax are simple, whereas those on the meso- and metathorax bear well-developed bristles. The posterior ventral papillae on the first seven abdominal segments are simple or setiform. The body-segments are devoid of any setiform structures (fig. 59,3), not counting the usual papillae. . . . . . . . . . . . . . . . . . . . . . . . . *Aphidoletes* Kieffer (p. 201)

28(27) The four ventral papillae on the eighth abdominal segment bear well-developed bristles. The sublateral papillae on all of the thoracic, as well as the posterior ventral papillae on the first seven abdominal segments, are setiform. The dorsal and lateral surfaces of the body-segments (fig. 61,3) bear groups of well-developed bristles, not counting the usual papillae.
*Monobremia* Kieffer (p. 206)

29(26) The spinule fields are weakly pronounced or absent. Prolegs or large, oval outgrowths (fig. 15,3; 64,5) occur instead of the spinule fields. The antennae are narrow, long, and pointed. The second antennal segment is more than five times as long as wide.

30(31) The body is bare. The head is weakly sclerotized; the internal tentorial arms are massively developed and visible through the thin cover (fig. 7,2). The length of the arms, which protrude beyond the head capsule, does not surpass the length of the capsule. The spatula is lacking. Long, conical prolegs (fig. 62,4) are found on the ventral surface of the meso- and metathoracic, and the first seven abdominal segments. The terminal segment bears six setiform terminal papillae (fig. 63,4).
*Lestodiplosis* Kieffer (p. 208)

31(30) The body is densely covered with cuticular structures. The head is heavily sclerotized; the exocephalic arms, which originate on the posterior edge of the head capsule, are highly developed (fig. 7,1) and more than 1.5 times as long as the head capsule. The one-toothed spatula is well-pronounced. The ventral surface of the metathorax and the first seven abdominal segments is provided with large, oval outgrowths, which bear on their tip the sternal (on the metathorax) and the anterior ventral (on the abdominal segments) papillae. The terminal segment (fig. 64,6) bears two large stick-like, as well as four small setiform papillae. . . . . . . . . . . . . . . . . . . . . . . *Arthrocnodax* Rübsaamen (p. 211)

32(21) The sublateral papillae on the meso- and metathorax, as well as the other papillae on the ventral surface of the body (not counting the intercalary papillae), are simple.

33(58) All or a part of the dorsal papillae, at least those on the prothorax, are simple.

34(41) The dorsal papillae on the eighth abdominal segment are in all cases setiform, those on the other segments may be setiform or simple.

35(38) The dorsal papillae on all or a part of the thoracic segments are simple, whereas those on the abdominal segments bear well-developed bristles.

36(37) The body is transversely banded. The dorsal papillae on the prothorax are simple, the other ones are setiform. The spatula (fig. 65,1) is well-developed and bears large teeth. . . . . . . . . . . . . . . . . . . . . . . . . *Xylodiplosis* Kieffer (p. 213)

37(36) The body is smooth. The dorsal papillae on all thoracic segments are simple, whereas those on the abdominal segments are setiform. The spatula (fig. 65,3) is small; the teeth are hardly pronounced. . . . . . . *Massalongia* Kieffer (p. 215)

38(35) The dorsal papillae are only setiform on the eighth abdominal segment; on the other segments they are simple.

39(40) The lateral surfaces of the first seven abdominal segments bear short transverse rows of spinules (fig. 66,3). The manubrium of the spatula (fig. 66,6) is long and narrow (considerably less wide than the base of a tooth). *Tricholaba* Rübsaamen (p. 215)

40(39) The lateral surfaces of the segments do not bear transverse rows of spinules. The manubrium of the spatula (fig. 67,2) is massive and wide (considerably wider than the base of a tooth). . . . . . . . . . . . . *Triommatomyia* Mamaev (p. 218)

41(34) On all segments the dorsal papillae are simple and without bristles.

42(43) The terminal segment (fig. 68,9) is rounded, and does not bear clear outgrowths. The large spatula (fig. 68,2) is long; the teeth and the head are split by a deep incision, which reaches far beyond the base of the teeth. The very small second antennal segment is apically truncated. . . . . . . . . . *Buhriella* Stelter (p. 220)

43(42) The terminal segment (fig. 69,2) bears two or more processes or teeth. The antennae have the usual shape.

44(45) There is no spatula. Six of the eight terminal papillae are simple, while the other two, which appear as blunt hooks, stand on a common base behind the anal slit (fig. 69,2). . . . . . . . . . . . . . . . . . . . . . . . *Thecodiplosis* Kieffer (p. 221)

45(44) The spatula is well-developed. Most terminal papillae are verruca-like or setiform; there are no paired terminal papillae behind the anal slit.

46(49) The spatula bears no teeth (fig. 70,1).

47(48) Spinule fields occur on the dorsal as well as on the ventral surface of the segments. The terminal segment bears eight small verrucae of equal dimensions (fig. 70,10). . . . . . . . . . . . . . . . . . . . *Haplodiplosis* Rübsaamen (p. 223)

48(47) Spinule fields occur only on the ventral surface of the segments. The terminal segment (fig. 69,4) bears eight setiform terminal papillae, the bristles of six of which are thick, and of the other two – thin. . . *Endopsylla* De Meijere (p. 225)

49(46) The spatula bears two well-pronounced teeth.

50(51) All eight terminal papillae have the appearance of well-developed verrucae with a short, conical appendage at the tip (fig. 69,6). . . *Harmandia* Kieffer (p. 226)

51(50) Two of the terminal papillae have the appearance of well-developed sclerotized verrucae. Either all or some of the remaining six bear short bristles. Those without short bristles are simple.

52(53) The second antennal segment is as long as wide. The terminal segment is small; the penultimate segment is at its posterior edge more than three times as wide as the terminal segment. . . . . . . . . . . . . . . . . . . *Agevillea* Hubault (p. 227)

53(52) The second antennal segment is nearly twice as long as wide. The terminal segment is only slightly smaller than the preceding one; the penultimate segment is at its posterior edge noticeably less than 1.5 times as wide as the terminal one.

54(55) The teeth of the spatula (fig. 71,2) are separated at the very base; the interstice is larger than the width of a tooth. The second antennal segment is no more than 1.5 times as long as wide. Two of the eight terminal papillae have the appearance of sclerotized verrucae, four bear short bristles, and two are simple. *Anisostephus* Rübsaamen (p. 228)

55(54) The teeth of the spatula (fig. 72,2) are separated by a comparatively shallow emargination; their bases are contiguous. The second antennal segment is twice as long as wide. Two of the eight terminal papillae have the appearance of sclerotized verrucae, the other six bear a short bristle.

56(57) The teeth of the spatula (fig. 72,2) are triangular and sharply pointed. The terminal segment is, on its dorsal and lateral surfaces as well as along its posterior edge, covered with small, elongated platelets of irregular shape; along the anterior edge occur protuberances. The manubrium of the spatula is wide (as wide as the base of a tooth). . . . . . . . . . . . *Syndiplosis* Rübsaamen (p. 229)

57(56) The teeth of the spatula (fig. 73,2) are rounded. The terminal segment is covered with large, weakly outlined platelets, which are more clearly outlined on the lateral surfaces. The dorsal surface of the other segments is bare or bears clearly discernable pointed protuberances on its middle part. The posterior and lateral margins of the segments are bare. The manubrium of the spatula is slightly narrower than in the preceding genus (less wide than the base of a tooth).
*Contarinia* Rondani (p. 230)

58(33) The dorsal papillae on the thoracic and the first eight abdominal segments bear well-developed bristles.

59(60) The prothoracic as well as the terminal pair of stigmata is shifted to the dorsal side. The central dorsal papillae on the prothorax are setiform and extremely long (as long as the segment). The dorsal papillae on the meso- and metathoracic, and the first seven abdominal segments have the appearance of spherical outgrowths (fig. 75,6). The head capsule (fig. 75,1) is strikingly elongated; its dorsal side is not reduced. The ventral side of the head capsule is twice as short as the dorsal side. . . . . . . . . . . . . . . . . . . . . . . . . . . . *Dichaetia* Mamaev (p. 234)

60(59) The prothoracic stigmata are found on the lateral sides of the segment. On all segments the dorsal papillae are setiform. The ventral side of the head capsule is as long as the dorsal side.

61(68) The antennae are long and gradually taper off from the base to the apex. The second antennal segment is no less than four times as long as wide.

62(65) The anal slit is situated on the dorsal side of the terminal segment. The second antennal segment (fig. 76,3) is ten to eleven times as long as wide. There is no spatula. The terminal segment bears four long, rod-like and four short terminal papillae.

63(64) The terminal papillae of the anterior pair bear a short bristle.
*Feltiella* Rübsaamen (p. 236)

64(63) The terminal papillae of the anterior pair have the appearance of small verrucae without bristles. . . . . . . . . . . . . . . . . *Therodiplosis* Kieffer (p. 238)

65(62) The anal slit is situated on the ventral side of the terminal segment. The second antennal segment (fig. 77,1) is four to six times as long as wide. The spatula is well-developed.

66(67) The dorsal papillae 'b' (fig. 1) bear on most segments, in particular on the prothorax, a markedly shorter bristle than the papillae 'a' and 'c'.
*Isodiplosis* Rübsaamen (p. 238)

67(66) The dorsal papillae are of approximately equal sizes on most segments.
*Acaroletes* Kieffer (p. 241)

68(61) The antennae are relatively thick and short, and only apically narrowed and rounded. The second antennal segment is no more than three times as long as wide.

69(72) The second antennal segment is as long as, or only slightly longer, than wide.

70(71) The head of the spatula (fig. 78,3) is not clearly demarcated; the manubrium is wide and short. The terminal segment is bare and bears eight conical terminal papillae. . . . . . . . . . . . . . . . . . . . . . . . . . *Plemeliella* Seitner (p. 242)

71(70) The spatula(fig. 78,5) is very small and has no manubrium. The terminal segment bears eight setiform terminal papillae. A few rows of tiny spinules occur on the anterior part of the dorsal surface of the terminal segment.
*Atylodiplosis* Rübsaamen (p. 243)

72(69) The second antennal segment is no less than 1.5 times as long as wide.

73(78) The spatula bears no teeth; the anterior edge of the head is slightly concave.

74(75) The head of the spatula (fig. 79,1) is wide; the manubrium is not clearly delimited by the head and gradually narrows to the posterior end. Spinule fields are absent. The exocephalic rods are short (no longer than the width of the base of the capsule). . . . . . . . . . . . . . . . . . . . . . . . . . . *Loewiola* Kieffer (p. 243)

75(74) The head of the spatula (fig. 80,2) is strongly widened and spindle-like; the manubrium is comparatively narrow. Spinule fields are present. The exocephalic rods are massive and long (more than 2.5 times as long as the width of the base of the capsule). The stigmata of the terminal pair are closely drawn towards each other and are situated on outgrowths of the penultimate segment.

76(77) The stigmata of the terminal pair (fig. 80,6) are located on comparatively short outgrowths, which are no longer than the width. The outgrowths do not protrude beyond the posterior edge of the terminal segment. . . *Itonida* Meigen (p. 245)

77(76) The stigmata of the terminal pair (fig. 81,7) are located on long outgrowths, which are noticeably longer than the width. The outgrowths protrude beyond the posterior edge of the terminal segment, which bears similar outgrowths at its tip.
*Stelechodiplosis* Möhn (p. 248)

78(73) The spatula bears two well-developed teeth.

79(84) The head capsule bears short exocephalic rods, which are as long as, or slightly shorter than, half the width of the head capsule. The body is densely covered with various cuticular structures.

80(81) The terminal segment is rounded and does not bear clear processes. The terminal papillae are long and setiform. The lateral surfaces of the segments bear thin, hair-like structures (fig. 82,4). The antennae have a peculiar shape: the apical part, which bears the sensorial unit, is no longer than the length of the base (fig. 82,1). The dorsal papillae are long; their bristles are no less than half as long as the segments. . . . . . . . . . . . . . . . . . . . . . . *Echinella* Mamaev (p. 250)

81(80) The terminal segment bears clearly discernable, conical outgrowths with sclerotized teeth at the tip. The terminal papillae are small and bear small bristles; they are found on the outgrowths. The lateral surfaces of the segments do not bear hair-like structures. The antennae (fig. 83,1) have the usual shape: the apical part, which bears the sensorial unit, is considerably longer than the length of the base. The dorsal papillae are relatively short; their bristles are noticeably less than half as long as the segments.

82(83) The antennae are long (nearly three times as long as wide). The terminal segment is small, clearly demarcated, and considerably narrower than the penultimate one. The terminal pair of stigmata is situated at the level of the anterior end of the anal slit. . . . . . . . . . . . . . . . . . . . . . . . . *Thomasiniana* Strand (p. 252)

83(82) The antennae are short (at most twice as long as wide). The terminal segment is noticeably smaller than the penultimate one and not clearly demarcated by the latter. The terminal pair of stigmata is shifted to the end of the body and situated at the level of, or behind, the posterior end of the anal slit. *Profeltiella* Kieffer (p. 255)

84(79) The head capsule bears well-developed exocephalic rods, which are never shorter than the width of the base of the head capsule.

85(86) The terminal segment bears at its tip two weakly delimited, rounded lateral lobes, which resemble a rectangle. All terminal papillae are located directly on the surface of the segment, not on verrucae. Four of the terminal papillae bear well-developed bristles of varying lengths, while the other four bear small, thick, conical appendages (fig. 85,3). . . . . . . . . *Mycodiplosis* Rübsaamen (p. 256)

86(85) The terminal segment bears clearly discernable outgrowths at its tip. All or a part of the terminal papillae have the appearance of sturdy verrucae, which apically bear bristles, conical appendages, etc.

87(92) The terminal segment bears six verruca-shaped and two setiform papillae.

88(89) The bristle of the outer terminal papilla (fig. 86,2) is noticeably shorter than the largest verruca-shaped terminal papilla. . . . . . . *Sitodiplosis* Kieffer (p. 258)

89(88) The bristle of the outer terminal papilla is as long as, or longer than, the largest verruca-shaped terminal papilla.

90(91) The six verruca-shaped terminal papillae (fig. 90,4) are not all the same size. The dorsal papillae 'b' (fig. 1) are strongly reduced. The dorsal surface of the terminal segment is either bare, or bears a few transverse rows of protuberances. *Clinodiplosis* Kieffer (p. 259)

91(90) At least four of the six verruca-shaped terminal papillae (fig. 91,2) are of the same size. The dorsal papillae 'b' are not reduced, their bristle is as long as that of the papillae 'c'; the outer papillae are somewhat larger than the other ones. The dorsal surface of the terminal segment bears small verrucae that are not arranged into clear rows. . . . . . . . . . . . . . . . . *Ametrodiplosis* Rübsaamen (p. 265)

92(87) The terminal segment bears eight verruca-shaped papillae, which apically bear various cuticular structures.

93(94) All terminal papillae (fig. 92,8) apically bear thick, pointed bristles of the same type. . . . . . . . . . . . . . . . . . . . . . . . . . . . *Karschomyia* Felt (p. 269)

94(93) The terminal papillae (fig. 93,6) apically bear structures of different morphological types.

95(96) Two of the eight terminal papillae have well-pronounced bristles. The spatula (fig. 93,2) is massive; the teeth, though not demarcated by the head, are separated by a deep, narrow incision. . . . . . . . . . . . . . . . *Putoniella* Kieffer (p. 271)

96(95) Four of the eight terminal papillae have well-pronounced bristles. The teeth of the spatula (fig. 94,2) are clearly demarcated by the head and are separated by a shallow, oval emargination. . . . . . . . . . . *Macrodiplosis* Kieffer (p. 273)

1. Genus **Dichrona** Rübsaamen (fig. 55,1-2)

*Morphology*. The second antennal segment is two or three times as long as wide. The one-toothed spatula (fig. 55,1) has a well-developed manubrium.

The bare dorsal surface is provided with large, pointed protuberances. All ten dorsal papillae on the thoracic segments are simple; a part of the dorsal papillae on the

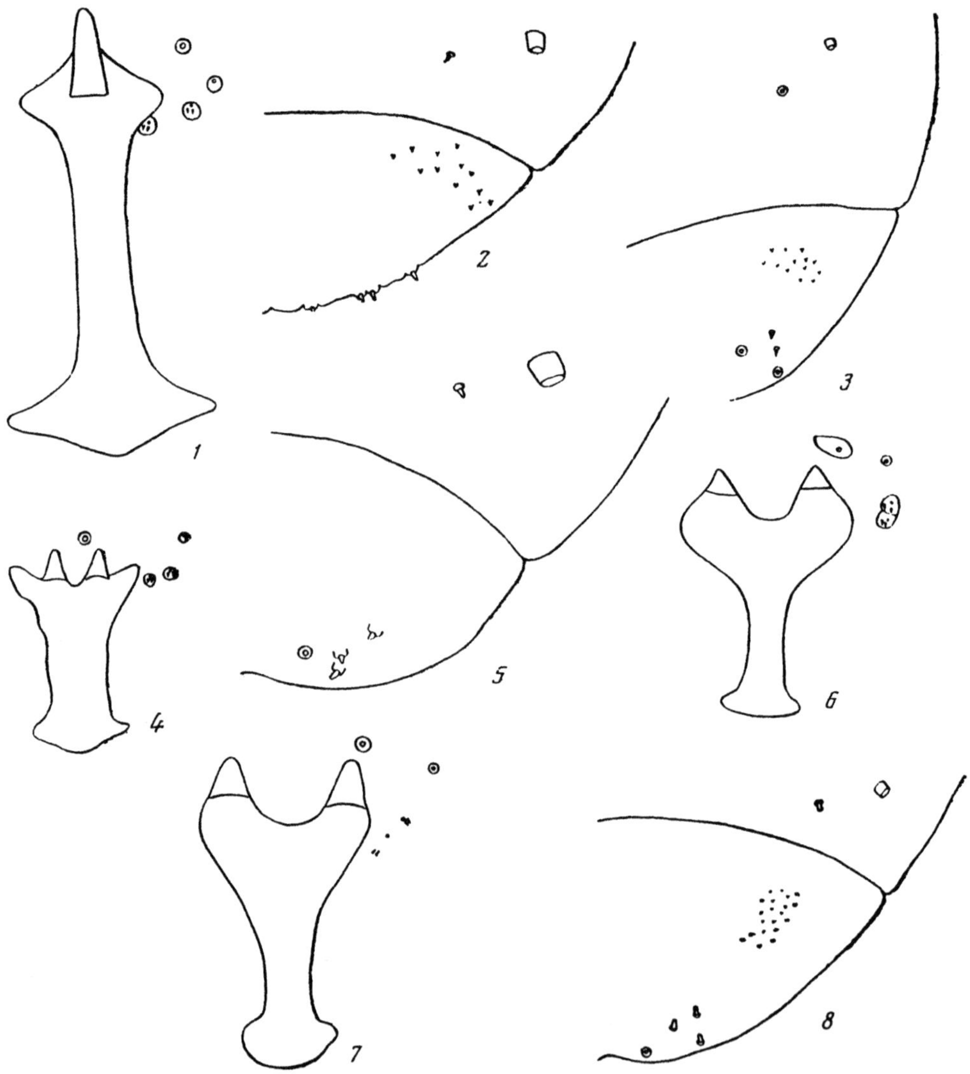


Figure 55. Larvae of the genera *Dichrona* Rübsaamen, *Pseudhormomyia* Kieffer, *Dyodiplosis* Rübsaamen, and *Proshormomyia* Kieffer
*Dichrona gallarum* Rübsaamen: 1 – spatula and papillae; 2 – terminal segment, dorsal view; *Pseudhormomyia caricis* (Rübsaamen): 3 – terminal segment, dorsal view; 4 – spatula and papillae; *Dyodiplosis arenariae* (Rübsaamen): 5 – terminal segment; 6 – spatula and papillae; *Proshormomyia fischeri* (Frauenfeld): 7 – spatula and papillae; 8 – terminal segment, dorsal view (after Möhn 1955b)

abdominal segments is simple, the other ones bear short bristles. The two dorsal papillae on the eighth abdominal segment are short and setiform. Two simple sternal papillae, four groups of normal intercalary papillae, and two simple sublateral papillae occur on the ventral surface of the thoracic segments. Four simple anterior and two simple posterior ventral papillae are found on each of the first seven abdominal segments. Four simple ventral papillae occur on the eighth abdominal segment. The terminal segment is provided with eight terminal papillae, which bear short bristles.

Spinule fields do not occur on the ventral surface of the segments; in their place, as well as on the remaining part of the surface of the body, pointed protuberances are found.

The terminal segment (fig. 55,2) bears on its dorsal surface isolated, tiny, pointed protuberances, while its ventral surface is bare. The anal slit is surrounded by longitudinal rows of spinules. All six anal papillae are simple.

*Biology*. The larvae develop in the leaf sheaths of various sedges.

**Dichrona gallarum** Rübsaamen

*Morphology*. All segments of the body bear pointed protuberances. Two of the eight terminal papillae have longer bristles than the other ones. Four or five longitudinal rows of spinules are found along both sides of the anal slit, whereas four or five transverse rows occur in front of the anal slit.

*Comment*. The authors made use of the description by Möhn (1955b).

2. Genus **Pseudhormomyia** Kieffer (fig. 55,3-4)

*Morphology*. The second antennal segment is as long as wide. The manubrium of the spatula (fig. 55,4) is distinct; the teeth are comparatively small.

The naked body bears isolated rows of pointed protuberances in the vicinity of the dorsal papillae. All dorsal papillae on the thoracic and the first seven abdominal segments are simple, as are the two dorsal papillae on the eighth abdominal segment. Two simple sternal, the usual set of intercalary, and two simple sublateral papillae occur on the ventral surface of each of the thoracic segments. Four simple anterior as well as two simple posterior ventral papillae are found on the first seven abdominal segments. The eighth abdominal segment bears four simple ventral papillae. The terminal segment is provided with eight terminal papillae. Of these, four have short bristles, and the other four are simple. The lateral papillae are simple. Just a few transverse rows of spinules are found on the ventral surface of the segments.

The dorsal surface of the terminal segment (fig. 55,3) bears pointed protuberances, its ventral surface is bare, and its lateral surfaces bear only a few isolated protuberances. Some longitudinal rows of spinules are found around the anal slit. The six anal papillae are simple.

*Biology*. The larvae are bound to several species of sedges.

**Pseudhormomyia caricis** (Rübsaamen)

*Morphology*. The body bears, in the vicinity of the dorsal papillae, six to eight transverse rows of protuberances. As many as eight to ten indistinct rows of protuberances are found on the terminal segment. The anterior ventral papillae stand partly between, partly behind, transverse rows of spinules. Twelve to fourteen longitudinal rows of spinules occur around the anal slit; eight to ten rows are found in front of the slit.

*Biology*. The larvae cause the formation of small galls on the leaves of sedges.

*Comment*. The authors made use of the description by Möhn (1955b).

3. Genus **Dyodiplosis** Rübsaamen (fig. 55,5-6)

*Morphology*. The second segment of the antennae is short. The spatula has a distinct manubrium; the small teeth of the spatula stand wide apart.

The dorsal surface of the body bears isolated, pointed protuberances. All dorsal papillae on the prothorax lack bristles; the dorsal papillae on the other segments bear short, conical bristles, or are (more rarely) simple. The two dorsal papillae on the eighth abdominal segment also bear short bristles. On the ventral surface of each of the thoracic segments are found: two simple sternal papillae, four groups of ordinary intercalary papillae, and two simple sublateral papillae. Four simple anterior and two posterior papillae occur on the ventral surface of each of the first seven abdominal segments. The ventral papillae on the eighth abdominal segment are simple. The terminal segment is provided with eight terminal papillae, six of which bear a short, conical bristle, while two are simple. The lateral papillae bear short conical bristles.

The ventral surface lacks spinule fields; they are replaced by rows of pointed protuberances.

The dorsal as well as the ventral surface of the terminal segment (fig. 55,5) is bare. Some rows of spinules are found around the anal slit. All six anal papillae are simple.

*Biology*. The larvae develop on various species of sedges.

**Dyodiplosis arenariae** (Rübsaamen)

*Morphology*. The anterior part of the spatula (fig. 55,6) is broad; its comparatively small teeth are separated by a rounded emargination. The anterior ventral papillae are located immediately behind rows of pointed protuberances. Six to eight longitudinal rows of spinules occur around the anal slit.

*Biology*. The larvae develop in small galls that frequently occur on the subterranean parts of the leaves of sedges.

*Comment*. The authors made use of the description by Möhn (1955b).

4. Genus **Proshormomyia** Kieffer (fig. 55,7-8)

*Morphology*. The second antennal segment is as long as it is wide. The spatula (fig. 55,7) is two-toothed; the well-developed, broad manubrium has a widened base.

The body is smooth; from the mesothorax onwards the segments each bear transverse rows of toothed platelets at the level of the dorsal papillae. Ten dorsal papillae, partly with blunt bristles and partly simple, occur on the thoracic and the first seven abdominal segments. The two dorsal papillae on the eighth abdominal segment bear blunt bristles. On the ventral surface of each of the thoracic segments are located: two simple sternal papillae, four groups of intercalary papillae (each group consisting of two setiform and one simple papilla), and two simple sublateral papillae. The ventral surface of the first seven abdominal segments bears two anterior as well as two posterior ventral papillae. The ventral papillae on the eighth abdominal segment are also simple. The terminal segment is provided with eight terminal papillae, six of which bear a blunt bristle, while two are simple. Spinule fields are distributed on the ventral surface of the segments.

The terminal segment (fig. 55,8) bears isolated pointed protuberances on its dorsal surface; the ventral surface is bare. The anal slit is surrounded by longitudinal rows of spinules. All six anal papillae are simple, without bristles.

*Biology.* The species of this genus develop on sedges.

**Proshormomyia fischeri** (Frauenfeld)
*Morphology.* The rounded teeth of the spatula (fig. 55,7) are separated by a wide emargination.

Eight to ten rows of toothed platelets occur on the dorsal surface of the body segments. The spinule fields consist of twelve to fourteen rows. On the abdominal segments the anterior ventral papillae are located between the fifth and seventh rows (counted from the posterior edge) of the spinule fields.

*Biology.* The larvae cause the formation of elongated thickenings on the leaf sheaths of various species of sedges.

*Comment.* The authors made use of the description by Möhn (1955b).

5. Genus **Planetella** Westwood (fig. 56,1-2)

*Morphology.* The second antennal segment is as long as its width. The strongly broadened spatula (fig. 56,1) has no distinct manubrium. The long, thin teeth are separated by a deep emargination.

The body bears oval platelets with small points. All ten dorsal papillae have moderately long bristles; but the dorsal papillae on the prothorax are simple. The two dorsal papillae on the eighth abdominal segment also bear bristles. The sternal papillae on the ventral surface of the thoracic segments are simple, without bristles. The intercalary papillae on the prothorax are grouped in pairs which stand on a common base; on the other segments their distribution is as usual. Four simple anterior as well as two simple posterior ventral papillae are found on each of the first seven abdominal segments. The four ventral papillae on the eighth abdominal segment are simple. The terminal segment bears eight simple terminal papillae. The lateral papillae are setiform. The ventral surface of the segments is provided with oval verrucae, which bear weakly developed teeth. Spinule fields do not occur. In their place rows of oval verrucae with well-developed, sometimes double, teeth are found on the anterior part of the segments.

The dorsal and ventral surfaces of the terminal segment (fig. 56,2) are covered with oval, toothed platelets. The anal slit is surrounded by longitudinal rows of spinules. The six anal papillae are simple.

*Biology.* The larvae develop on sedges.

*Comment.* The authors made use of the description by Möhn (1955b).

**Planetella tuberifica** (Rübsaamen)
*Morphology.* The second antennal segment is slightly wider than its length. The anterior ventral papillae are located behind sixteen to eighteen rows of oval, toothed verrucae.

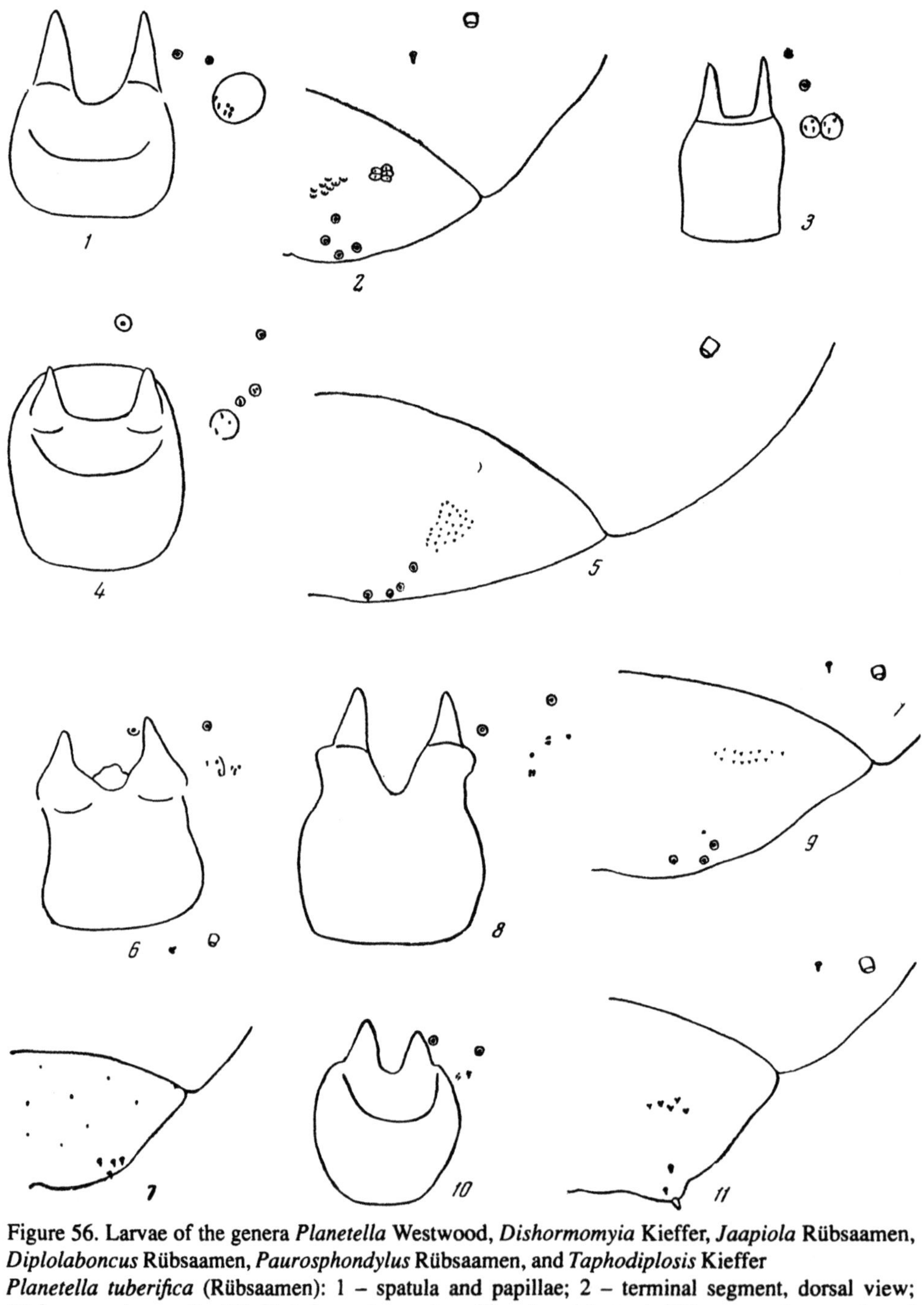


Figure 56. Larvae of the genera *Planetella* Westwood, *Dishormomyia* Kieffer, *Jaapiola* Rübsaamen, *Diplolaboncus* Rübsaamen, *Paurosphondylus* Rübsaamen, and *Taphodiplosis* Kieffer
*Planetella tuberifica* (Rübsaamen): 1 – spatula and papillae; 2 – terminal segment, dorsal view; *Dishormomyia cornifex* (Kieffer): 3 – spatula and papillae; *Jaapiola tarda* Rübsaamen: 4 – spatula and papillae; 5 – terminal segment, dorsal view; *Diplolaboncus* (Rübsaamen): 6 – spatula and papillae; 7 – terminal segment, dorsal view; *Paurosphondylus* (Rübsaamen): 8 – spatula and papillae; 9 – terminal segment, dorsal view; *Taphodiplosis subterranea* (Kieffer et Trotter): 10 – spatula and papillae; 11 – terminal segment, dorsal view (after Möhn 1955b)

The anal slit is surrounded by ten to twelve longitudinal rows of spinules. All six anal papillae are simple.

*Biology*. The larvae cause blister-like calluses on the leaf sheaths of sedges.

*Comment*. The authors made use of the description by Möhn (1955b).

## 6. Genus **Dishormomyia** Kieffer (fig. 56,3)

*Morphology*. The second antennal segment is short. The two-toothed, strongly widened spatula (fig. 56,3) has no distinct manubrium. The body bears oval, toothed verrucae.

All ten dorsal papillae bear short, thick bristles. The two dorsal papillae on the eighth abdominal segment also bear thick bristles. The sternal papillae on the ventral surface of the thoracic segments are simple, while the intercalary papillae are of the ordinary type. Four simple anterior ventral papillae occur on the ventral surface of the first seven abdominal segments. The two posterior ventral papillae on the first seven abdominal segments, as well as the four ventral papillae on the eighth abdominal segment, are simple. The terminal segment bears eight setiform terminal papillae. The lateral papillae have short, though thick, bristles. Spinule fields are found on the ventral surface of the body segments.

The terminal segment bears rounded, toothed verrucae on its dorsal surface; the ventral surface is bare.

*Biology*. The larvae develop on sedges.

**Dishormomyia cornifex** (Kieffer)

*Morphology*. The narrow, pointed teeth of the spatula (fig. 56,3) are separated by a wide, rectangular emargination. The dorsal surface of the body bears isolated, oval, toothed verrucae. The anterior ventral papillae are located within the spinule fields.

*Biology*. The larvae cause deformations of the lateral runners of various species of sedges.

*Comment*. The authors made use of the description by Möhn (1955b).

## 7. Genus **Jaapiola** Rübsaamen (fig. 56,4-5)

*Morphology*. The second antennal segment is twice as long as its width. The strongly widened spatula (fig. 56,4) has no clearly pronounced manubrium. The teeth of the spatula are short and pointed.

The bare dorsal surface of the body is, in the vicinity of the dorsal papillae, provided with rows of pointed protuberances. While some of the ten dorsal papillae on the abdominal segments bear short bristles, some of them are simple. The two dorsal papillae on the eighth abdominal segment have short bristles. On the dorsal surface each of the thoracic segments bears: two simple sternal papillae, four groups of ordinary intercalary papillae, and two simple sublateral papillae. Four simple anterior, as well as two simple posterior ventral papillae occur on the ventral surface of each of the first seven abdominal segments. Four simple ventral papillae are found on the eighth abdominal segment. The terminal segment is provided with eight terminal papillae, four of which bear a bristle and four of which are simple. The lateral papillae are partly setiform, partly simple. The spinule fields are well-pronounced.

The smooth dorsal surface of the terminal segment (fig. 56,5) bears tiny, pointed protuberances, which do not form distinct rows. The ventral surface of the segment is also smooth. A large number of longitudinal rows of spinules is found around the anal slit. There are just four anal papillae.

*Biology*. The larvae are found on sedges.

**Jaapiola tarda** Rübsaamen

*Morphology*. The emargination between the teeth of the spatula (fig. 56,4) is rounded. The dorsal surface of the segments, from the prothoracic onwards, bears ten to twelve transverse rows of pointed protuberances in the vicinity of the dorsal papillae. All ten dorsal papillae on the thoracic segments are simple; on the abdominal segments they are partly simple, and partly bear short bristles.

The spinule fields on the ventral surface of the body comprise 30-32 rows. The anterior ventral papillae on the first seven abdominal segments are located between the fifth and seventh rows of spinules, reckoned from the posterior edge of the field. Around the anal slit the terminal segment bears 28-30 longitudinal rows of spinules, and in front of the slit 24-26 rows.

*Biology*. The larvae develop within grain-like calluses on the leaf bases of *Carex vesicaria*.

*Comment*. The authors made use of the description by Möhn (1955b).

8. Genus **Diplolaboncus** Rübsaamen (fig. 56,6-7)

*Morphology*. The second antennal segment is as long as its width. The two-toothed and considerably widened spatula (fig. 56,6) has no distinct manubrium.

The bare dorsal surface of the body bears, from the mesothorax onwards, indistinct, pointed protuberances in the vicinity of the dorsal papillae. Ten dorsal papillae (partly with well-developed bristles, partly simple) occur on each of the thoracic and the first seven abdominal segments. Two papillae with short bristles are found on the eighth abdominal segment. On their ventral surface the thoracic segments each bear two simple sternal and sublateral papillae as well as four groups of ordinary intercalary papillae. The first seven abdominal segments each bear on their ventral surface four simple anterior and two simple posterior ventral papillae. Four simple ventral papillae occur on the eighth abdominal segment. The terminal segment bears eight terminal papillae with short bristles. The lateral papillae have short bristles. The bare ventral surface of the body is provided with rows of sturdy protuberances, some of which are many-toothed.

The bare dorsal surface of the terminal segment (fig. 56,7) bears only a few isolated protuberances. On the bare ventral surface some rows of protuberances are found in front of the anal slit. Large longitudinal rows of spinules are found around the anal slit. None of the six anal papillae bears a bristle.

*Biology*. The larvae develop on various species of sedges.

**Diplolaboncus tumorificus** (Rübsaamen)
*Morphology*. The second antennal segment is slightly wider than the length. The teeth of the spatula (fig. 56,6) are rather sharp. An irregular protuberance is found on the bottom of the emargination separating the teeth.

Fourteen to sixteen rows of indistinct protuberances occur on the dorsal surface of the body-segments in the vicinity of the dorsal papillae. Eighteen to twenty rows of protuberances are found on the ventral surface of the segments. The anterior ventral papillae are located between the last two rows of protuberances. The anal slit is surrounded by eighteen to twenty longitudinal rows of spinules.

*Biology*. The larvae inhabit blister-like swellings on the leaf bases of various species of sedges.

*Comment*. The authors made use of the description by Möhn (1955b).

9. Genus **Paurosphondylus** Rübsaamen (fig. 56,8-9)

*Morphology*. The second antennal segment is nearly as long as wide. The noticeably widened spatula (fig. 56,8) has no clearly distinct manubrium. The teeth of the spatula are pointed.

The smooth body bears only a few rows of pointed protuberances at the level of the dorsal papillae. The thoracic and the first seven abdominal segments each bear ten dorsal papillae with short bristles (just a few of the papillae lack a bristle). The two dorsal papillae on the eighth abdominal segment have short bristles. The ventral surface of each of the thoracic segments bears: two simple sternal papillae, four groups of intercalary papillae, each group of which consists of one simple and two setiform papillae, and two simple sublateral papillae. Four simple anterior and two simple posterior ventral papillae are found on the ventral surface of each of the first seven abdominal segments. Just four ventral papillae occur on the eighth abdominal segment. The terminal segment bears six terminal papillae, two of which are provided with a short bristle. Spinule fields occur on the ventral surface of the segments.

The rounded terminal segment (fig. 56,9) carries on its dorsal surface some very tiny, weakly pronounced, pointed protuberances; the ventral surface of the segment is smooth. The anal slit is surrounded by longitudinal rows of tiny spinules. All six anal papillae are simple.

*Biology*. The species of this genus develop in the leaf sheaths of sedges.

**Paurosphondylus rosenhaueri** (Rübsaamen)
*Morphology*. The second antennal segment is widened; its width slightly surpasses its length. The teeth of the spatula (fig. 56,8) are divided by a deep, rounded emargination.

All anterior ventral papillae on the abdominal segments are located behind the spinule fields. The spinule fields on the ventral surface of the body-segments consist of 16-18 rows, on the terminal segment of 20-22 rows. The anal slit is surrounded by 16-18 longitudinal rows of spinules. All terminal papillae (fig. 56,9) are found on the dorsal surface of the terminal segment.

*Biology*. The larvae cause swellings on the leaf sheaths of *Carex acuta*.

*Comment*. The authors made use of the description by Möhn (1955b).

## 10. Genus **Taphodiplosis** Kieffer (fig. 56,10-11)

*Morphology*. The second antennal segment is as long as wide. The strongly widened spatula (fig. 56,10) has no distinct manubrium; the teeth of the spatula are pointed.

The dorsal surface of the body is provided with pointed protuberances. In most cases all ten dorsal papillae on the thoracic and the first seven abdominal segments have short bristles; the dorsal papillae are rarely simple. The two dorsal papillae on the eighth abdominal segment bear short bristles. The ventral surface of each of the thoracic segments bears: two simple sternal papillae, four groups of ordinary intercalary papillae and two simple sublateral papillae. Four simple anterior as well as two simple posterior ventral papillae occur on the first seven abdominal segments. Four simple ventral papillae are found on the eighth abdominal segment. The terminal segment bears six terminal papillae, four of which are provided with very short bristles and two with a light coloured, conical apical part. On each of the lateral surfaces of the segments two setiform lateral papillae are found. The ventral surface of the body bears transverse rows of large spinules.

The dorsal surface of the terminal segment (fig. 56,11) is provided with pointed protuberances, while the ventral surface is bare. Longitudinal rows of spinules surround the anal slit. All anal papillae are simple.

*Biology*. The larvae develop on sedges.

**Taphodiplosis subterranea** (Kieffer et Trotter)

*Morphology*. The ventral surface of the body is bare. Pointed protuberances are found only along the lateral surface of the segments and near the posterior ventral papillae. The four anterior ventral papillae are located behind the spinule fields, which consist of fourteen to fifteen rows. Twelve to fourteen longitudinal rows of spinules occur around the anal slit; in front of the slit six or seven transverse rows are found.

*Biology*. The larvae live in calluses on the subterranean parts of the leaves of *Carex divulsa*.

*Comment*. The authors made use of the description by Möhn (1955b).

## 11. Genus **Endaphis** Kieffer (fig. 57)

*Morphology*. The smooth larvae are of moderate size. The spindle-shaped body bears at the tip two acicular processes.

The small, rounded head capsule (fig. 57,1) is as long as, or slightly longer than, the width at the base. The well-developed, albeit small, exocephalic rods are approximately as long as the width of the head capsule. The rods are apically slightly widened. The antennae are well-developed. The second antennal segment is nearly triangular and pointed at the tip; it is at least four times as long as its width at the base.

The spatula (fig. 57,2) is two-toothed.

The body is bare. The majority of the papillae is provided with a short bristle. Six setiform papillae (fig. 57,5) are distributed on the dorsal surface of each of the thoracic and the first seven abdominal segments. Just two papillae occur on the eighth abdominal segment. On the ventral surface of the thoracic segments the usual set of papillae is found: two simple sternal and two sublateral papillae with short bristles, as well as four groups of intercalary papillae, each consisting of one simple and two setiform papillae. The ventral surface of the first seven abdominal segments bears four anterior ventral papillae with short bristles (fig. 57,4), the two central ones of which are located within, the two marginal ones behind the spinule fields. The two posterior ventral papillae are also setiform. Two short lateral papillae with short bristles are found on each lateral surface of the thoracic and the first seven abdominal segments. The terminal segment (fig. 57,3) bears setiform terminal and anal papillae.

Some rows of short, transverse folds are found along the anterior margin of the dorsal surface of the segments. On the ventral surface the folds pass into more clearly demarcated and longer rows of extremely tiny spinules.

The terminal segment (fig. 57,3) is apically provided with two conical outgrowths which are separated by a deep, rounded emargination. The tip of each outgrowth bears a single long spine. The ventral surface of the segment is covered with numerous tiny spinules which form a compact field (fig. 57,3). The anal slit is shifted to the dorsal surface and is not surrounded by longitudinal rows of spinules.

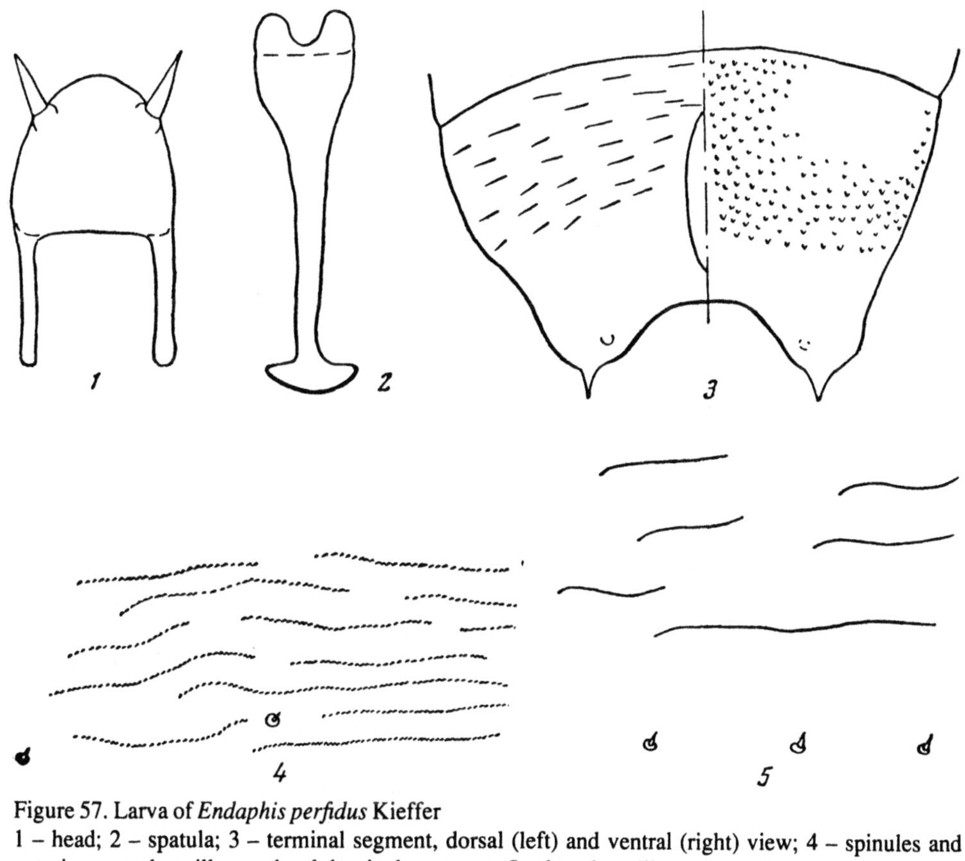


Figure 57. Larva of *Endaphis perfidus* Kieffer
1 – head; 2 – spatula; 3 – terminal segment, dorsal (left) and ventral (right) view; 4 – spinules and anterior ventral papillae on the abdominal segments; 5 – dorsal papillae (orig.)

The stigmata are small; the terminal pair is shifted to the dorsal surface of the eighth abdominal segment.

*Biology.* The larvae are endoparasites of aphids.

**Endaphis perfidus** Kieffer

*Morphology.* The blunt, rounded teeth of the spatula are separated by a deep emargination. The well-pronounced manubrium has a widened base. The meso- and metathoracic, and all abdominal segments bear on the dorsal surface of the otherwise bare body straight, transverse folds and on the ventral surface fifteen to twenty rows of extremely tiny spinules.

*Biology.* The larvae are unspecialized endoparasites of aphids.

*Material.* Pavlova Sloboda, Moskovskaya oblast, 1 July 1962, from aphids on *Sorbus*, three larvae.

12. Genus **Monarthropalpus** Rübsaamen (fig. 58)

*Morphology.* The body of the larvae is densely covered with sclerotized verrucae.

The comparatively small, nearly square head capsule (fig. 58,1) is as long as its width at the base. The relatively thick and short exocephalic rods are no longer than the width of the base of the head capsule. The short antennae are apically rounded; the second segment is 2.5-3 times as long as it is wide.

The two-toothed spatula (fig. 58,2) bears large, conical teeth. The manubrium is slightly sclerotized.

The body is densely covered with sclerotized verrucae (fig. 58,3). The verruca-shaped papillae may be simple or setiform. Six dorsal papillae (fig. 58,3) are found on the dorsal surface of each of the thoracic and the first seven abdominal segments. The apical part of these papillae is represented by a thick, conical bristle. The eighth abdominal segment bears only two dorsal papillae. On the ventral surface of the thoracic segments occur: two sternal and two sublateral papillae, as well as two groups of intercalary papillae, each group of which consists of one simple and two setiform papillae. The sternal and sublateral papillae bear a small conical bristle. Two setiform anterior and posterior ventral papillae (fig. 58,4) are located on the ventral surface of the first seven abdominal segments. The two ventral papillae on the eighth abdominal segment look like small, broad verrucae. Two of the four terminal papillae are setiform, whereas the other two are conical. The anal papillae are verruca-like. Two setiform lateral papillae (fig. 58,5) occur on each of the lateral surfaces of the segments. There are no spinule fields.

The terminal segment (fig. 58,6) is well-developed. It is apically divided into two conical outgrowths which bear the terminal papillae. The dorsal surface of the terminal segment is covered with oval verrucae (fig. 58,7), the ventral surface with rounded and pointed protuberances of various shapes (fig. 58,8-10). The anal slit, which is found on the ventral surface of the segment, is surrounded by longitudinal rows of spinules (fig. 58,6).

The stigmata are comparatively small; the terminal pair is not shifted to the dorsal surface of the segment.

*Biology.* The larvae develop within galls of the parenchyma.

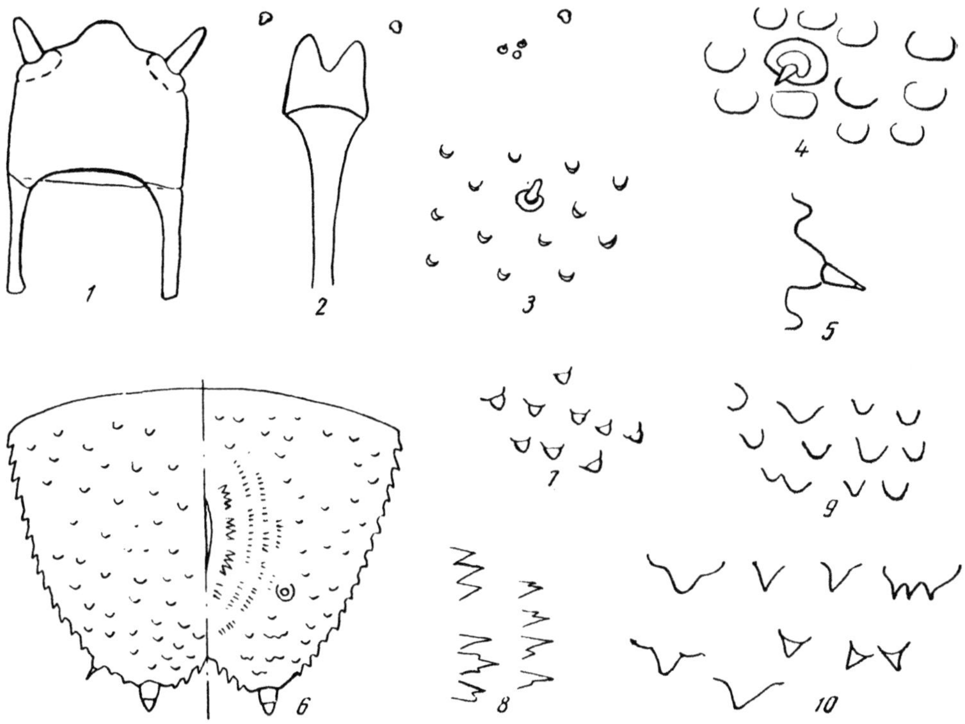


Figure 58. Larva of *Monarthropalpus buxi* (Laboulbène)
1 – head; 2 – spatula and papillae; 3 – cuticular structures and dorsal papilla on the abdominal segments; 4 – cuticular structures and posterior ventral papilla on the abdominal segments; 5 – lateral papillae; 6 – terminal segment, dorsal (left) and ventral (right) view; 7 – cuticular structures on the terminal segment, dorsal view; 8-10 – spinules and cuticular structures on the terminal segment, ventral view (orig.)

**Monarthropalpus buxi** (Laboulbène)
*Morphology.* The conical papillae are short and thick. The entire dorsal surface of the body segments is uniformly covered with small, oval verrucae (fig. 58,3). The larger, slightly pointed verrucae on the ventral surface are more densely distributed along the middle line of the body.

*Biology.* The larvae damage *Buxus* shrubs. They develop within parenchymatous galls on the leaves, causing their premature shedding.

*Material.* Prague, 1 November 1963, five larvae (M. Skuhravá and B. Mamaev).

13. Genus **Aphidoletes** Kieffer (fig. 59, 60)

*Morphology.* The larvae are slightly dorso-ventrally flattened. The head is conspicuously dark brown. The posterior edge of the heavily sclerotized head capsule (fig. 59,1) is smooth. The head capsule is as long as, or somewhat longer than, the width of its base. The long, straight, heavily sclerotized exocephalic rods originate on the posterior edge of the head capsule. They are nearly 1.5 times as long as the head capsule. The first segment of the well-developed, pointed antennae is short. The long second antennal segment is four to five times as long as the width of its base.

The two-toothed spatula (fig. 59,2; 60,4) has heavily sclerotized teeth and a broad manubrium, which, however, may be weakly pronounced.

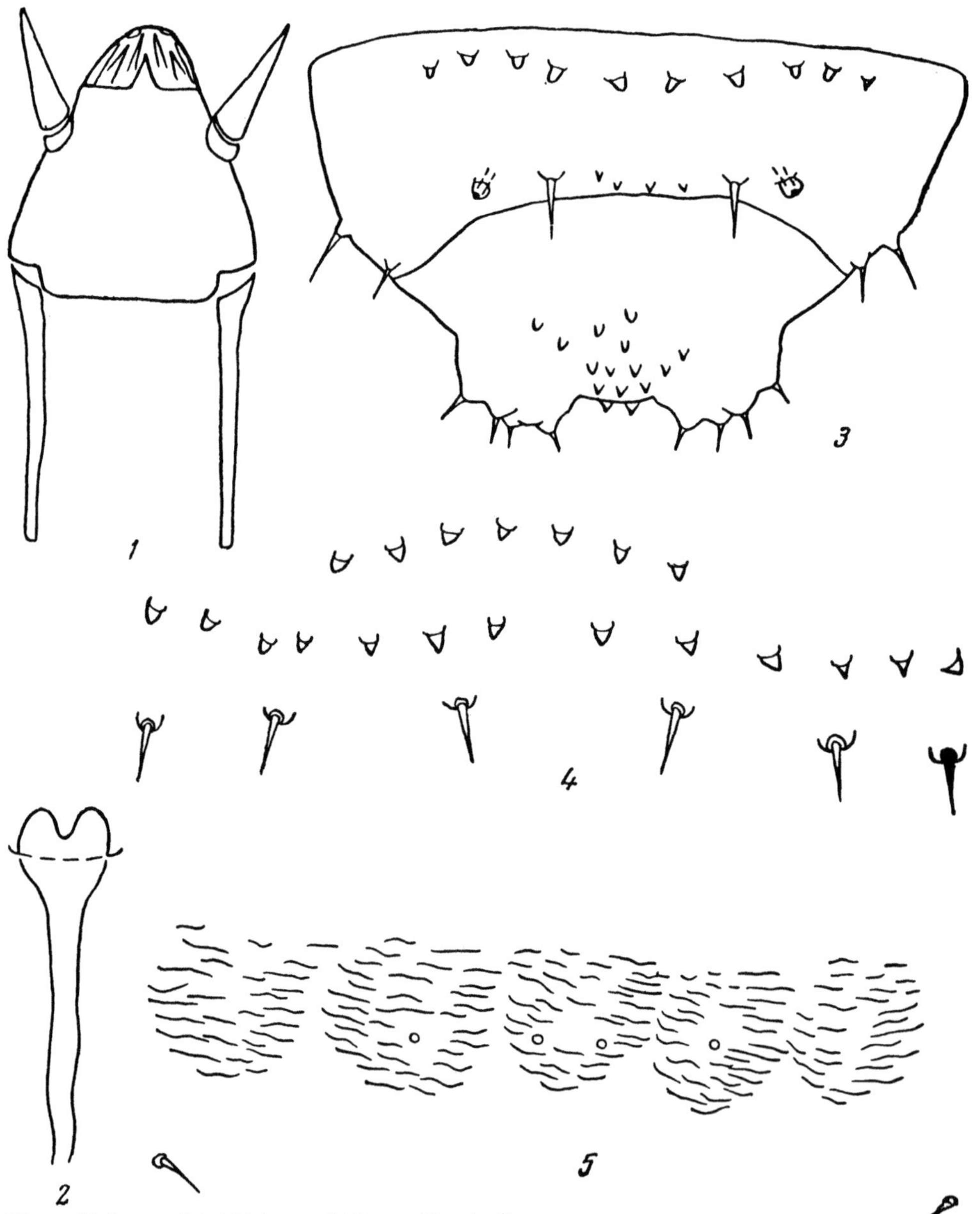


**Figure 59.** Larva of *Aphidoletes aphidimyza* (Rondani)
1 – head; 2 – spatula; 3 – last two segments, dorsal view; 4 – cuticular structures and dorsal papillae on the third abdominal segment; 5 – spinule fields and ventral papillae on the abdominal segments (orig.)

The body is bare. In addition to the setiform papillae sparsely distributed sclerotized verrucae (fig. 59,4) may be found. Six setiform dorsal papillae occur on the dorsal surface of each of the thoracic and the first seven abdominal segments. The dimensions of the central and marginal dorsal papillae are approximately equal.

On the ventral surface of each of the thoracic segments are found: two simple sternal and two sublateral papillae, and four groups of intercalary papillae, each group of which consists of one simple and two setiform papillae. The sublateral papillae on the prothorax are simple; on the meso- and metathorax they may be either simple or setiform. Four simple anterior ventral papillae (fig. 60,6), situated within the spinule fields, as well as two posterior ventral papillae (either simple or setiform) occur on the first seven abdominal segments. On the eighth abdominal segment just four simple ventral papillae occur. Two setiform lateral papillae are found on each of the lateral surfaces of the segments. The terminal segment bears only eight setiform terminal papillae, four of which have longer bristles than the other ones.

The spinule fields, which occur on the ventral surface of the meso- and metathoracic, and all abdominal segments, contain spinules of various shapes. The spinules in the anterior half of the fields are wide and blunt, and not contiguous; in the posterior half of the fields the spinules are narrower and more densely distributed.

The comparatively small and short terminal segment (fig. 59.3) is apically provided with two wide, rounded lobes, which bear the easily visible setiform terminal papillae. The anal slit is surrounded by some longitudinal rows of spinules.

The stigmata appear as sturdy, sclerotized verrucae. The terminal pair is shifted to the dorsal surface of the eighth abdominal segment.

*Biology*. The larvae live in colonies of various species of aphids.

*Comment*. We consider the genera *Phaenobremia* Kieffer and *Isobremia* Kieffer as synonyms of the genus *Aphidoletes* Kieffer.

Key to the species of *Aphidoletes* Kieffer

1(2) The posterior ventral papillae on the first seven abdominal (fig. 59,5) and the sublateral papillae on the meso- and metathoracic segments bear long, well-developed hairs. On its anterior part the dorsal surface of the abdominal segments is provided with clearly visible, large, sclerotized verrucae (fig. 59,4) which form a single row of seven or eight verrucae. In some cases a second row is also found, consisting of six to fourteen verrucae. This second row is divided into two symmetrical parts by the midline. . . . . . . . . . . . . . . *A. aphidimyza* (Rondani)

2(1) The posterior ventral papillae on the first seven abdominal and the sublateral papillae on the meso- and metathoracic segments are simple, without bristles. No large, sclerotized verrucae are pronounced on the dorsal surface of any abdominal segment, except the terminal one.

3(4) The eighth abdominal segment has the usual shape, without processes. The lateral papillae are located directly on the lateral surfaces of the segments. The dorsal surface of the terminal segment (fig. 60,1) bears an extensive field of large, sclerotized verrucae. . . . . . . . . . . . . . . . . . . . . . . . *A. urticariae* (Kieffer)

4(3) The eighth abdominal segment is provided with conspicuous processes, which bear the lateral papillae. The dorsal surface of the terminal segment is bare, without conspicuous, large, sclerotized verrucae. . . . . . . . . . . . . *A. abietis* (Kieffer)

**Aphidoletes aphidimyza** (Rondani)

*Morphology.* The elongated head capsule (fig. 59,1) is markedly longer than the width of its base. The comparatively long exocephalic rods are nearly twice as long as the width of the base of the head capsule. The second antennal segment is five times as long as the width of its base. The segment is pointed and narrows considerably towards the tip. The body is almost bare. Just one or two rows of sclerotized verrucae occur in front of the dorsal papillae (fig. 59,4) on the anterior part of the dorsal surface of the segment. On the metathorax a row of four verrucae can be seen. One or two rows of verrucae, divided into two parts by the midline, occur on the first five abdominal segments. The first row consists of seven or eight, the second one of six to fourteen verrucae. On the last two segments the number of verrucae per row may be lowered to respectively five to six and twelve. On the eighth abdominal segment a single row of ten verrucae is found in front of the dorsal papillae, while four to six papillae are distributed between the bristles. A field of fourteen or fifteen sclerotized verrucae occurs on the dorsal surface of the terminal segment. All verrucae are large, triangular, and apically rounded.

The sublateral papillae on the ventral surface of the thoracic segments are not all the same shape: they are simple on the prothorax, but they posses long, well-developed bristles on the meso- and metathorax. On the first seven abdominal segments both posterior ventral papillae are provided with well-developed bristles (fig. 59,5).

The spinule fields on the middle segments of the body comprise 20-28 rows. The anal slit is surrounded by longitudinal rows of spinules.

*Biology.* The larvae are unspecialized predators in colonies of various species of aphids.

*Material.* Kuntsevo, Moskovskaya oblast, 4 July 1955, from aphids on willow twigs, five larvae; Stavropol'skiy region, 24 July 1955, from aphids on *Urtica*, nine larvae; from the collection of W. Nijveldt (Institute for Phytopathological Research), Wageningen, Netherlands, five larvae.

*Comment.* A detailed description has been given by Roberti (1946).

**Aphidoletes urticariae** (Kieffer)

*Morphology.* The structure of the head is the same as in the preceding species. The dorsal surface of the bare body is not provided with large, sclerotized verrucae. Only on the abdominal segments, does one find two rows of five to seven tiny spinules (fig. 60,2), symmetrical with respect to the midline. These rows are most clearly pronounced on the first five abdominal segments. All sublateral papillae on the ventral surface of the thoracic segments are simple, without bristles. On the first seven abdominal segments both posterior ventral papillae are simple, without bristles, as well.

The terminal segment dorsally bears an extensive field consisting of 25-37 pointed verrucae and covering a considerable portion of the surface of the segment (fig. 60,1).

The anal slit is surrounded by rows of spinules.

*Biology.* The larvae are unspecialized predators in colonies of various species of aphids.

*Material.* Pavlova Sloboda, Moskovskaya oblast, 15 July 1962, from aphids on willow, two larvae; idem, 1 July 1962, from aphids on *Sorbus*, four larvae; idem, 27 June 1963, from aphids on birch, two larvae; Khust, Zakarpatskaya oblast, Ukr. S.S.R., 15 June 1963, from aphids on thistle, three larvae.

*Comment.* An accurate diagnosis of the species is given for the first time.

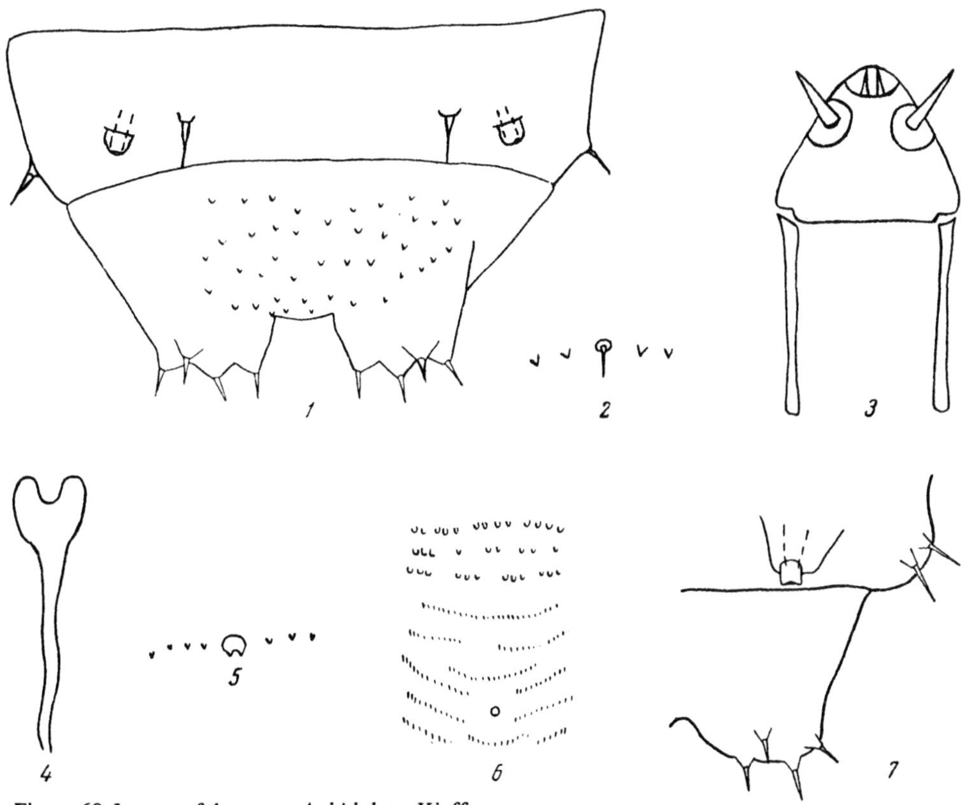


Figure 60. Larvae of the genus *Aphidoletes* Kieffer
*A. urticariae* (Kieffer): 1 – last two segments, dorsal view; 2 – cuticular structures on the abdominal segments, dorsal view; *A. abietis* (Kieffer): 3 – head; 4 – spatula; 5 – cuticular structures on the abdominal segments, dorsal view; 6 – spinules and anterior ventral papilla; 7 – last two segments, dorsal view (orig.)

**Aphidoletes abietis** (Kieffer)

*Morphology*. The comparatively short and wide head capsule (fig. 60,3) is as long as the width of its base. The small exocephalic rods are less than 1.5 times as long as the width of the base of the head capsule. The antennae are comparatively short. The second (longest) segment is four times as long as its width at the base and broadly rounded at the tip.

The body is bare but for two dorsal rows. These two rows are symmetrical with respect to the midline, and each consists of a single shiny pore and five to seven tiny spinules (fig. 60,5). The rows on the first five abdominal segments are the most clearly pronounced ones. The sublateral papillae on the ventral surface of the thoracic segments are all alike, simple, without bristles. Both posterior ventral papillae on the first seven abdominal segments are also simple. The anterior ventral papillae on the first seven abdominal segments are located within the spinule fields (fig. 60,6).

The bare terminal segment has no spinules on its dorsal surface (fig. 60,7). The anal slit is surrounded by rows of tiny spinules. The terminal pair of stigmata (fig. 60,7) is situated on small, conical processes of the eighth abdominal segment. Two similar processes, which bear the lateral papillae, are found on the lateral surfaces of the segment.

*Biology.* The larvae live as predators within galls of pine and spruce aphids.

*Material.* Pavlova Sloboda, Moskovskaya oblast, 8 July 1962, from galls of *Chermes (Adelges) abietis*, four larvae.

*Comment.* According to Möhn (1955b), the anterior ventral papillae on the first seven abdominal segments are located outside the spinule fields, which is at variance with our findings (see above).

## 14. Genus **Monobremia** Kieffer (fig. 61)

*Morphology.* The elongated-oval body of the larvae is densely covered with setiform papillae.

The comparatively short head capsule (fig. 61,1) is only slightly longer than its width at the base. Its straight posterior edge is not emarginate. The well-developed exocephalic rods are slightly widened at the tip, and markedly longer than the width of the head capsule. The apically pointed second antennal segment is more than four times as long as the width of its base.

The two teeth of the spatula (fig. 61,2) are wide, blunt, and apically rounded; they are separated by a rounded emargination. The base of the well-developed manubrium is not widened.

The dorsal and lateral surfaces of the bare body are provided with numerous setiform papillae (fig. 61,5-6). The papillae on the dorsal surface of the thoracic and the first seven abdominal segments are arranged into a few transverse rows. The first row consists of eight papillae, the second one (which is not found on the thoracic segments) of sixteen to eighteen. These are followed by a row of six dorsal papillae behind which another four or six small papillae are located. On the anterior half of the eighth abdominal segment (fig. 61,3) a single row of papillae occurs, as well as two large and three or four tiny papillae, which are situated between the stigmata. The anterior half of the terminal segment bears on either side of the midline two or three papillae; while three or four papillae are found between the processes. On the meso- and metathorax the lateral surfaces are respectively provided with seven and ten papillae. The lateral surfaces of the first seven abdominal segments bear ten to fourteen, and those of the eighth abdominal segment nine papillae. The number of papillae on the lateral surfaces of most segments, and on the dorsal surface of the last two segments is not constant and varies as well between the segments of a single larva as between different larvae.

The ventral surface of the body lacks well-developed setiform papillae. Two simple sternal and two sublateral papillae (consisting of a short bristle on a well-pronounced verruca) occur on each of the thoracic segments.

Moreover, four groups of three short, setiform intercalary papillae are found on each segment. The first seven abdominal segments each bear four simple anterior ventral papillae (fig. 61,7), located within the spinule fields, as well as two posterior ventral papillae consisting of a verruca with a short apical bristle.

All four ventral papillae on the eighth abdominal segment have well-developed bristles. The terminal segment bears eight long, setiform terminal papillae and six anal papillae with a short, though easily visible, bristle. Large spinule fields (fig. 61,7) are situated on the meso- and metathoracic, as well as on the first seven and the ninth abdominal segments. The spinule fields are compact. A few anterior rows consist of rather large spinules. The next rows consist of very tiny spinules (fig. 61,4).

The terminal segment (fig. 61,3) is apically provided with two rounded processes, each bearing four setiform terminal papillae. The anal slit is surrounded by longitudinal rows of tiny spinules.

The stigmata appear as well-developed sclerotized verrucae. The terminal pair is shifted to the dorsal surface of the segment.

Figure 61. Larva of *Monobremia subterranea* (Kieffer)
1 – head; 2 – spatula; 3 – last two segments, dorsal view; 4 – rows of spinules on the abdominal segments, ventral view; 5 – sixth abdominal segment, dorsal view; 6 – lateral papillae on the seventh abdominal segment; 7 – spinule field and anterior ventral papillae on the abdominal segments (orig.)

*Biology.* The larvae live in colonies of various species of aphids, often together with larvae of other aphidophagous gall midges.

*Comment.* An accurate diagnosis of the genus is given for the first time. The diagnosis of the genus as given by Möhn (1955b) is not corroborated by our material.

**Monobremia subterranea** (Kieffer)

*Morphology.* The compact spinule fields consist of 17-27 rows on the thoracic and of 19-23 on the abdominal segments; on the seventh abdominal and the terminal segment the fields comprise respectively 16-17 and 18-20 rows. The two or three anterior rows contain comparatively large, blunt spinules, whereas the spinules in the remaining rows are extremely tiny and hardly discernable.

*Biology.* The larvae are unspecialized predators of various aphid species.

*Material.* Pavlova Sloboda, Moskovskaya oblast, 22 July 1962, from a colony of aphids on birch twigs, five larvae; idem, from aphids on *Urtica*, three larvae.

*Comment.* Möhn (1955b) erroneously considered the above described larvae of *M. subterranea* (Kieffer) as a pathological form of *Aphidoletes urticariae* (Kieffer).

15. Genus **Lestodiplosis** Kieffer (fig. 7,2; 9,3; 15,3; 62; 63)

*Morphology.* The elongated-oval larvae have a well-developed head capsule and six setiform terminal papillae.

The elongated head capsule (fig. 7,2) is noticeably longer than it is wide. Its dorsal side has a small emargination, whereas the ventral side is deeply emarginate. The lateral sides bear wide, lobe-shaped outgrowths. The long, apically somewhat widened, posterior tentorial arms are well-developed and partly protrude beyond the head capsule. The long and slender antennae have a pointed tip. The second antennal segment is no less than four times as long as its width at the base. The spatula is not pronounced.

The body is bare. The papillae are predominantly setiform. The dorsal surface of each of the thoracic and the first seven abdominal segments bears six setiform dorsal papillae (fig. 62,2; 63,2). The papillae 'a' and 'c' (fig. 1) are of approximately equal size, whereas the papillae 'b' are noticeably shorter. Just two setiform papillae occur on the eighth abdominal segment. Two simple, sternal papillae (fig. 62,1) are found only on the prothorax. Each thoracic segment is provided with four groups of intercalary papillae, each group comprising two simple and one well-developed setiform papillae. The sublateral papillae have a long bristle. The first seven abdominal segments bear two well-developed setiform posterior ventral papillae on their ventral surface. Four simple ventral papillae are found on the eighth abdominal segment.

Instead of the two sternal papillae two prolegs (fig. 63,1) are developed on the ventral surface of the meso- and metathorax. On the first seven abdominal segments three prolegs (fig. 63,5) are found instead of the four anterior ventral papillae. Some rows of tiny spinules (fig. 63,3) surround the bases of the prolegs. Two well-developed, setiform lateral papillae are located on each of the lateral surfaces of the segments. The wide, nearly square terminal segment (fig. 63,4) bears six well-developed setiform terminal papillae. The very small anal slit is located on the ventral side, near the anterior margin of the segment. Some rows of tiny spinules occur in front of the anal slit. A few more rows

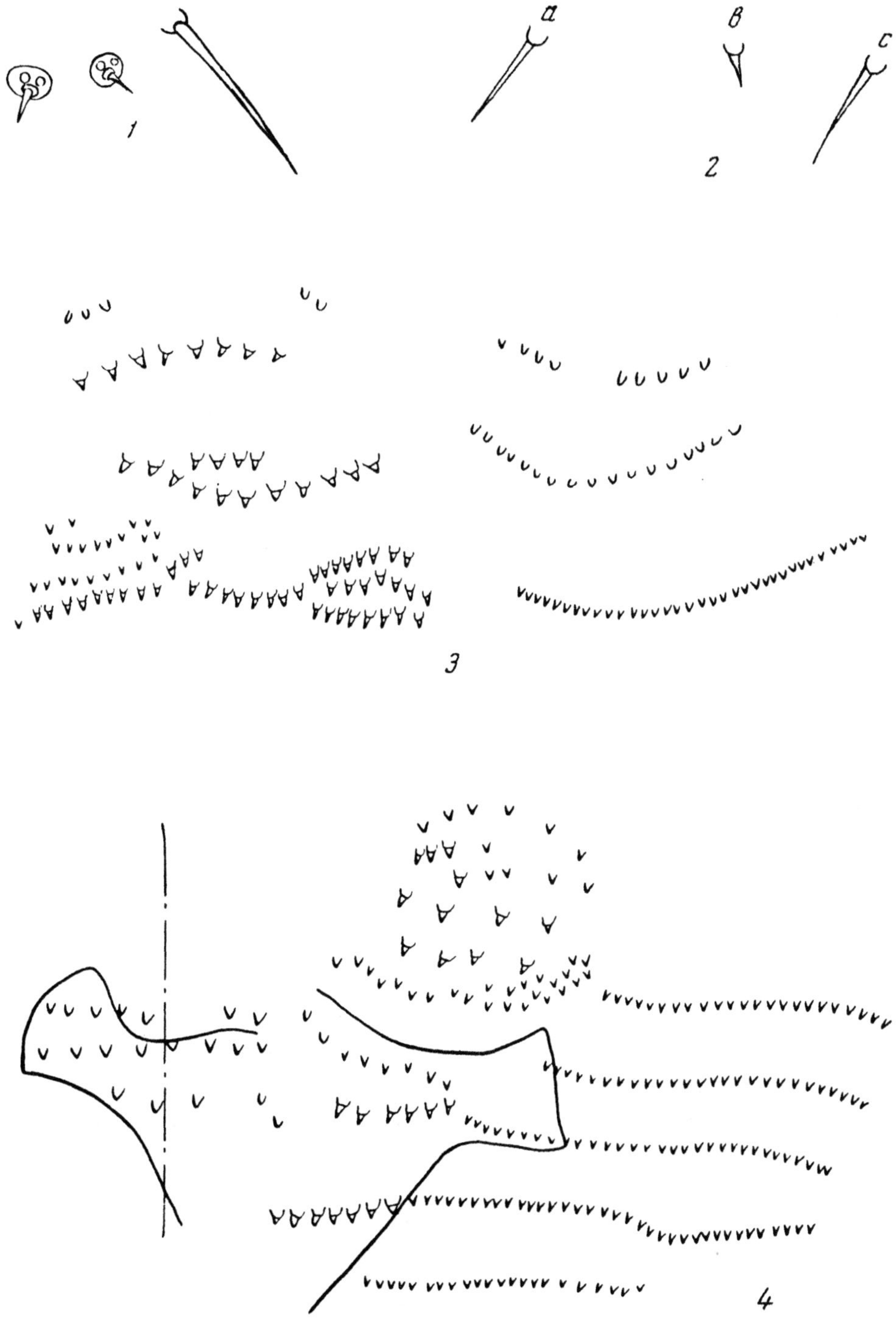


Figure 62. Larva of *Lestodiplosis pallidicornis* Kieffer
1 – papillae on the thoracic segments, ventral view; 2 – dorsal papillae on the abdominal segments; 3 – rows of spinules on the mesothorax, ventral view; 4 – rows of spinules and 'false legs' on the abdominal segments, ventral view (orig.)

are distributed on the central part of the segment, behind the anal slit. The well-developed stigmata look like verrucae. The posterior pair is shifted to the dorsal surface of the segment.

*Biology.* The larvae prey on the larvae of various species of gall midges (mycetophages as well as phytophages), and may be found in litter, fruiting bodies of fungi, galls of other gall midges, and flowers.

Key to the species of *Lestodiplosis* Kieffer

1(2) The comparatively short antennae are no more than five times as long as the width of their base. The length of the very short dorsal papillae 'b' (fig. 1) is no more than one third of the length of the papillae 'c'. The spinules on the ventral surface of the segments (fig. 62,3) are short and no longer than the width at the base. *L. pallidicornis* Kieffer

2(1) The comparatively long antennae are no less than eight times as long as their width at the base. The length of the relatively long dorsal papillae 'b' is no less than two thirds of the length of the papillae 'c'. The spinules on the ventral surface of the segments (fig. 63,3) are comparatively long; noticeably longer than their width at the base. . . . . . . . . . . . . . . . . . . . . . . . . . . . . . . *L. polypori* (H. Loew)

**Lestodiplosis pallidicornis** Kieffer

*Morphology.* The comparatively short antennae are no more than five times as long as their width at the base. The dimensions of the dorsal papillae vary. The papillae 'c' (fig. 1) are longest, the papillae 'b' shortest (no longer than one third the length of the papillae 'c'). The papillae 'a' are slightly shorter than the papillae 'c', their length is no more than two thirds that of the papillae 'c'. The interstice between the papillae 'a' – 'a' is twice the length of each of these papillae. The spinules form some transverse rows on the ventral surface of the segments around the bases of the prolegs. Between the rows spots with irregularly distributed spinules (fig. 62,3-4) occur. All spinules are pointed and short; they are no longer than their width at the base.

*Biology.* The larvae develop in the flower heads of red clover. Their narrow specialization is, however, not proven.

*Material.* Pavlova Sloboda, Moskovskaya oblast, 22 June 1963, ten larvae.

*Comment.* Milne (1960) has given a fragmentary description of the larvae.

**Lestodiplosis polypori** (H. Loew)

*Morphology.* The comparatively long antennae are no less than eight times as long as the width of their base. The dorsal papillae are well-developed. The sizes of the papillae 'a' and 'c' (fig. 1) are approximately equal, whereas the papillae 'b' are shorter (about two thirds the length of the neighbouring papillae). The distance between the central papillae exceeds their length by one third. Some transverse rows of spinules are found on the ventral surface of the segments, around the bases of the prolegs (fig. 63,3,5). All spinules are slender and considerably longer than their width at the base.

*Biology.* The larvae inhabit the fruiting bodies of woody fungi.

*Material.* Pavlova Sloboda, Moskovskaya oblast, 10 September 1962, from the fruiting bodies of polyporaceous fungi on *Betula*, four larvae.

*Comment.* An accurate diagnosis of the larvae is given for the first time.

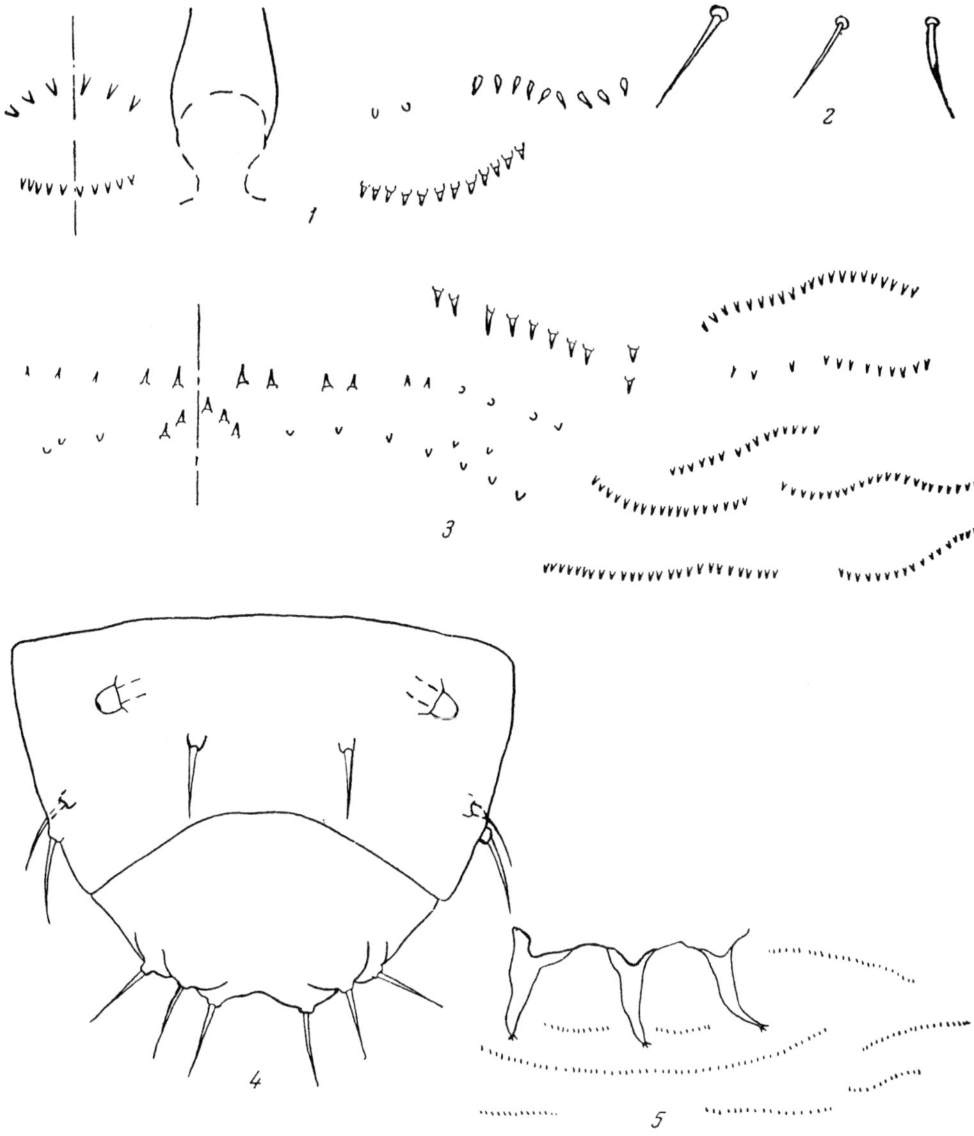


Figure 63. Larva of *Lestodiplosis polypori* (H. Loew)
1 – rows of spinules and 'false leg' on the mesothorax, ventral view; 2 – dorsal papillae on the abdominal segments; 3 – rows of spinules on the abdominal segments; 4 – last two segments, dorsal view; 5 – rows of spinules and 'false legs' on the abdominal segments (orig.)

## 16. Genus **Arthrocnodax** Rübsaamen (fig. 7,1; 64)

*Morphology*. The larvae have a strongly elongated head capsule, and elongated supernumerary and anterior body segments.

The heavily sclerotized, elongated-oval head capsule (fig. 7,1) has a smooth posterior edge. The capsule is 2.5 times as long as its width at the base. The exocephalic rods are extremely long and over their whole length equally wide. They are more than 1.5 times as long as the head capsule. The long, slender, apically pointed antennae are almost half as long as the head capsule.

The spatula (fig. 64,4) is one-toothed. The base of the well-pronounced manubrium is not widened.

Barring the supernumerary segment and the prothorax, the dorsal and ventral surface of the body (fig. 64,3) are densely covered with platelets and protuberances. The bare supernumerary and prothoracic segments are longitudinally striated. The dorsal surface of the thoracic and the first seven abdominal segments each bear six setiform dorsal papillae (fig. 64,2). The papillae 'a' and 'c' (fig. 1) are markedly longer than the papillae 'b'. On the eighth abdominal segment only two setiform papillae occur. The ventral

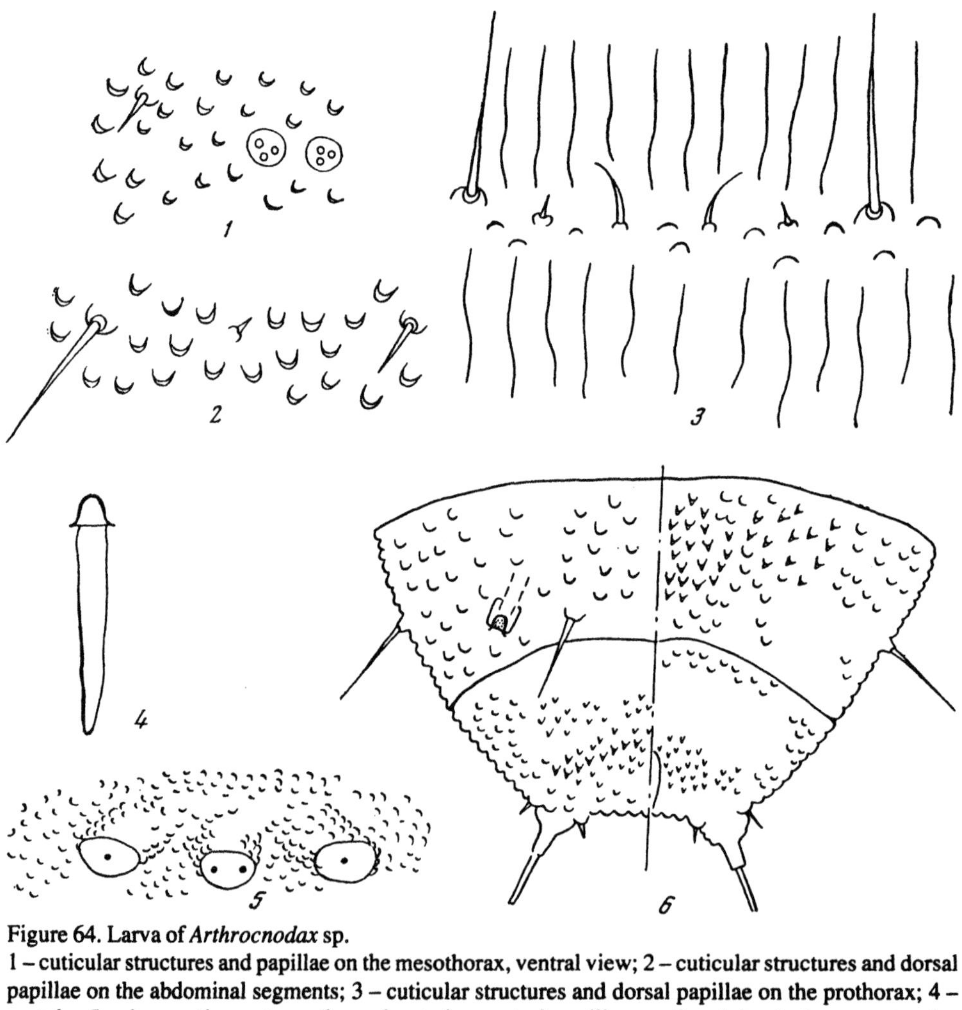


Figure 64. Larva of *Arthrocnodax* sp.
1 – cuticular structures and papillae on the mesothorax, ventral view; 2 – cuticular structures and dorsal papillae on the abdominal segments; 3 – cuticular structures and dorsal papillae on the prothorax; 4 – spatula; 5 – locomotive outgrowths and anterior ventral papillae on the abdominal segments; 6 – terminal segment, dorsal (left) and ventral (right) view (orig.)

surface of each thoracic segment is provided with two simple sternal papillae, four groups of intercalary papillae (fig. 64,1) (each group consisting of one simple and two setiform papillae), and two sublateral papillae. The latter are simple on the prothorax, but setiform on the meso- and metathorax. In addition to two setiform posterior ventral papillae each of the first seven abdominal segments bears on its ventral surface (fig. 64,5) four simple anterior ventral papillae which are distributed over three oval outgrowths (two on the central and a single one on each lateral outgrowth). The four simple ventral papillae on the eighth abdominal segment stand in a single transverse row. The terminal segment bears only six terminal papillae, four of which are provided with a short bristle, and two with a long stick-like appendage. Two setiform lateral papillae are found on each lateral surface of the segments.

There are no spinule fields. In their place the ventral surface of the metathoracic, the first seven abdominal, and the terminal segments are provided with three large, oval outgrowths. The outgrowths are small on the mesothorax and absent from the eighth abdominal segment.

The trapezium-shaped terminal segment (fig. 64,6) is well-developed. The anterior half of its dorsal surface is bare, the posterior half is densely covered with tiny, pointed spinules. A considerable portion of the ventral surface bears small, oval platelets. The anal slit, which is found on the ventral side of the segment, is surrounded by longitudinal rows of spinules.

The stigmata appear as elongated, cylindrical verrucae; the terminal pair is shifted to the dorsal surface.

*Biology*. The larvae prey especially inside mite galls.

*Comment*. Nearly twenty species, which differ only biologically (inhabiting galls of various plant mites) have been described and incorporated into the genus *Arthrocnodax* Rübsaamen. Their biological differences are, evidently, apparent. In reality, there are a few unspecialized, predatory species which feed just as happily on various species of mites.

**Arthrocnodax** sp.

*Morphology*. The narrow second segment is distinctly elongated and pointed. It is eight times as long as the width of its base. The entire surface of the body is covered with small, oval platelets and protuberances. Only on the terminal segment the protuberances are slightly pointed. The very small papillae 'b' (fig. 1) bear a short bristle. The papillae 'c' are longest, whereas the papillae 'a' are approximately one third the length of 'c'.

*Biology*. The larvae have been found in mite galls on *Thymus*, *Ulmus*, *Salix*, and *Tilia*.

*Material*. Stavropol'skiy kray, 15 August 1955, from mite galls on *Thymus*, three larvae; Pavlova Sloboda, Moskovskaya oblast, 8 July 1962, from mite galls on *Ulmus*, three larvae.

## 17. Genus **Xylodiplosis** Kieffer (fig. 65,1-2)

*Morphology*. The body is eight to nine times as long as its diameter. The second antennal segment is twice as long as its width. The teeth of the two-toothed spatula (fig. 65,1) are rounded. The dorsal and ventral surface of the body are transversely striated (up to 130 stripes may be counted per segment). The six dorsal papillae on the prothorax are simple. The dorsal papillae on the other segments (including the two dorsal papillae on the eighth abdominal segment) bear a short bristle. Two simple sternal and sublateral as well as four groups of ordinary intercalary papillae are found on the ventral surface of each thoracic segment (fig. 65,1). Four simple anterior ventral papillae, which are located behind the rows of spinules, and two simple posterior ventral papillae occur on the ventral surface of the first seven abdominal segments. On the eighth abdominal segment four simple ventral papillae are found. Four of the eight terminal papillae on the terminal segment bear a short bristle, two of which have a big, light brown verruca, whereas the other two have a small conical appendage at the tip. Two of the three lateral papillae, which occur on each lateral surface of the segments, bear a short bristle, while the third one is simple. Spinule fields are present.

The dorsal as well as the ventral surface of the terminal segment (fig. 65,2) is bare. The transverse striation is only visible on the lateral surfaces of the segment. Some longitudinal rows of spinules occur around the anal slit. In front of the anal slit a few transverse rows are found. All six anal papillae are simple.

*Biology*. The larvae are inhabitants of holes in woody stems.

**Xylodiplosis** sp.

*Morphology*. The spinule fields consist of 20-22 rows of spinules. Eight to ten longitudinal rows of spinules are found around the anal slit while as many transverse rows occur in front of the slit.

*Biology*. The larvae have been observed in holes in a felled *Fraxinus* trunk.

*Comment*. The authors made use of the description by Möhn (1955b).

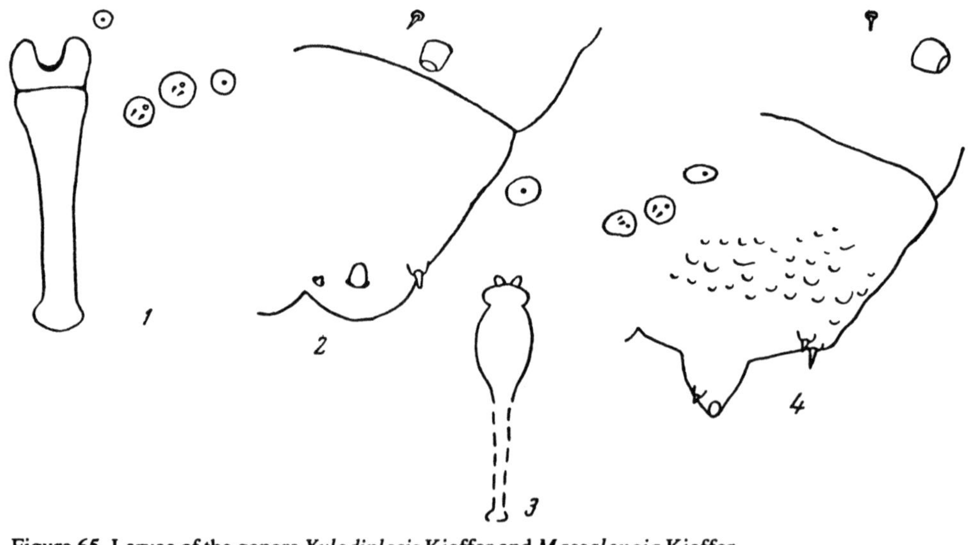


Figure 65. Larvae of the genera *Xylodiplosis* Kieffer and *Massalongia* Kieffer
*Xylodiplosis* sp.: 1 – spatula and papillae; 2 – terminal segment, dorsal view; *M. rubra* (Kieffer): 3 – spatula and papillae; 4 – terminal segment, dorsal view (after Möhn 1955b)

## 18. Genus **Massalongia** Kieffer (fig. 65,3-4)

*Morphology*. The second antennal segment is twice as long as its width. The spatula (fig. 65,3) is hardly developed; the teeth on the very small head are weakly pronounced.

In addition to the dorsal papillae the prothorax bears on its dorsal surface some verrucae. The other segments bear rows of large spines, which cover two thirds of the dorsal surface. On the lateral surfaces rounded verrucae occur. Six simple dorsal papillae are found on the dorsal surface of the thoracic segments, whereas six setiform dorsal papillae occur on each of the first seven abdominal segments. The dorsal papillae on the eighth abdominal segment are also setiform. The thoracic segments bear on the ventral surface (fig. 65,3): two simple sternal papillae, four groups of ordinary intercalary papillae, and two simple sublateral papillae. Four simple anterior ventral papillae, which are located within the spinule fields, and two simple posterior ventral papillae are found on the first seven abdominal segments. Four simple ventral papillae occur on the eighth abdominal segment. The terminal segment bears eight terminal papillae, four of which are provided with a short bristle and four of which are conical. Two setiform lateral papillae are found on each of the lateral surfaces of the segments. The ventral surface of the body bears spinule fields.

The anterior half of the dorsal surface of the terminal segment (fig. 65,4) is bare, but on its posterior half and the lateral surfaces indistinctly outlined, rounded verrucae occur. The ventral surface is bare. Around the anal slit numerous longitudinal rows of spinules are found. All six anal papillae are simple.

*Biology*. The species of this genus develop in parenchymatous galls of *Betula* leaves.

### **Massalongia rubra** (Kieffer)

*Morphology*. Only a few oval verrucae occur on the prothorax in the vicinity of the dorsal papillae. Twelve to fourteen transverse rows of large spinules are found on the other segments. The lateral surfaces of the body are provided with rounded platelets. The four anterior ventral papillae are located between the third and fourth posterior rows of spinules. The spinule fields consist of fourteen to sixteen rows. Thirteen to fifteen longitudinal rows of spinules occur around the anal slit.

*Biology*. The larvae live within parenchymatous galls which are connected to the veins of *Betula* leaves.

*Comment*. The authors made use of the description by Möhn (1955b).

## 19. Genus **Tricholaba** Rübsaamen (fig. 17,2; 66)

*Morphology*. The elongated-oval larvae have a well-pronounced, two-toothed spatula and eight small terminal papillae, which look like verrucae.

The head capsule (fig. 66,1) is small and comparatively short. Its length is slightly less than the width at the base. The ventral side of the posterior edge is straight, the dorsal side has an oval emargination. The well-developed exocephalic rods are not widened at the tip. The antennae are short and apically rounded. They are more than twice as long as the width at the base.

The well-developed spatula (fig. 66,6) is two-toothed. The teeth are broad and apically slightly pointed. The heavily sclerotized manubrium has a widened base.

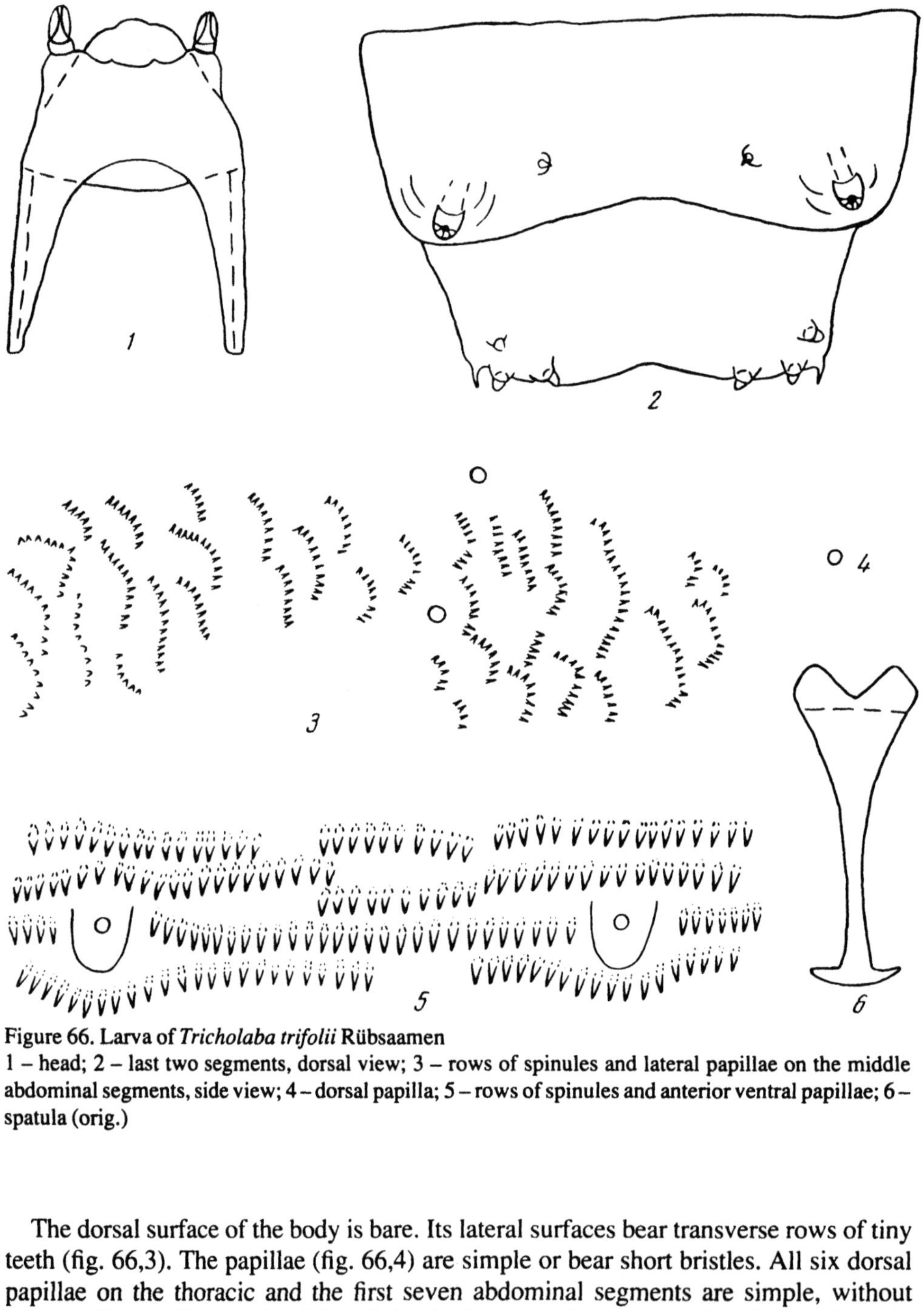

Figure 66. Larva of *Tricholaba trifolii* Rübsaamen
1 – head; 2 – last two segments, dorsal view; 3 – rows of spinules and lateral papillae on the middle abdominal segments, side view; 4 – dorsal papilla; 5 – rows of spinules and anterior ventral papillae; 6 – spatula (orig.)

The dorsal surface of the body is bare. Its lateral surfaces bear transverse rows of tiny teeth (fig. 66,3). The papillae (fig. 66,4) are simple or bear short bristles. All six dorsal papillae on the thoracic and the first seven abdominal segments are simple, without bristles. The papillae on the eighth abdominal segment have a short bristle. The thoracic segments bear on the ventral surface: two simple sternal papillae, four groups of intercalary papillae (each group of which consists of one simple and two setiform papillae), and two simple sublateral papillae. Four simple anterior ventral papillae are located between the last second and third row of spinules (counted from the posterior edge) on the ventral surface of the first seven abdominal segments (fig. 66,5). The two posterior ventral papillae are also simple. On the eighth abdominal segment four simple ventral papillae occur. The terminal segment bears eight terminal papillae, two of which are provided with a short, thick bristle, while the other ones have a rounded, stick-shaped appendage (fig. 17,2). Two weakly pronounced, simple or setiform papillae are found on each lateral surface of the segments.

The ventral surface of the meso- and metathorax, and the first seven abdominal segments bear large, clearly outlined spinule fields. The fields on the first seven abdominal segments each contain six oval spots the four central ones of which are close together.

The bare, trapezium-shaped, terminal segment (fig. 66,2) is somewhat shorter and narrower than the penultimate one. The anal slit is surrounded by clearly pronounced, longitudinal rows of spinules. All six anal papillae are simple.

The terminal pair of stigmata is shifted to the dorsal side of the eighth segment.

*Biology*. The larvae are co-inhabitants in the galls of other gall midges. They may, however, form galls of their own.

Key to the species of *Tricholaba* Rübsaamen

1(2) The dorsal papillae on the eigth abdominal segment bear a short bristle. The spinule fields on the middle segments consist of 20-21 rows. . . . . *T. trifolii* Rübsaamen

2(1) The dorsal papillae on the eighth abdominal segment are simple. The spinule fields on the middle segments consist of 26-28 rows. . . . . . . . . *T. similis* Rübsaamen

**Tricholaba trifolii** Rübsaamen

*Morphology*. The dorsal papillae are simple (except on the eighth abdominal segment where they bear a short bristle). In most cases the lateral papillae have a short, hardly discernable bristle; on the eighth abdominal segment the bristles, though short, are well-pronounced. The lateral surfaces of the thoracic segments are bare. On the abdominal segments they bear, ventrally to the stigmata, transverse rows of short teeth, which border weakly pronounced polygonal platelets. Twelve to fourteen rows of teeth are found on the first, seventh, and eighth abdominal segments. On the other segments the number of rows rises to 20-29.

The large spinule fields on the ventral surface cover the anterior part of the segments. The spinules are over the whole extension of the fields uniformly elongated and thin. The field on the mesothorax consists of eight rows, the fields on the metathorax and the first abdominal segment consist of 18 rows, and on the next three segments of 20-21. On the abdominal segments from the fifth onwards the fields comprise respectively 19, 18, 17, 13, and 7 rows.

*Biology*. The larvae inhabit the flower heads of *Trifolium pratense* and, as inquilines, the galls of *Dasyneura trifolii* (F. Loew).

*Material*. Volkovo, Yaroslavskaya oblast, 18 July 1961, from flower heads of clover, ten larvae.

*Comment*. Milne (1960) has described gall midges from clover heads under the name *T. barnesi* Milne. The authors share the opinion of Nijveldt (1963) that *T. barnesi* Milne is a synonym of *T. trifolii* Rübsaamen.

**Tricholaba similis** Rübsaamen

*Morphology*. The dorsal papillae are simple on most segments, including the eighth abdominal one. The lateral papillae bear a short bristle only on the eighth abdominal segment. The lateral surfaces of the first seven abdominal segments are provided with up to 28-30 transverse rows of teeth. The lateral surfaces on the first and seventh abdominal segments bear 13-14 rows. The spinule fields consist of 26-28 rows of spinules.

*Biology*. The larvae are known as co-inhabitants in the galls of *Dasyneura viciae* (Kieffer). They can, however, form galls of their own on the leaves of *Vicia*.

*Material*. Pavlova Sloboda, Moskovskaya oblast, 13 June 1962, from galls of *Dasyneura viciae* (Kieffer), three larvae.

*Comment*. The recently described species *T. viciarum* Stelter (Stelter 1963) cannot be morphologically distinguished from the specimens of *T. similis* Rübsaamen that we have at our disposition.

20. Genus **Triommatomyia** Mamaev (fig. 67)

*Morphology*. The elongated-oval larva has a weakly sclerotized head. Eight clearly visible, conical protuberances are found on the end of the body.

The comparatively short head capsule (fig. 67,1) is slightly longer than the width of its base. The straight posterior edge of the head capsule is not emarginate. The long

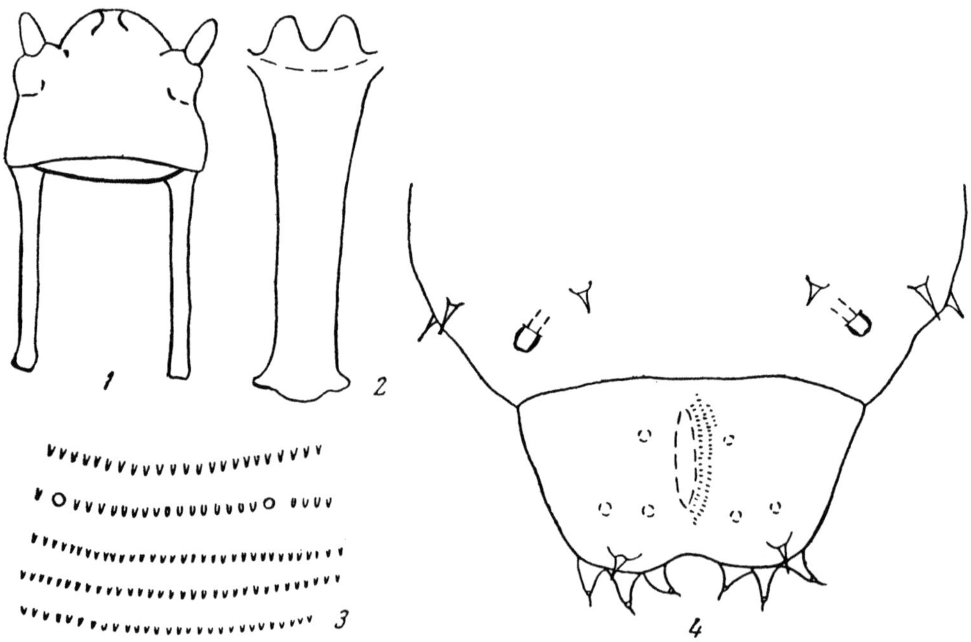


Figure 67. Larva of *Triommatomyia barnesi* Mamaev
1 – head; 2 – spatula; 3 – rows of spinules and anterior ventral papillae; 4 – last two segments, dorsal view (orig.)

exocephalic rods are somewhat longer than the head capsule. The antennae are relatively short. The apically rounded second antennal segment is approximately twice as long as wide.

The spatula (fig. 67,2) is two-toothed. The emargination between the blunt teeth is rounded. The teeth are narrower than the comparatively broad manubrium. The more weakly sclerotized base of the manubrium is slightly widened.

The bare body bears neither sclerotized platelets nor broad protuberances. Six hardly discernable papillae are found on the dorsal surface of the thoracic and the first seven abdominal segments. On the eighth abdominal segment the normal set of two setiform dorsal papillae occurs. The well-pronounced bristles are almost as long as the diameter of the stigmata.

The ventral surface of the thoracic segments bears: two simple sternal papillae, four groups of intercalary papillae (each group consisting of one simple and two setiform papillae), and two simple sublateral papillae. Four simple anterior ventral papillae, which are located within the spinule fields (fig. 67,3), as well as two simple posterior ventral papillae are found on the first seven abdominal segments. Four simple ventral papillae occur on the eighth abdominal segment. The terminal segment (fig. 67,4) bears eight terminal papillae, which appear as broad, conical verrucae with an apical bristle. Two setiform lateral papillae occur on each of the lateral surfaces of the segments.

Spinule fields are found on the ventral surface of the meso- and metathoracic, and the first seven abdominal segments. They are absent from the eighth abdominal segment. The spinule fields on the thoracic segments are compact. On the first seven abdominal segments they contain four oval bare spots, which are located along the posterior margin of the fields.

The terminal segment (fig. 67,4) is bare. The anal slit is surrounded by longitudinal rows of thin hairs.

Only the terminal pair of stigmata is well-pronounced. It is found on the dorsal side of the eighth abdominal segment. The small stigmata have the appearance of oval verrucae.

*Biology.* The larvae are inhabitants of damp, decayed wood.

*Comment.* A diagnosis of the genus is given for the first time.

**Triommatomyia barnesi** Mamaev

*Morphology.* The spatula (fig. 67,2) has a well-pronounced, wide manubrium, but its head is not demarcated.

The spinule fields (fig. 67,3) consist of uniform, tiny, pointed spinules, which are arranged into twelve or thirteen rows on the thoracic, and into 20-23 rows on the first seven abdominal segments. On the terminal segment just seven to eight transverse rows are found (in front of the anal slit). The anterior ventral papillae are situated within the spinule fields, at the level of the third to fourth row from the posterior edge. Four or five dense rows of hairs occur around the anal slit.

*Biology.* The larvae have been found in light coloured, strongly decomposed, damp wood of a *Fagus* stub.

*Material.* Novo-Prokhladnoe, Krasnodarskiy kray, 29 May 1959, four larvae.

*Comment.* A description of the larvae of this species is given for the first time.

## 21. Genus **Buhriella** Stelter (fig. 68)

*Morphology.* The very characteristic, large, bare larvae have a powerful spatula and a small head.

The short, oval head capsule (fig. 68,1) is no longer than the width of its base. The well-developed exocephalic rods are thick and short. They are slightly longer than half the width of the head capsule at its base. The short, broad antennae have a bluntly truncated tip. The second antennal segment is no longer than its width.

The two-toothed spatula (fig. 68,2) is sturdy. Its pointed teeth stand wide apart and are separated by a deep and wide emargination. The manubrium is long and broad.

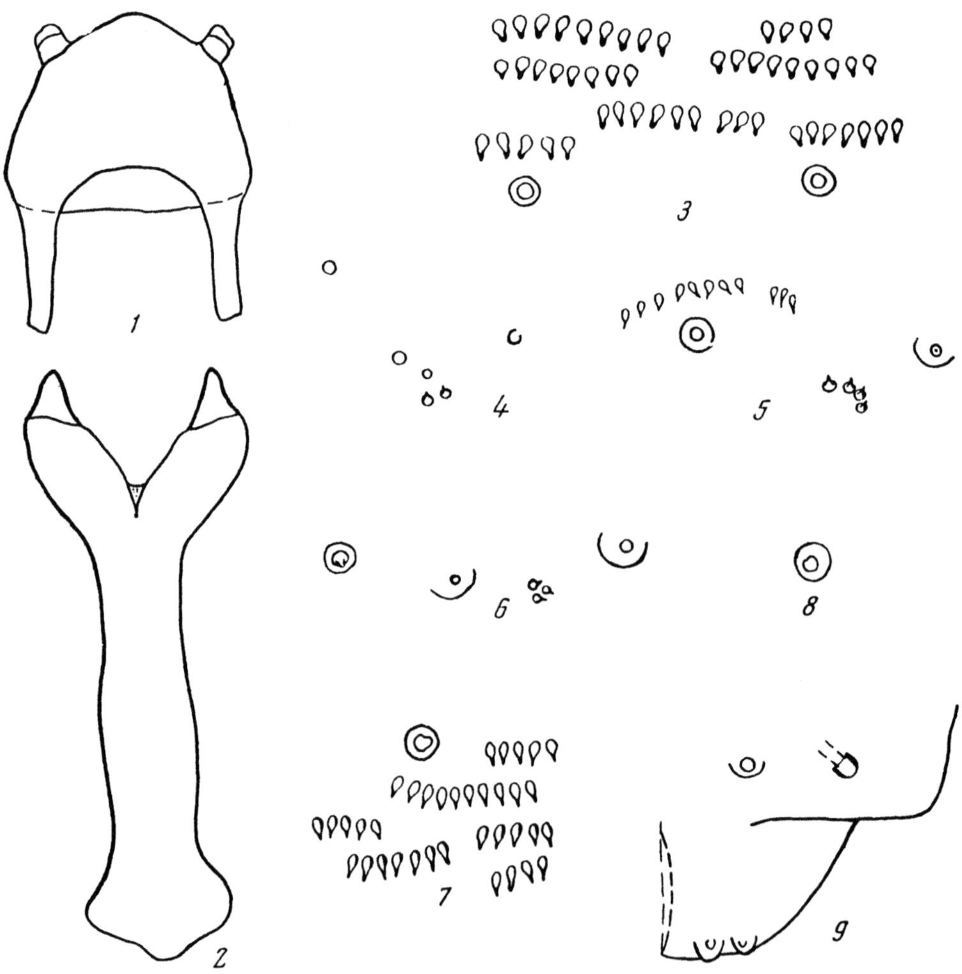


Figure 68. Larva of *Buhriella rubicola* Stelter
1 – head; 2 – spatula; 3 – rows of spinules and anterior ventral papillae on the first five abdominal segments; 4-6 – papillae on the pro-, meso-, and metathorax, ventral view; 7 – posterior ventral papillae and cuticular structures on the seventh abdominal segment; 8 – dorsal papilla; 9 – last two segments, dorsal view (orig.)

The dorsal as well as the ventral surface of the body is bare, except for the spinule fields. The thoracic and the first seven abdominal segments each bear six small, simple dorsal papillae. Just two simple dorsal papillae occur on the eighth abdominal segment. The ventral surface of the thoracic segments (fig. 68,4-6) bears: two simple sternal papillae, two sublateral papillae, and six to eight intercalary papillae, which are not arranged into clear groups. Most of the intercalary papillae are setiform. Four simple anterior ventral papillae are located on the first five abdominal segments, behind the spinule fields. On the next two segments the ventral papillae stand within the fields. The posterior ventral papillae are simple (fig. 68,7). Four simple ventral papillae are found on the eighth abdominal segment. The terminal segment (fig. 68,9) bears four terminal papillae. Two lateral papillae occur on each of the lateral surfaces of the segments. With a few exceptions, all papillae on the ventral surface of the thoracic segments look superficially alike (fig. 68,8).

The ventral surface of the meso- and metathorax, and of the first eight abdominal segments bears well-pronounced spinule fields.

The penultimate segment is provided with conspicuous, wide, lobe-shaped lateral outgrowths. The oval terminal segment (fig. 68,9) is small and short. The anal slit is bare. The stigmata are of the usual type. The terminal pair is shifted to the dorsal side of the segment.

*Biology*. The larvae damage *Rubus* species.

*Comment*. A detailed diagnosis of the genus has been drafted for the first time.

**Buhriella rubicola** Stelter

*Morphology*. The extremely long manubrium of the spatula (fig. 68,2) is more than twice as long as the head, including the teeth. The anterior ventral papillae (fig. 68,3) are located either immediately behind the spinule fields or between the last two to four rows.

The spinules that constitute the fields (fig. 68,3) are arranged into short, transverse rows. The fields on the thoracic and the eighth abdominal segments are compact, the fields on the other segments contain four oval bare spots. The fields consist of 10-18 rows on all thoracic, and the first, seventh and eighth abdominal segments; on the other segments they comprise 20-24 rows.

*Biology*. The larvae develop in galls on the leaves of *Rubus*.

*Material*. Two larvae, which H. Stelter (Institute for Plant Cultivation, Grosz-Lüsewitz, G. D. R.) had sent to the authors.

*Comment*. Stelter (1960) has given a short description of the larvae.

22. Genus **Thecodiplosis** Kieffer (fig. 69,1-2)

*Morphology*. The second antennal segment is twice as long as wide. The spatula is absent.

The body is bare. The dorsal surface of the segments, from the mesothorax onwards, bears transverse rows of spinules. Six simple dorsal papillae are found on the thoracic and the first seven abdominal segments; on the eighth abdominal segment two simple dorsal papillae occur. The ventral surface of the thoracic segments (fig. 69,1) bears two simple sternal and sublateral papillae, as well as four groups of ordinary intercalary papillae. Four simple anterior ventral papillae, which are located within the spinule fields, and two simple posterior ventral papillae are found on the first seven abdominal segments. On the eighth abdominal segment four simple ventral papillae occur. Six of the eight terminal papillae on the terminal segment are simple. The other two are conical, and resemble the hooks that are found on a common base on the central part of the segment, behind the anal slit. The lateral papillae are simple. Spinule fields are present.

Some transverse rows of spinules occur on the anterior half of the otherwise bare dorsal surface of the terminal segment. The ventral surface of the segment (fig. 69,2) is also bare. Around the anal slit longitudinal rows of spinules occur; in front of the slit some transverse rows are found as well. All anal papillae are simple.

*Biology*. The larvae develop in the base of the needles of coniferous trees.

**Thecodiplosis brachyntera** (Schwägrichen)
*Morphology*. The rows of spinules that occur on the dorsal surface of the segments are united with the spinule fields on the ventral surface to form a complete ring. On the meso- and metathorax 13-15 dorsal rows of spinules occur, whereas on the abdominal segments

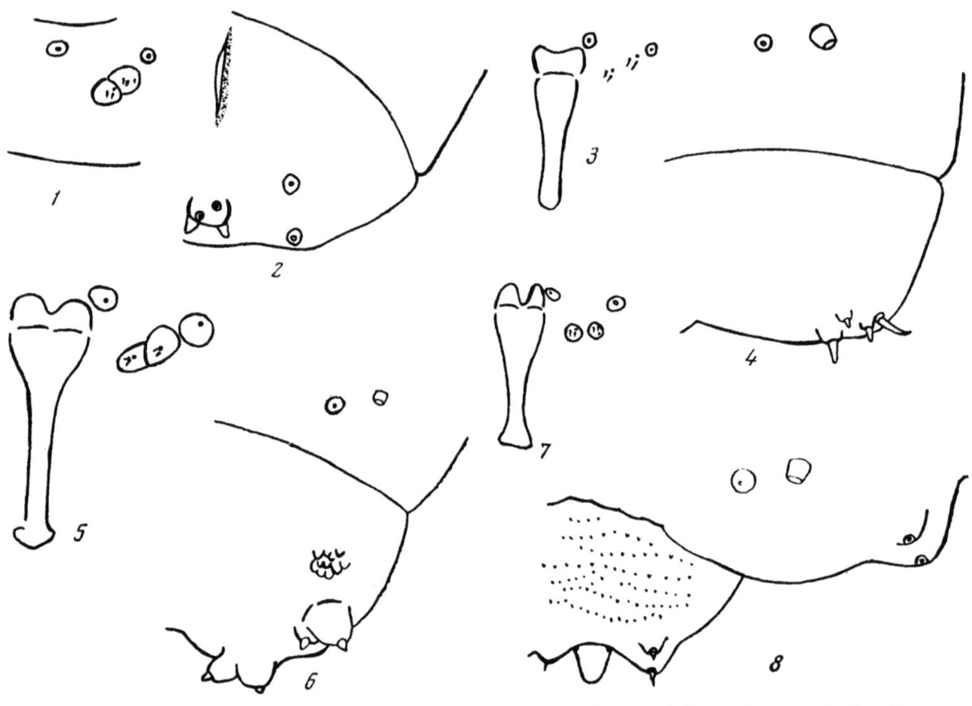


Figure 69. Larvae of the genera *Thecodiplosis* Kieffer, *Endopsylla* De Meijere, *Harmandia* Kieffer, and *Agevillea*
*Th. brachyntera* (Schwägrichen): 1 – papillae on the thoracic segments, ventral view; 2 – terminal segment, ventral view; *E. agilis* De Meijere: 3 – spatula and papillae; 4 – last two segments, dorsal view; *H. loewi* (Rübsaamen): 5 – spatula and papillae; 6 – terminal segment, dorsal view; *A. abietis* Hubault: 7 – spatula and papillae; 8 – last two segments, dorsal view (after Möhn 1955b)

28-32 occur. The terminal segment (fig. 69,2) bears ten to twelve transverse rows of spinules. The spinule fields on the ventral surface of the body consist of 30-35 rows. The anterior ventral papillae are located between the seventh and eighth row of spinules (as counted from the posterior edge). Four to six longitudinal rows of spinules are found around the anal slit; three to four transverse rows occur in front of the slit.

*Biology*. The larvae develop on the common base between the two needles of *Pinus*, and cause a stunting of the needles. In case of massive reproduction, severe damage may result.

*Comment*. The authors made use of the description by Möhn (1955b).

## 23. Genus **Haplodiplosis** Rübsaamen (fig. 70)

*Morphology*. The slightly dorso-ventrally flattened, large larvae can be up to five mm. long.

The massive, broad head capsule (fig. 70,2) is almost square. The head capsule has a small process on the dorsal side of the posterior edge and a 'V'-shaped emargination at the ventral side. The exocephalic rods are short (somewhat shorter than the width of the base of the head capsule), but sturdy. They are noticeably widened at the tip. The antennae are wide and short. The cylindrical second antennal segment is apically rounded. It is twice as long as the width of its base.

The broad, massive spatula (fig. 70,1) bears no teeth. Its anterior edge is smooth. The base of the manubrium is widened.

The body is covered with large cuticular protuberances. The dorsal surface of the thoracic and the first seven abdominal segments bears six simple dorsal papillae. Two simple dorsal papillae are found on the eighth abdominal segment. On the ventral surface of the thoracic segments occur: two simple sternal papillae, four groups of intercalary papillae (each group consisting of two setiform and one simple papilla), and two simple sublateral papillae. The ventral surface of the first seven abdominal segments bears four simple anterior ventral papillae, which are located behind the spinule fields, and two posterior ventral papillae. On the eighth abdominal segment four simple, hardly visible papillae occur.

Small spinule fields (fig. 70,3-5) are distributed on the dorsal surface of the metathorax and the first four or five abdominal segments. On the ventral surface of the body, fields (fig. 70,6-7) are found on the meso- and metathoracic, and all abdominal segments. Moreover, groups of verrucae (fig. 70,8-9) occur on the lateral surfaces of the middle abdominal segments.

The broad and comparatively short terminal segment (fig. 70,10) is slightly wider than its length. The rounded tip bears eight simple terminal papillae, which are situated on small swellings. The anal slit is surrounded by numerous longitudinal rows of setiform structures.

The well-developed stigmata look like cylindrical, apically rounded verrucae. They are slightly longer than wide. The first as well as the terminal pair of stigmata is noticeably shifted to the dorsal side of its segment.

*Biology*. The larvae damage grass stems.

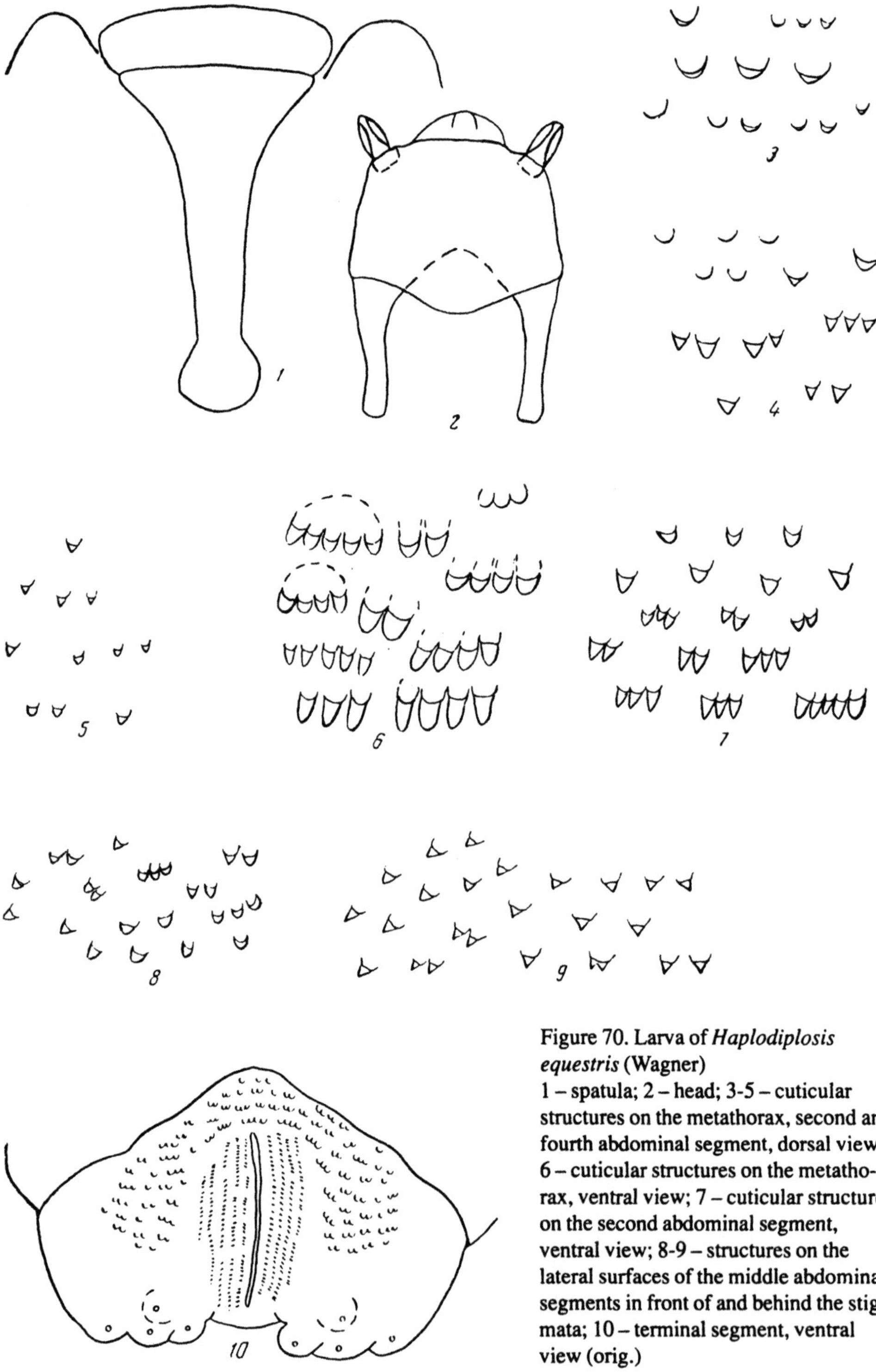


Figure 70. Larva of *Haplodiplosis equestris* (Wagner)
**1 – spatula; 2 – head; 3-5 – cuticular structures on the metathorax, second and fourth abdominal segment, dorsal view; 6 – cuticular structures on the metathorax, ventral view; 7 – cuticular structures on the second abdominal segment, ventral view; 8-9 – structures on the lateral surfaces of the middle abdominal segments in front of and behind the stigmata; 10 – terminal segment, ventral view (orig.)**

**Haplodiplosis equestris** (Wagner)
*Morphology.* The wide head of the spatula (fig. 70,1) is indistinctly demarcated by the manubrium. At each side of the head a single, broad process with a sclerotized tip occurs.

The body is over a considerable extension bare, not counting the spinule fields. The spinule fields (fig. 70,3-5) on the dorsal surface of the segments are small. The fields on the meso- and metathorax are divided along the midline; on the abdominal segments the fields are compact. All spinules on the fields on the thoracic segments are blunt and short, resembling protuberances, whereas the spinules on the abdominal segments are in part pointed (fig. 70,5). The spinules do not form clear rows, but are randomly distributed (fig. 70,5); in some cases the fifth abdominal segment has no spinules. The anterior half of the spinule field on the ventral surface of the thoracic segments consists of blunt and pointed protuberances, which in the posterior half of the field pass into long, toothlike structures with a rounded tip (fig. 70,6). The protuberances on the abdominal segments have a similar shape, but are sharper (fig. 70,7).

The protuberances on the ventral surface of the body do not form clear rows either. The fields on the thoracic and the first seven abdominal segments consist of approximately six to ten rows of blunt and short, and six rows of elongated protuberances. Just six rows of protuberances are found on the eighth abdominal segment. On the terminal segment occur four rows of short, and six rows of elongated protuberances. The lateral surfaces of the thoracic and the first abdominal segment are bare. The next segments bear two groups of protuberances on each lateral surface. The first group consists of blunt protuberances (fig. 70,8), which are located on the anterior part of the segments, in front of the stigmata. The second group is made up of pointed protuberances (fig. 70,9) and is found on the posterior half of the segments, behind the stigmata. The number of protuberances decreases towards the end of the body, so that on the seventh abdominal segment only a few spinules can be found on the posterior half of the segment behind the stigmata.

The anal slit is surrounded by dense longitudinal rows of contiguous setiform structures.

*Biology.* The larvae develop in subepidermal cavities in the stems of rye, wheat and barley. They are a serious pest.

*Material.* Five larvae, which M. Skuhravá and V. Skuhravý (Czechoslovakia, Prague, Entomological Institute of the Czechoslovakian Academy of Sciences) had placed at our disposal.

24. Genus **Endopsylla** De Meijere (fig. 69,3-4)

*Morphology.* The second antennal segment is four times as long as its width. The two-toothed spatula (fig. 69,3) is comparatively small, its incision is shallow and smooth.

The body is bare. The thoracic and the first seven abdominal segments each bear six simple dorsal papillae, the eighth abdominal segment only two. On the ventral surface of the thoracic segments (fig. 69,3) one finds: two simple sternal papillae, four groups of ordinary intercalary papillae, and two simple sublateral papillae. The first seven abdominal segments each bear, behind the spinule fields, four simple anterior ventral papillae, and two simple posterior ventral papillae. Four simple ventral papillae are found on the eighth abdominal segment. The terminal segment is provided with eight terminal papillae, six of which bear a thick, conical bristle, and two a thin, pointed one. The lateral papillae are simple. Spinule fields are present.

The tip of the terminal segment (fig. 69,4) is divided into two lobes by a deep emargination. Its dorsal and ventral surfaces are bare. The anal slit is bare; but in front of the slit transverse rows of spinules occur. All six anal papillae are simple.

*Biology.* The larvae are endoparasites of various insects.

**Endopsylla agilis** De Meijere

*Morphology.* The spinule fields consist of twelve to fourteen rows. Fourteen to fifteen transverse rows of spinules are found in front of the anal slit. The terminal papillae have various sizes (fig. 69,4).

*Biology.* The larvae are known as endoparasites of species of the genus *Psylla.*

*Comment.* The authors made use of the description by Möhn (1955b).

### 25. Genus **Harmandia** Kieffer (fig. 69,5-6)

*Morphology.* The second antennal segment is twice as long as wide. The spatula (fig. 69,5) is provided with two rounded teeth.

The body is bare. The papillae are predominantly simple. Each segment bears six simple dorsal papillae, except for the eighth abdominal segment, on which two simple dorsal papillae are found. On the ventral surface of each of the thoracic segments one finds: two simple sternal papillae, two simple sublateral papillae, and the usual set of intercalary papillae (fig. 69,5). Four simple anterior ventral papillae, which are located behind the spinule fields, and two simple posterior ventral papillae are found on the first seven abdominal segments. The eighth abdominal segment bears four simple ventral papillae. The lateral papillae are simple. Spinule fields are present.

The terminal segment (fig. 69,6) is provided with eight well-pronounced, conical terminal papillae. Its dorsal surface bears verruca-like protuberances, the ventral surface is bare. Around the anal slit occur longitudinal rows of spinules; in front of the slit transverse rows are found. The six anal papillae are mostly devoid of bristles. The distribution of the stigmata is typical.

*Biology.* The larvae develop in galls on the leaves of *Populus.*

Key to the species of *Harmandia* Kieffer

1(2) The lateral papillae on the eighth abdominal segment are simple. The spinule fields comprise no more than 22-24 rows. . . . . . . . . . . . . . *H. globuli* (Rübsaamen)

2(1) One pair of lateral papillae on the eighth abdominal segment bears short bristles. The spinule fields comprise 26-28 rows. . . . . . . . . . . . *H. loewi* (Rübsaamen)

**Harmandia globuli** (Rübsaamen)

*Morphology.* All lateral papillae are simple. The spinule fields consist of 22-24 rows. The anal slit is surrounded by six to eight longitudinal rows of spinules; in front of the slit nine to ten transverse rows occur. All lateral papillae are simple.

*Biology.* The larvae develop within reddish, thick-walled galls, which project from the upper surface of the leaves of *Populus tremula.*

*Comment.* The authors made use of the description by Möhn (1955b).

**Harmandia loewi** (Rübsaamen)

*Morphology.* The anterior lateral papillae on the eighth abdominal segment bear a very short bristle. The spinule fields consist of 26-28 rows. The anal slit is surrounded by four to six longitudinal rows of spinules; in front of the slit twelve to fourteen rows occur. Two of the six anal papillae bear a short bristle.

*Biology.* The larvae develop within rounded, thick-walled galls, which are found on the upper surface of the leaves of *Populus tremula.*

*Comment.* The authors made use of the description by Möhn (1955b).

26. Genus **Agevillea** Hubault (fig. 69,7-8)

*Morphology.* The second antennal segment is as long as it is wide. The teeth of the two-toothed spatula (fig. 69,7) are rounded and separated by a deep incision.

The body is bare. Only on the posterior margin of the eighth abdominal segment are some transverse rows of spinules found. The anterior third part of the abdominal segment bears some transverse rows of spinules. All six dorsal papillae on the thoracic and the first seven abdominal segments are simple, without bristles. The two dorsal papillae on the eighth abdominal segment also lack bristles. The ventral surface of the thoracic segments (fig. 69,7) bears two simple sternal, two simple sublateral, and four groups of ordinary intercalary papillae. Four simple anterior ventral papillae, located behind the spinule fields, and two simple posterior ventral papillae are found on the first seven abdominal segments. On the eighth abdominal segment just four simple ventral papillae occur.

The terminal segment bears eight terminal papillae, six of which have a short bristle, and two a light brown, conical apical part. In some cases the two outer terminal papillae are simple. The lateral papillae are simple. On the eighth abdominal segment they are located on verrucae. Spinule fields are present.

The dorsal surface of the terminal segment (fig. 69,8) is provided with transverse rows of blunt verrucae; the ventral surface is bare. The anal slit is surrounded by some longitudinal rows of spinules. In front of the slit a few transverse rows are found. All six anal papillae are simple.

*Biology.* The larvae develop on coniferous species.

**Agevillea abietis** Hubault

*Morphology.* The eighth abdominal segment bears on the posterior margin of the dorsal surface four or five rows of blunt protuberances. On the terminal segment (fig. 69,8) seven or eight transverse rows of blunt protuberances occur. The spinule fields consist of 20-22 rows. The anal slit is surrounded by four to six longitudinal rows of spinules. Ten to twelve transverse rows are found in front of the slit.

*Biology.* The larvae develop within parenchymatous galls on the needles of *Abies alba.*

*Comment.* The authors made use of the description by Möhn (1955b).

27. Genus **Anisostephus** Rübsaamen (fig. 71)

*Morphology*. The small, oval head capsule (fig. 71,1) is slightly longer than its width. The tip of the well-developed exocephalic rods is somewhat broadened. The rods are at least as long as the width of the base of the capsule. The antennae are short and wide. The second antennal segment is no more than 1.5 times as long as width.

The spatula (fig. 71,2) is two-toothed. The apically rounded teeth are triangular. The teeth are separated by a deep, oval emargination, so that the bases of the teeth are not united. The interstice between the teeth is no less than half the width of a tooth. The well-pronounced manubrium has a widened base.

The body is bare. The papillae are predominantly simple. The thoracic and the first seven abdominal segments bear on the dorsal surface six simple dorsal papillae, on the eighth abdominal segment two simple dorsal papillae occur. The ventral surface of the thoracic segments bears: two simple sternal papillae, four groups of ordinary intercalary papillae, and two simple sublateral papillae. Four simple anterior ventral papillae, which are located behind the spinule fields (fig. 71,3), and two simple posterior ventral papillae are found on the first seven abdominal segments. On the eighth abdominal segment four simple ventral papillae (fig. 71,4) occur. The terminal segment is provided with eight terminal papillae, two of which are large and conical, four bear short bristles, and two are simple. The lateral papillae are simple.

The spinule fields are distributed on the ventral surface of the meso- and metathoracic, and all abdominal segments. The fields on the first seven abdominal segments contain four oval bare spots.

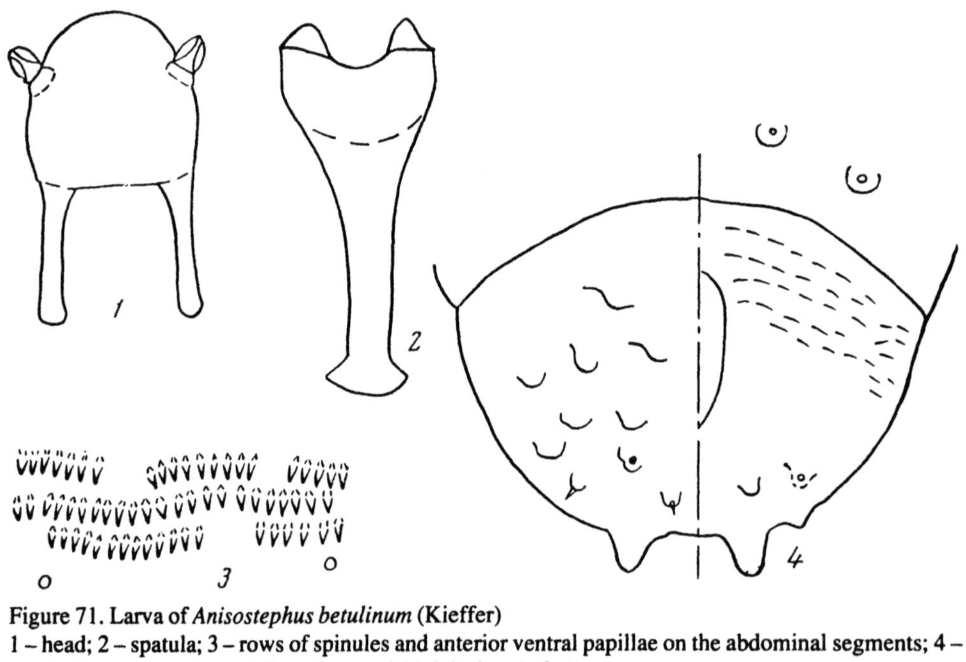


Figure 71. Larva of *Anisostephus betulinum* (Kieffer)
1 – head; 2 – spatula; 3 – rows of spinules and anterior ventral papillae on the abdominal segments; 4 – terminal segment, dorsal (left) end ventral (right) view (orig.)

The terminal segment (fig. 71,4) is bare. Its dorsal surface bears only hardly discernable oval and polygonal platelets. The anal slit is not surrounded by longitudinal rows of spinules. The anal papillae lack bristles. The stigmata are of the usual type. The terminal pair is shifted to the dorsal side of the segment.

*Biology*. The larvae live in parenchymatous galls.

**Anisostephus betulinum** (Kieffer)

*Morphology*. The naked body bears, in particular on the last three segments, weakly outlined oval and polygonal platelets along the lateral surfaces of the segments. Similar structures occur on the dorsal surface of the terminal segment. The spinule fields comprise 17-19 rows on the first eight abdominal segments, and 7-8 rows on the terminal segment.

*Biology*. The larvae cause blister-like parenchymatous galls on the leaves of *Betula*.

*Material*. Bykovo, Moskovskaya oblast, 10 June 1960, five larvae.

28. Genus **Syndiplosis** Rübsaamen (fig. 72)

*Morphology*. The short, heavily sclerotized head capsule (fig. 72,1) is considerably shorter than the width of its base. The posterior edge of the head capsule has on the dorsal side a small emargination; the ventral side is almost straight. The relatively short, well-developed exocephalic rods are wide and sturdy. They are noticeably shorter than the width of the capsule. The antennae are short. The second antennal segment is slightly longer than twice its width at the base.

The spatula (fig. 72,2) is two-toothed. The teeth have a wide base and a pointed tip. They are separated by a wide, oval emargination. As a rule, the manubrium is well-pronounced and broad (as broad as the width of a tooth).

The dorsal surface of the body, not counting the terminal segment, is bare. A single row, or a few sparse rows, of tiny spinules (fig. 72,4) occur on the dorsal surface of the anterior part of the first seven abdominal segments. The eighth abdominal segment bears small oval platelets at the level of the dorsal papillae. A complete set of dorsal and lateral papillae is found on the thoracic and the first seven abdominal segments. The small, hardly discernable papillae are provided with a short bristle. Two setiform dorsal papillae occur on the eighth abdominal segment. The ventral surface of the thoracic segment bears: two simple sternal papillae, four groups of intercalary papillae, and two simple sublateral papillae. Four simple anterior ventral papillae as well as two simple posterior ones are found behind the spinule fields (fig. 72,5) on the first seven abdominal segments. On the eighth abdominal segment four simple ventral papillae occur. The terminal segment (fig. 72,3) bears eight terminal papillae, six of which are setiform and of various sizes, whereas the remaining two look like blunt, sclerotized verrucae.

Well-developed spinule fields are found on the ventral surface of the meso- and metathoracic, and all abdominal segments.

The dorsal surface of the terminal segment (fig. 72,3) is densely covered with numerous verrucae; the ventral surface is bare. The anal slit is surrounded by longitudinal rows of spinules. The terminal pair of stigmata is shifted to the dorsal side of the segment.

*Biology*. The species of this genus are mostly associated with the genus *Populus*.

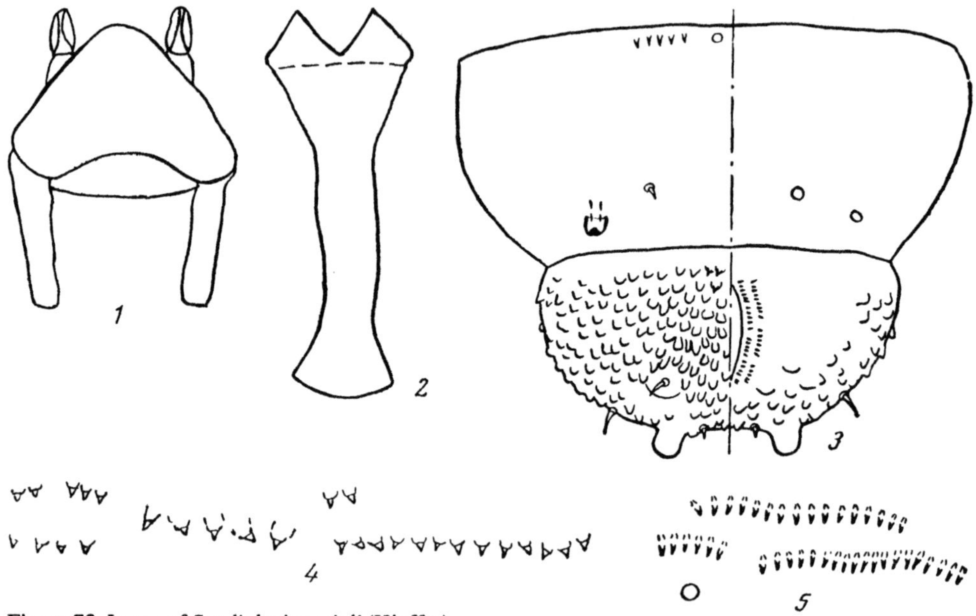


Figure 72. Larva of *Syndiplosis petioli* (Kieffer)
1 – head; 2 – spatula; 3 – last two segments, dorsal (left) and ventral (right) view; 4 – rows of spinules on the abdominal segments, dorsal view; 5 – rows of spinules and anterior ventral papilla on the abdominal segments (orig.)

**Syndiplosis petioli** (Kieffer)
*Morphology.* The dorsal surface of the thoracic and the first eight abdominal segments is provided with isolated rows of wide, pointed verrucae. The number of rows increases towards the end of the body.

The spinule fields on the thoracic segments are compact; on the first seven abdominal segments the fields contain six indistinct bare spots, the two central ones of which are paired. The fields on the first two abdominal segments consist of 18 and 21 rows, on the next four segments of 25, on the seventh and eighth segments of 17, and on the terminal segment of 8-9 rows.

*Biology.* The larvae develop in galls on the petioles of *Populus tremula*.

*Material.* Pavlova Sloboda, Moskovskaya oblast, 23 June 1962, five larvae.

29. Genus **Contarinia** Rondani (fig. 73, 74)

*Morphology.* The larvae have a small, weakly pronounced head capsule. The terminal segment is provided with short processes.

The short, almost triangular head capsule (fig. 73,5) is slightly shorter than the width of its base. The posterior edge of the head capsule is straight. The length of the exocephalic rods is somewhat less than the width of the capsule. The antennae are short and wide. The apically rounded second antennal segment is twice as long as its width at the base.

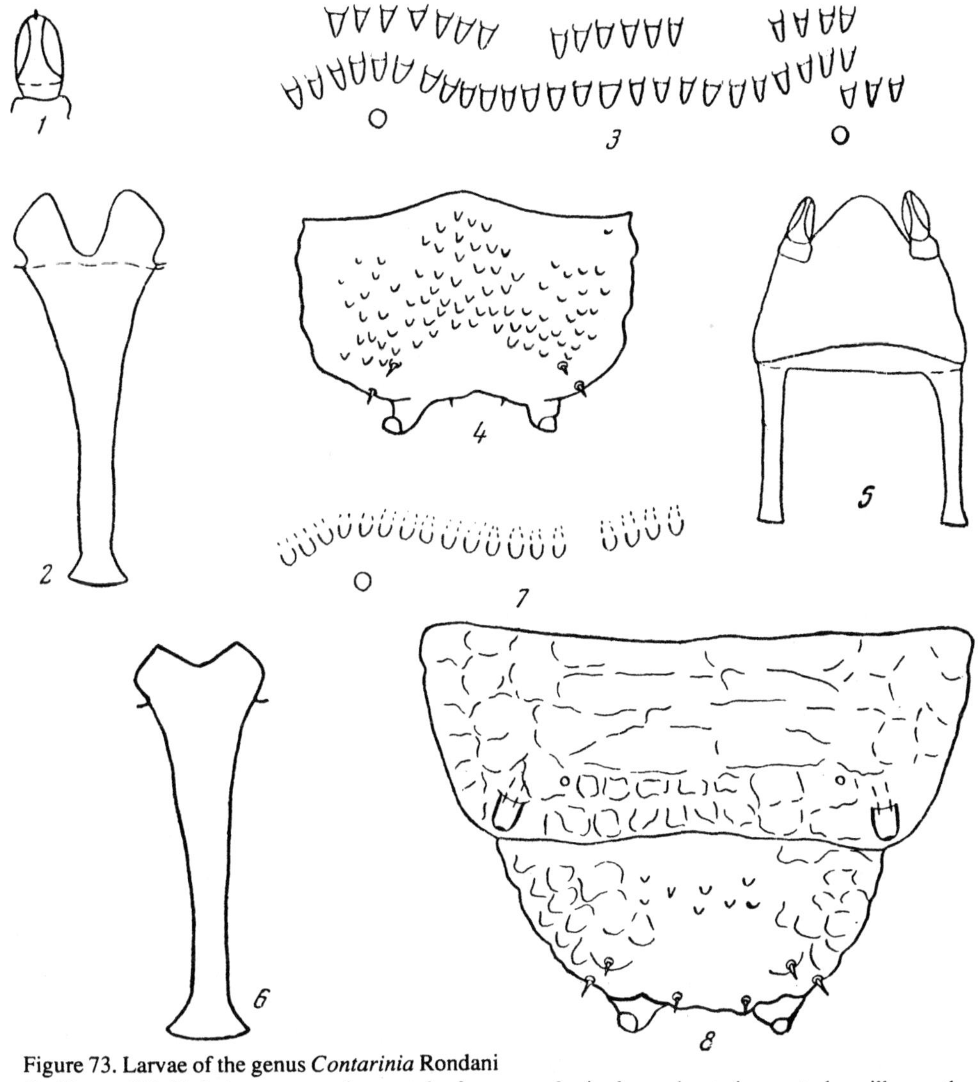


Figure 73. Larvae of the genus *Contarinia* Rondani
*C. tiliarum* (Kieffer): 1 – antenna; 2 – spatula; 3 – rows of spinules and anterior ventral papillae on the abdominal segments; 4 – terminal segment, dorsal view; *C. nasturtii* (Kieffer): 5 – head; 6 – spatula; 7 – rows of spinules and anterior ventral papilla on the abdominal segments; 8 – last two segments, dorsal view (orig.)

The spatula (fig. 73,2) is two-toothed. The short, apically rounded teeth are separated by a shallow emargination, which does not reach the base of the teeth. The base of the well-pronounced manubrium is slightly widened.

The body is bare. The papillae are predominantly simple. The thoracic and the first seven abdominal segments bear six simple dorsal papillae. On the eighth abdominal segment two simple dorsal papillae occur. The ventral surface of the thoracic segments is provided with two simple sternal papillae, four groups of ordinary intercalary papillae, and two simple sublateral papillae. Four simple anterior ventral papillae, as well as two simple posterior ventral papillae are found behind the spinule fields on the first seven abdominal segments (fig. 73,7). On the eighth abdominal segment four simple ventral papillae occur. The terminal segment (fig. 73,4) bears eight terminal papillae, six of which have a short bristle, and the remaining two of which resemble conical outgrowths with a blunt tip. The lateral surfaces of the segments are provided with simple lateral papillae.

Clearly pronounced spinule fields occur on the ventral surface of the meso- and metathoracic, and all abdominal segments. In the species that we have at our disposal the spinule fields are not compact, but contain two to six oval bare spots.

The penultimate segment is bare. In some cases it bears extremely weakly outlined polygonal platelets. The terminal segment (fig. 73,8) bears on its lateral surfaces quite indistinct polygonal platelets and in its central part a few protuberances; the remaining part of the segment is bare. There are six simple anal papillae. The anal slit is surrounded by some longitudinal rows of spinules. The stigmata are well-pronounced. The terminal pair is shifted to the dorsal side of the eighth abdominal segment.

*Biology*. The larvae are typical gall formers. They develop in flowers, fruits, leaves and offshoots of various plants.

*Comment*. The genus comprises approximately one hundred species, based on European material only, many of which cannot be morphologically distinguished from one another. The identification key that is given below allows the recognition of several groups of larvae.

Key to the species of *Contarinia* Rondani

1(2) The dorsal surface of the terminal segment (fig. 73,4) is densely covered with small oval protuberances. The spinule fields on the first seven abdominal segments contain four oval bare spots. . . . . . . . . . . . . . . . . . . . . . *C. tiliarum* (Kieffer)

2(1) The terminal segment is bare or bears on its dorsal surface isolated, scattered, pointed protuberances. The spinule fields on the first seven abdominal segments contain four or six oval bare spots.

3(4) The spinule fields on the first seven abdominal segments contain four oval bare spots. The teeth of the spatula (fig. 73,6) are separated by a wide, shallow emargination. . . . . . . . . . . . . . . . . . . . . . . . . . . . *C. nasturtii* (Kieffer)

4(3) The spinule fields on the first seven abdominal segments contain six small, but distinct, oval bare spots, the four central ones of which are paired. The teeth of the spatula (fig. 74,4) are separated by a comparatively deep, oval emargination.

5(6) The tip of the exocephalic rods is widened. The spinule fields on the middle abdominal segments consist of 15-19 rows. . . . . . . . . . *C. tanaceti* Rübsaamen

6(5) The tip of the exocephalic rods is not widened. The spinule fields on the middle abdominal segments consist of 22-24 rows. . . . . . . *C. dipsacearum* Rübsaamen

**Contarinia tiliarum** (Kieffer)

*Morphology*. The head capsule is provided with well-developed, apically widened exocephalic rods, which are as long as the width of the base of the head capsule. The second antennal segment (fig. 73,1) is cask-shaped.

The well-pronounced teeth of the spatula (fig. 73,2) are separated by a deep emargination, which reaches almost to the base of the teeth. The spinule fields on the first seven abdominal segments contain four indistinct bare spots. The fields consist of uniform, large spinules with a pointed tip (fig. 73,3). The fields comprise on the mesothorax 15 rows, on the next eight segments 17-19, and on the last two segments respectively nine and six rows. The dorsal surface of the terminal segment (fig. 73,4) bears densely distributed oval protuberances.

*Biology.* The yellow larvae develop within bud galls on the leaves of *Tilia*.

*Material.* Pavlova Sloboda, Moskovskaya oblast, 26 June 1962, five larvae.

**Contarinia nasturtii** (Kieffer)

*Morphology.* The exocephalic rods are not apically widened. They are almost as long as the width of the head capsule at its base. The second antennal segment narrows towards the tip (fig. 73,5). The apically pointed teeth of the spatula (fig. 73,6) are separated by a shallow, rounded emargination. The spinule fields on the first seven abdominal segments contain four small, oval bare spots. The fields consist of large, blunt spinules (fig. 73,7), which form on the first two abdominal segments respectively 14 and 18 rows, on the next four segments 19-20, on the seventh segment 15, and on the last two segments 6-10 rows. As a rule, a few more rows occur on the terminal segment than on the penultimate one.

To a considerable extent, the dorsal surface of the terminal segment is bare, although in the central part (fig. 73,8) some small, pointed protuberances are found.

*Biology.* The larvae cause the deformation of the flowers or offshoots of many species of Cruciferae, and are often found together with other species of gall midges.

*Material.* Kamensk-Shakhtinskiy, Luganskaya oblast, 3 June 1953, in the flowers of *Nasturtium*, five larvae.

**Contarinia tanaceti** Rübsaamen

*Morphology.* The comparatively well-developed exocephalic rods are as long as the width of the head capsule. The tip of the rods is broadened. The second antennal segment tapers off at the tip (fig. 74,1). The teeth of the spatula (fig. 74,2) are rounded. The spinule fields on the first seven abdominal segments contain six small, bright oval spots the four central ones of which are paired. The fields are made up of large, blunt spinules (fig. 74,3). The spinules are arranged into 15-19 rows on the first seven abdominal segments, and into 6-7 rows on the last two segments. The dorsal surface of the terminal segment is provided with small, pointed spinules and irregularly polygonal platelets.

*Biology.* The larvae develop in the flowers of *Tanacetum*.

*Material.* Pavlova Sloboda, Moskovskaya oblast, 8 August 1963, five larvae.

*Comment.* An accurate description of the larvae has been drafted for the first time.

**Contarinia dipsacearum** Rübsaamen

*Morphology.* The head capsule bears short exocephalic rods which are not apically widened. The rods are markedly shorter than the width of the capsule at its base. The blunt teeth of the spatula (fig. 74,4) have a rounded tip. The spinule fields on the first seven abdominal segments contain six distinct, small bare spots, the four central ones of which are paired. The fields consist of uniform, rather tiny spinules (fig. 74,5). The spinules are distributed as follows: on the first two abdominal segments 14-18 rows

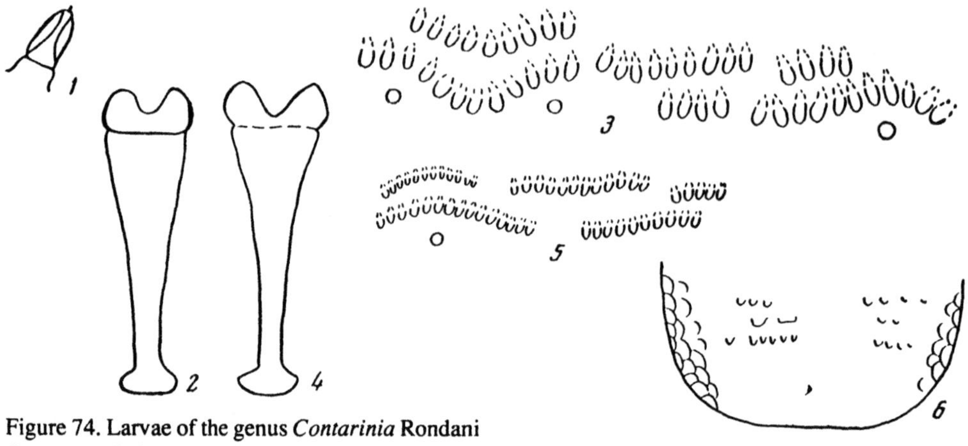


Figure 74. Larvae of the genus *Contarinia* Rondani
*C. tanaceti* Rübsaamen: 1 – antenna; 2 – spatula; 3 – rows of spinules and anterior ventral papilla on the abdominal segments; *C. dipsacearum* Rübsaamen: 4 – spatula; 5 – rows of spinules and anterior ventral papilla on the abdominal segments; 6 – cuticular structures on the terminal segment, dorsal view (orig.)

occur, on the next four segments 22-24 rows, on the seventh segment 19 rows, and on the last two segments 8-9 rows. The dorsal surface of the terminal segment bears some protuberances on its central part, and irregularly polygonal platelets (fig. 74,6) along the edges.

*Biology*. The larvae develop in the flowers of *Knautia*.

*Material*. Pavlova Sloboda, Moskovskaya oblast, 25 June 1962, five larvae.

*Comment*. An accurate description of the larvae has been drafted for the first time.

## 30. Genus **Dichaetia** Mamaev (fig. 75)

*Morphology*. The larvae are characterized by a large, elongated head capsule, characteristic, semi-spherical dorsal papillae, and a dense cover of hair-shaped structures.

The elongated head capsule (fig. 75,1) is at the dorsal side nearly twice as long as its width at the base. The ventral side is half as long as the dorsal side. The exocephalic rods are covered by the dorsal wing of the head capsule. The well-developed antennae are comparatively long. The second antennal segment is at least three times as long as it is wide.

The spatula (fig. 75,2) is two-toothed. The well-developed, conical teeth are separated by a deep emargination. The manubrium is not pronounced.

The dorsal surface of the body is covered with variously shaped, pointed protuberances (fig. 75,8), which, on the lateral surfaces, pass into hair-like structures. The papillae are setiform, simple, or spherical. On the dorsal surface of the prothorax one finds six setiform dorsal papillae, the two central ones of which are extremely large (almost as long as the segment itself (fig. 75,4)). Six spherical papillae (fig. 75,6), which bear at the tip a 'corolla' of spreading bristles, are found on the meso- and metathoracic, and the first seven abdominal segments. On the dorsal surface of the eighth abdominal

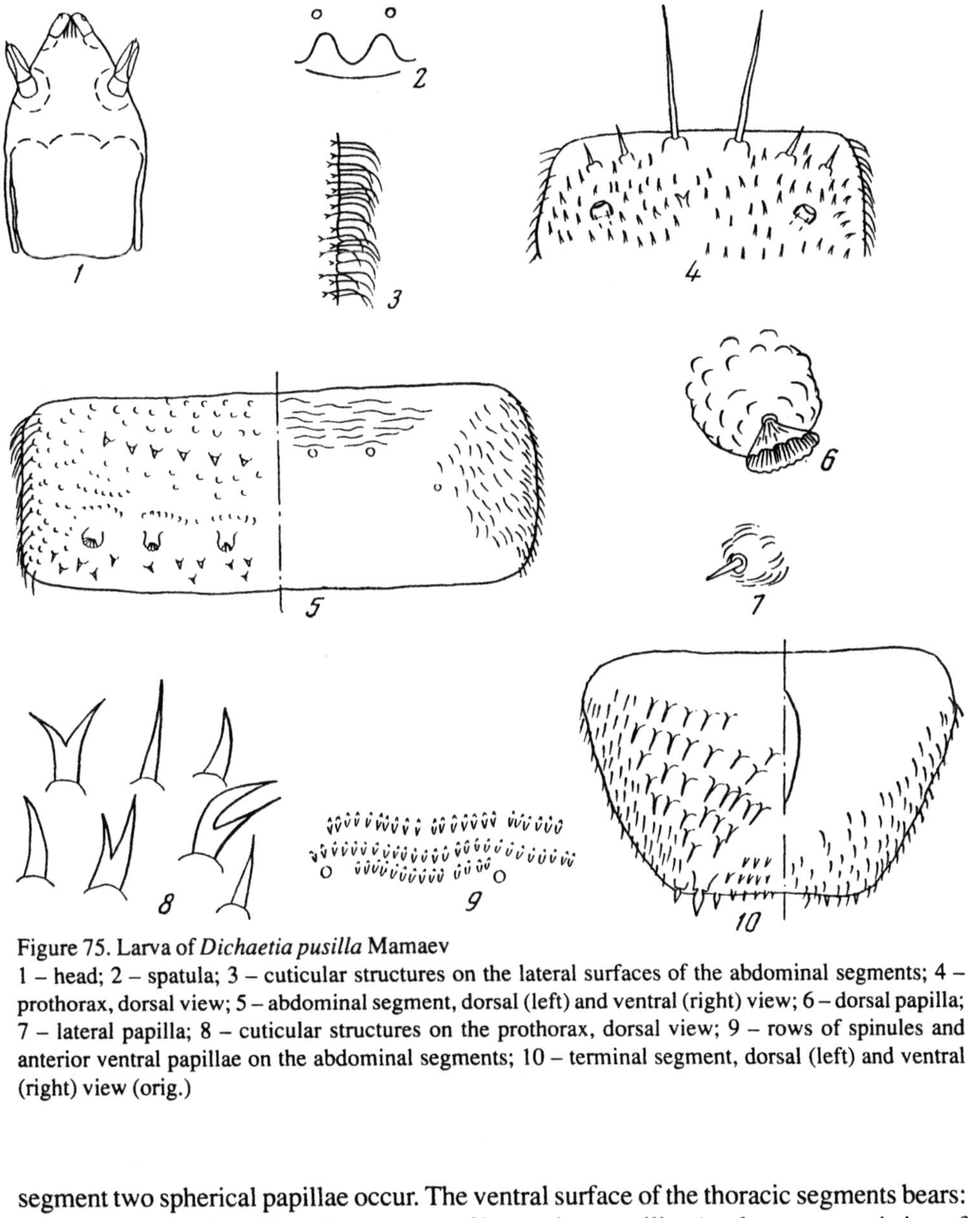


Figure 75. Larva of *Dichaetia pusilla* Mamaev
1 – head; 2 – spatula; 3 – cuticular structures on the lateral surfaces of the abdominal segments; 4 – prothorax, dorsal view; 5 – abdominal segment, dorsal (left) and ventral (right) view; 6 – dorsal papilla; 7 – lateral papilla; 8 – cuticular structures on the prothorax, dorsal view; 9 – rows of spinules and anterior ventral papillae on the abdominal segments; 10 – terminal segment, dorsal (left) and ventral (right) view (orig.)

segment two spherical papillae occur. The ventral surface of the thoracic segments bears: two simple sternal papillae, four groups of intercalary papillae (each group consisting of one simple and two setiform papillae), and two simple sublateral papillae. The first seven abdominal segments are on the ventral surface (fig. 75,5) provided with four simple anterior ventral papillae, which are located behind the spinule fields, and two simple posterior ventral papillae. Four simple ventral papillae occur on the eighth abdominal segment. The terminal segment (fig. 75,10) bears six setiform terminal papillae; two of which, being situated somewhat more dorsally, are larger and more distinct than the other ones. Two lateral papillae (fig. 75,7), which consist of a verruca and a well-pronounced bristle, are found on the lateral surfaces of the segments. The lateral and terminal papillae are hidden among hair-like structures, which cover the lateral surfaces of the segments (fig.75,3).

The well-pronounced, compact spinule fields (fig. 75,9) are distributed on the ventral surface of the metathoracic and all abdominal segments.

The small terminal segment (fig. 75,10) has a rounded tip. The dorsal surface of the segment is covered with long, pointed protuberances, which, on the lateral and ventral surfaces, pass into hair-like structures. The anal slit is surrounded by dense, longitudinal rows of hairs.

*Biology*. The larvae live on the surface of fragments of wood that are covered with mould.

*Comment*. The diagnosis of the genus is given for the first time.

**Dichaetia pusilla** Mamaev

*Morphology.* The dorsal surface of the body is covered with tiny, acute, sometimes split protuberances (fig. 75,8) which are arranged in transverse rows. Larger, acute protuberances are arranged in a transverse row in front of the dorsal papillae.

The dorsal papillae on the prothorax bear setae which are of unequal length (fig. 75,4). The medially situated setae are at least five times longer than the laterally situated ones. The dorsal papillae on the further segments (fig. 75,6) are globose and covered with little verrucae. A fan-shaped seta is present on top of each papilla. All papillae are situated equidistant from each other. Both pairs of papillae on the thorax and the eighth abdominal segment bear setae; the further segments bear one pair of simple lateral papillae and one pair of setose laterals. On the metathorax and the final body segments only 8-9 spinule rows are developed, whereas further segments bear 11-14 spinule rows.

*Biology.* The larvae have been collected from the surface of limewood which was covered with mycelium.

*Material.* Tulskye Zaseky, 2 June 1959, three larvae.

*Comment.* The diagnosis of the larvae is given for the first time.

31. Genus **Feltiella** Rübsaamen (fig. 76,1-2)

*Morphology*. The supernumerary segment and the first body segments are strongly elongated. The second antennal segment is ten times as long as it is wide. There is no spatula. Barring the eighth segment, the dorsal surface of the body is bare. On the dorsal surface of the eighth segment a single row of connivent oval verrucae is found. All six dorsal papillae bear well-developed bristles. The two dorsal papillae on the eighth abdominal segment are setiform as well.

The ventral surface of each of the thoracic segments (fig. 76,1) bears two simple sternal and sublateral papillae, and four groups of ordinary intercalary papillae. Four simple anterior and two simple posterior ventral papillae are found on the ventral surface of the first seven abdominal segments. On the eighth abdominal segment four simple ventral papillae occur. The terminal segment is provided with eight terminal papillae bearing thick bristles of varying lengths. Each lateral surface of the segment bears two setiform lateral papilllae. Instead of spinule fields some rows of coarse protuberances are found on the ventral surface of the segment.

The terminal segment (fig. 76,2) is bare at the dorsal as well as at the ventral surface. The anal slit is located on the dorsal side of the segment. Two of the six anal papillae are found alongside the anal slit, the other four stand on the ventral surface. The anal slit is surrounded by longitudinal rows of spinules.

*Biology.* The larvae prey on colonies of spider mites.

**Feltiella tetranychi** Rübsaamen

*Morphology.* The coarse protuberances on the ventral surface are arranged into six to nine rows. Four of the eight terminal papillae (fig. 76,2) bear very long bristles, the other four bear short bristles.

*Biology.* The larvae have been found in colonies of *Tetranychus* sp. on the leaves of *Humulus lupulus.*

*Comment.* The authors have made use of the description by Möhn (1955b).

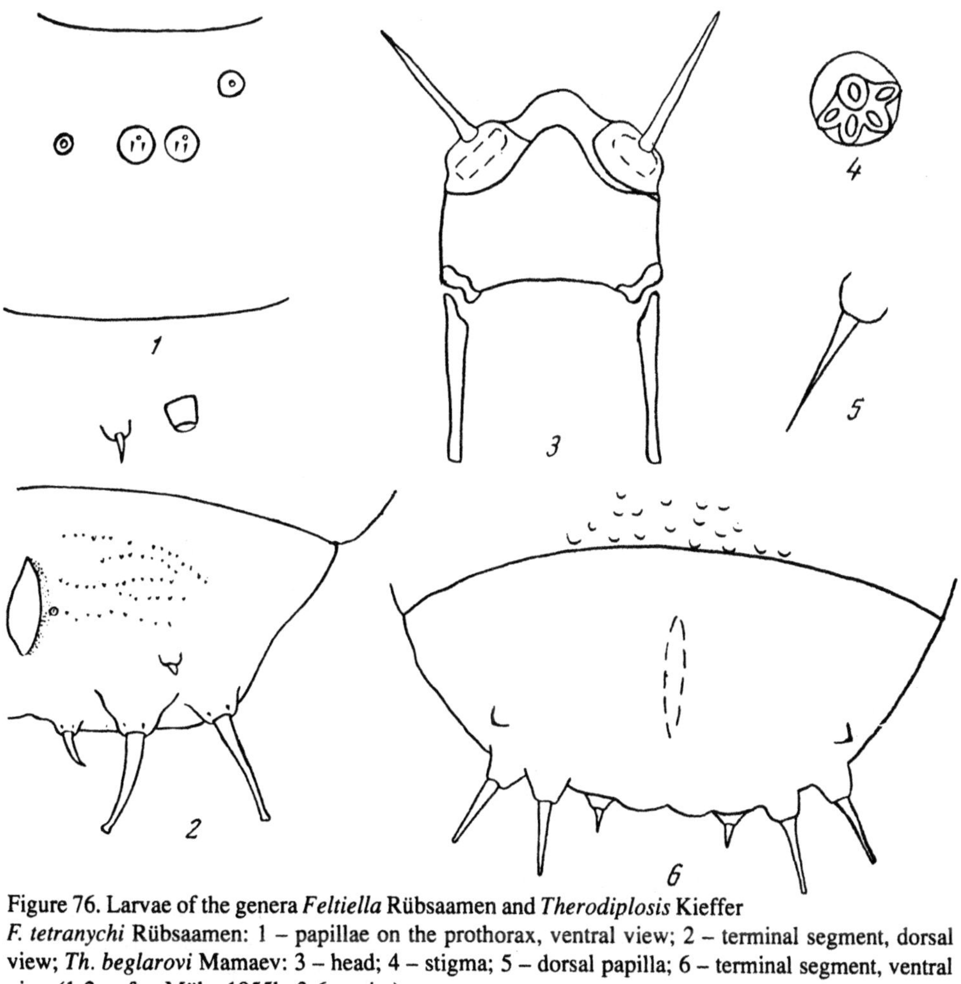


Figure 76. Larvae of the genera *Feltiella* Rübsaamen and *Therodiplosis* Kieffer
*F. tetranychi* Rübsaamen: 1 – papillae on the prothorax, ventral view; 2 – terminal segment, dorsal view; *Th. beglarovi* Mamaev: 3 – head; 4 – stigma; 5 – dorsal papilla; 6 – terminal segment, ventral view (1-2 – after Möhn 1955b; 3-6 – orig.)

## 32. Genus **Therodiplosis** Kieffer (fig. 76,3-6)

*Morphology.* The body of the elongated-oval larvae is in most cases bare.

The comparatively short, broad head capsule (fig. 76,3) is almost square; it is no longer than the width of its base. The relatively thin exocephalic rods have no widened tip. They articulate with the posterior edge of the capsule. The shape of the antennae is peculiar. The first antennal segment is short and wide, while the second one is thin and extremely long (ten or eleven times as long as the width of its base). There is no spatula.

On the otherwise bare body a small number of sparsely distributed platelets and verrucae of irregular shape is found. Six well-developed setiform dorsal papillae (fig. 76,5), which are situated on sturdy verrucae, occur on the dorsal surface of the thoracic and the first seven abdominal segments. Just two dorsal papillae are found on the eighth abdominal segment. On the ventral surface of the thoracic segments one finds: two simple sternal papillae, four groups of intercalary papillae (each of which consists of a single simple and two setiform papillae), and two simple sublateral papillae. Four simple anterior as well as two simple posterior ventral papillae are found on the ventral surface of the first seven abdominal segments. Just four simple ventral papillae occur on the eighth abdominal segment. The terminal segment bears eight terminal papillae, four of which have the appearance of verrucae with long rod-like appendages, two bear shorter appendages, and two have no appendages at all. On the lateral surfaces of the segments two setiform lateral papillae occur. The ventral surface of the segments sometimes bears a few rows of blunt verrucae.

The bare, rounded terminal segment (fig. 76,6) is broad and short. The anal slit is located on the dorsal side of the segment. It may be bare or surrounded by three or four longitudinal rows of spinules. The very small stigmata are found on verruca-shaped protuberances of the cuticle; the terminal pair of stigmata is shifted to the dorsal side of the segment.

*Biology.* The larvae live as predators in colonies of spider mites.

### **Therodiplosis beglarovi** Mamaev

*Morphology.* A field, consisting of three or four rows of wide, blunt verrucae, is found on the eighth segment along the posterior edge of the ventral surface (fig. 76,6).

*Biology.* The larvae prey in colonies of the mite *Schizotetranychus telarius.*

*Material.* The larvae were put at our disposal by G.A. Beglyarov.

*Comment.* The diagnosis of the species is given for the first time. Since we have no larvae of the second species of this genus, *Therodiplosis persicae* Kieffer, in our collection, we are not in a position to accurately define the differences between these species.

## 33. Genus **Isodiplosis** Rübsaamen (fig. 77)

*Morphology.* The tiny, orange larvae are elongated-oval.

The well-pronounced head capsule (fig. 77,1) is slightly elongated and somewhat longer than its width at the base. The sturdy exocephalic rods have a strikingly widened tip. Both the dorsal and ventral side of the posterior edge of the head capsule have a small emargination. The antennae are well-developed. The almost cylindrical second antennal segment is apically rounded. It is no less than 4.5 times as long as its width at the base.

The well-pronounced manubrium of the two-toothed spatula (fig. 77,2) has a somewhat widened base. The teeth are rounded and blunt.

The dorsal and lateral surfaces of the body are covered with variously shaped platelets and protuberances (fig. 77,4). The papillae are predominantly setiform. The thoracic and the first seven abdominal segments dorsally bear six setiform papillae, the papillae 'b' (fig. 1) of which are noticeably shorter than the papillae 'a' and 'c' (fig. 77,4-5). Just two setiform dorsal papillae are found on the eighth abdominal segment. The ventral surface of each of the thoracic segments (fig. 77,2) bears two simple sternal papillae, four groups of intercalary papillae (each group consisting of a simple papilla and two short setiform papillae), and two simple sublateral papillae. Four simple anterior as well as two simple posterior ventral papillae (fig. 77,3,6) are found behind the spinule fields on the ventral surface of the first seven abdominal segments. On the eighth abdominal segment a single row occurs, which consists of four simple anterior ventral papillae. The terminal segment (fig. 77,9) is provided with eight terminal papillae, four of which are setiform and four bear a short, conical apical part. On the lateral surfaces of the segment two simple lateral papillae (fig. 77,8) occur.

Except for the spinule fields (fig. 77,6), the ventral surface of the meso- and metathoracic, and all abdominal segments is bare. The compact spinule fields consist of large spinules.

The trapezium-shaped terminal segment (fig. 77,9) is markedly shorter and narrower than the penultimate one. Tiny, pointed protuberances are sparsely distributed on the dorsal surface of the segment. The ventral surface of the segment is over a considerable extension bare. Numerous rows of tiny spinules are found on the anterior half of the segment, in front of the anal slit. The anal slit is surrounded by some longitudinal rows of tiny spinules.

*Biology*. The larvae live as unspecialized mycetophages in decaying plant tissues.

**Isodiplosis involuta** Rübsaamen

*Morphology*. The second antennal segment (fig. 77,1) is 4.5 times as long as its width. The papillae 'b' (fig. 1) are, in particular, short on the thoracic segments (fig. 77,4); on the other segments they are somewhat longer. They are all found on well-developed verrucae. The sclerotized structures that cover the dorsal surface of the segments are rather sparsely distributed. Their shapes and sizes are quite variable. A few rows of distinct, rounded platelets are found on the prothorax (fig. 77,4), in front of the dorsal papillae. On the meso- and metathorax four rows of platelets are found in front of the dorsal papillae, while some rows are found behind them. Relatively large platelets of irregular shape as well as smaller, pointed protuberances are scattered over the entire surface of the first seven abdominal segments. The eighth abdominal segment bears only platelets, whereas on the terminal segment only protuberances are found.

Spinule fields (fig. 77,6) occur on the ventral surface of the segments; the fields consist of clearly outlined spinules. On most segments the fields comprise six to eight rows. The field on the eighth segment consists of only five rows, while the field on the terminal segment is comprised of fourteen rows.

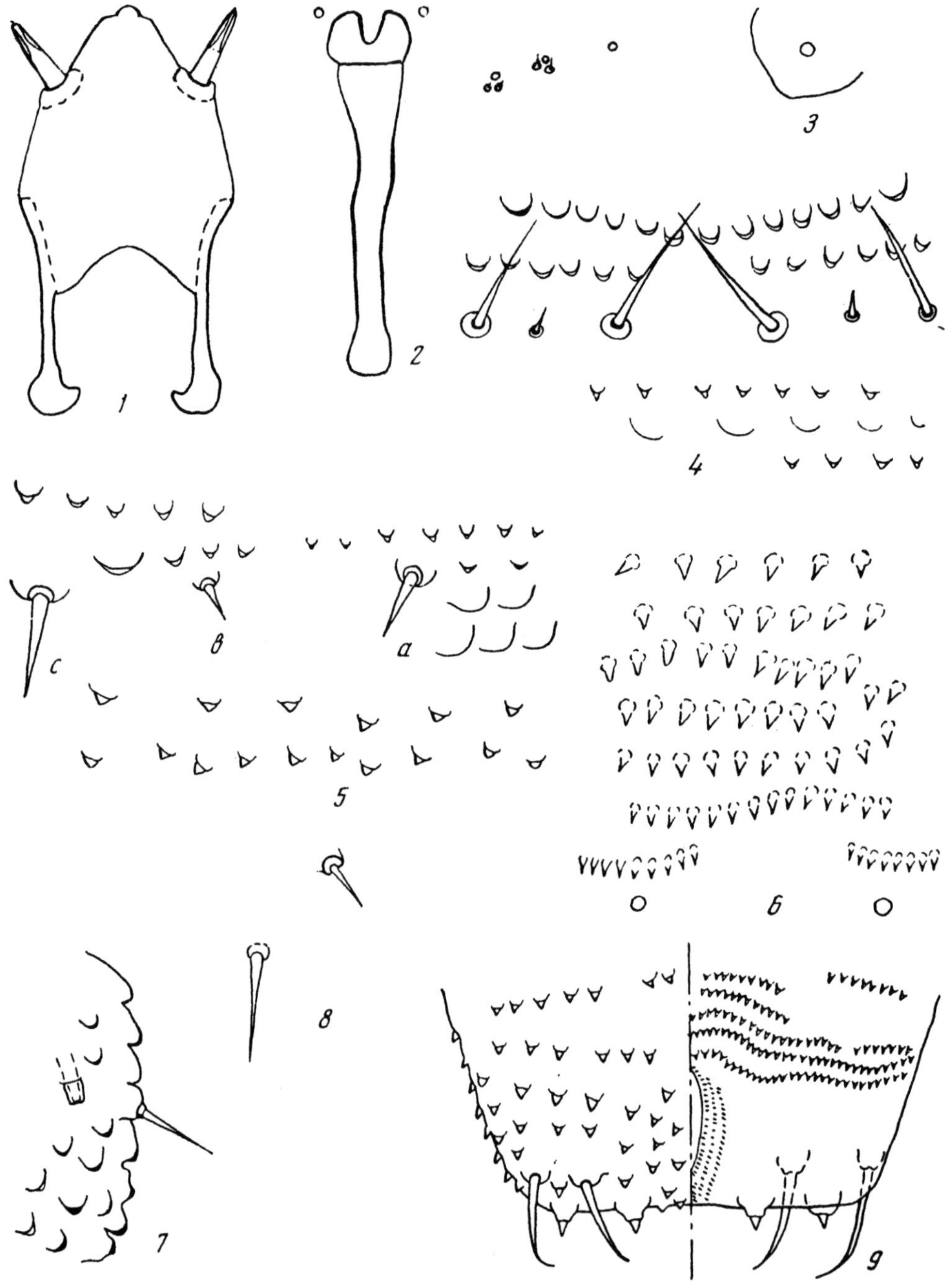


Figure 77. Larva of *Isodiplosis involuta* Rübsaamen
1 – head; 2 – spatula and papillae; 3 – posterior ventral papilla; 4 – cuticular structures and dorsal papillae on the prothorax; 5 – cuticular structures and dorsal papillae on the abdominal segments; 6 – rows of spinules and anterior papillae on the abdominal sepments; 7 – cuticular structures and lateral papillae on the abdominal segments; 8 – lateral papillae; 9 – terminal segment, dorsal (left) and ventral (right) view (orig.)

*Biology.* The larvae have been found in decaying inflorescences of *Polygonum* and in the flower heads of many Compositae.

*Material.* Pavlova Sloboda, Moskovskaya oblast, 14 September 1963, in decaying, shrivelled inflorescences of *Polygonum.*

*Comment.* The independence of the species *I. deutera* Milne, which was first described in England (Milne 1960), should be verified.

## 34. Genus **Acaroletes** Kieffer (fig. 78,1-2)

*Morphology.* The second antennal segment is five to six times as long as its width at the base. The spatula (fig. 78,1) is provided with two broad, rounded teeth.

The thoracic and the first seven abdominal segments bear six setiform dorsal papillae of approximately equal size (the central papillae may be slightly larger). Only two setiform dorsal papillae occur on the eighth abdominal segment. The ventral surface of the thoracic segments is provided with: two simple sternal papillae, four groups of intercalary papillae (each group consisting of one simple and two setiform papillae), and two simple sublateral papillae. Four simple anterior as well as two simple posterior ventral papillae are found on the first seven abdominal segments. The eighth abdominal segment bears four simple ventral papillae.

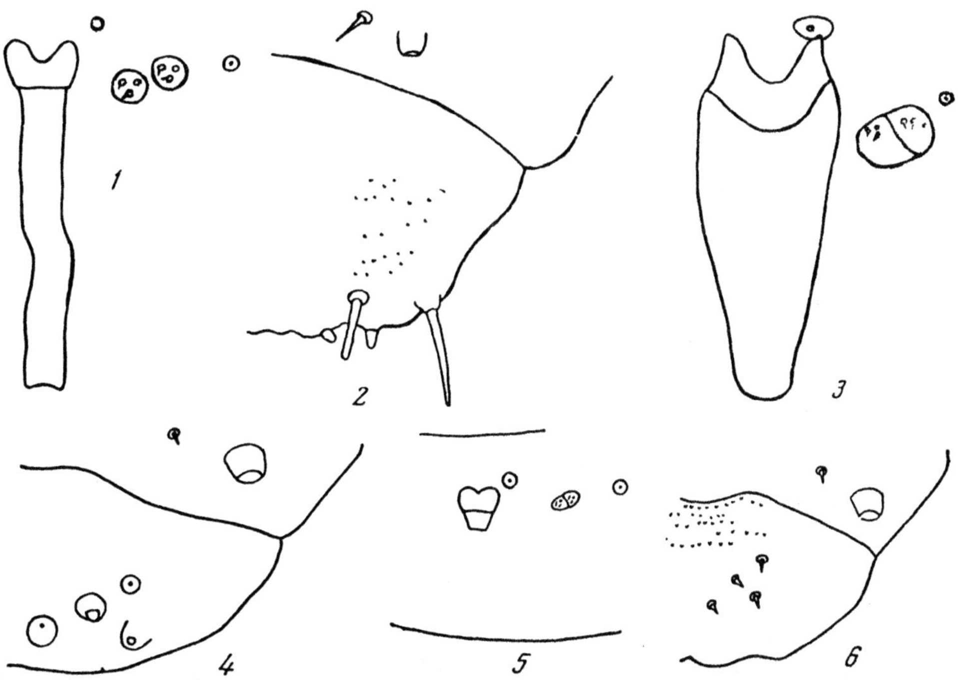


Figure 78. Larvae of the genera *Acaroletes* Kieffer, *Plemeliella* Seitner, and *Atylodiplosis* Kieffer *Acaroletes* sp.: 1 – spatula and papillae; 2 – last two segments, dorsal view; *Plemeliella abietina* Seitner: 3 – spatula and papillae; 4 – last two segments, dorsal view; *Atylodiplosis acetosellae* (Rübsaamen): 5 – spatula and papillae; 6 – last two segments, dorsal view (after Möhn 1955b)

The terminal segment (fig. 78,2) is provided with eight terminal papillae, four of which are setiform, whereas the four remaining ones bear short, rod-like appendages. On the lateral surfaces of each segment two setiform lateral papillae occur. The ventral surface of the body bears spinule fields and, locally, rounded verrucae.

The dorsal as well as the ventral surface of the terminal segment is partly covered with oval, pointed protuberances. The anal slit bears some longitudinal rows of spinules. The anal papillae are simple.

*Biology*. The larvae are specialized predators of the mite genus *Tetranychus*.

**Acaroletes** sp.
*Morphology*. The dorsal as well as the ventral surface of the body is densely covered with toothed, oval verrucae. The central dorsal papillae are slightly longer than the other ones. The anterior ventral papillae are located behind the spinule fields. The spinule fields consist of ten to twelve transverse rows of spinules. Two or three longitudinal rows of spinules are found around the anal slit.

*Biology*. The larvae prey on spider mites.

*Comment*. The authors made use of the description by Möhn (1955b).

35. Genus **Plemeliella** Seitner (fig. 78,3-4)

*Morphology*. The second antennal segment is as long as it is wide. The manubrium of the spatula (fig. 78,3) is broad and short. The teeth are pointed.

The dorsal surface of the body is bare. The lateral parts bear indistinctly outlined verrucae and protuberances. All six dorsal papillae bear a very short bristle. On the prothorax the four outer ones of the six dorsal papillae bear no or very short bristles which may be only just visible. The two dorsal papillae which occur on the eighth abdominal segment also bear short bristles. Two simple sternal and sublateral papillae as well as four groups of ordinary intercalary papillae (fig. 78,3) occur on the ventral surface of the thoracic segments. The ventral surface of the first seven abdominal segments bears four simple anterior ventral papillae (which are located behind the spinule fields) and two simple posterior ventral papillae. All four ventral papillae on the eighth abdominal segment are simple. The eight terminal papillae which are found on the terminal segment are conical. They are small and blunt, and have a rod-shaped apical part. Two setiform lateral papillae are found on each of the lateral surfaces of the segments. The ventral surface of the body bears spinule fields.

The dorsal as well as the ventral surface of the terminal segment (fig. 78,4) is bare. The anal slit is surrounded by a few longitudinal rows of spinules. No anal papillae bear bristles.

*Biology*. The larvae damage seeds.

**Plemeliella abietina** Seitner
*Morphology*. The dorsal surface of the body is bare. On the lateral parts some pointed protuberances and rounded, indistinctly outlined verrucae are found. The dorsal papillae bear very short bristles (only the central four dorsal papillae that are found on the prothorax, are simple). The four anterior ventral papillae are located behind the spinule

fields. The spinules in the fields are arranged into 20-22 rows. The anal slit is surrounded by five to seven longitudinal rows of spinules.

*Biology.* The larvae damage the seeds of *Picea*, and are recorded as pests.

*Comment.* The authors made use of the description that was drafted by Möhn (1955b).

36. Genus **Atylodiplosis** Rübsaamen (fig. 78,5-6)

*Morphology.* The second antennal segment is 1.5-2 times as long as its width. The manubrium of the spatula (fig. 78,5) is not developed.

The otherwise bare dorsal surface bears, from the mesothorax onwards, some transverse rows of spinules. The six dorsal papillae are provided with short bristles, as is the case with the two dorsal papillae on the eighth abdominal segment. The sternal and sublateral papillae on the ventral surface of the thoracic segments (fig. 78,5) are simple; the intercalary papillae are of the usual type. On the first seven abdominal segments four simple anterior ventral papillae as well as two simple posterior ones occur. The four ventral papillae which are found on the eighth abdominal segment are likewise simple. On the terminal segment eight terminal papillae bearing short bristles occur. The lateral papillae are also provided with short bristles. The ventral surface of the segments bears spinule fields.

The otherwise bare dorsal surface of the terminal segment (fig. 78,6) bears a few transverse rows of spinules on its anterior half. The anal slit is surrounded by longitudinal rows of spinules; in front of the slit some transverse rows are found.

*Biology.* The larvae deform flowers.

**Atylodiplosis acetosellae** (Rübsaamen)

*Morphology.* The length of the second antennal segment slightly surpasses its width. The rounded teeth of the spatula (fig. 78,5) are separated by a shallow emargination. From the mesothorax onwards the dorsal surface of the segments bears three or four rows of spinules. The anterior ventral papillae are found behind the spinule fields, which consist of seven to nine rows. Around the anal slit three or four longitudinal rows of spinules occur; in front of the slit three or four transverse rows are found.

*Biology.* The larvae deform the seeds of several species of *Rumex*.

*Comment.* The authors made use of the description by Möhn (1955b).

37. Genus **Loewiola** Kieffer (fig. 7,3; 79)

*Morphology.* The larvae have a characteristic spatula. The dorsal as well as the ventral surface of the body is entirely covered with platelets of various shapes.

The short and broad head capsule (fig. 7,3) is as long as, or slightly shorter than, its width at the base. The posterior edge of the head capsule has at the dorsal side a shallow emargination; the ventral side is straight. The short exocephalic rods are well-developed and about half as long as the width of the capsule. The antennae are comparatively short. The second antennal segment is approximately 2.5 times as long as it is wide. It bears a small, rod-like appendage at the tip.

Figure 79. Larva of *Loewiola centaureae* (F. Loew)
1 – spatula; 2 – papillae on the prothorax, ventral view; 3 – cuticular structures and dorsal papillae on the abdominal segments; 4 – cuticular structures on the lateral surfaces of the abdominal segments; 5 – last two segments, dorsal view; 6 – abdominal segment, ventral view; 7 – cuticular structures and anterior ventral papillae on the abdominal segments (orig.)

The spatula (fig. 79,1) bears no teeth. The broad manubrium is about half as wide as the head capsule. The base of the manubrium is widened.

The dorsal as well as the ventral surface of the body is densely covered with oval verrucae and platelets of various shapes. The papillae may be setiform or simple. The thoracic and the first seven abdominal segments each bear six setiform dorsal papillae (fig. 79,3) whose sizes increase from the central ones towards the outer ones. On the eighth abdominal segment two setiform dorsal papillae are found. The ventral surface of the segments (fig. 79,2) is provided with: two simple sternal papillae, four groups of intercalary papillae (each group comprising one simple and two setiform papillae), and two simple sublateral papillae. On the first seven abdominal segments four simple anterior (fig 79,6) and two posterior ventral papillae occur. All four ventral papillae on the eighth abdominal segment are simple and lack bristles. The terminal segment is provided with eight terminal papillae, four of which are setiform whereas the four remaining ones look like broad and blunt verrucae. Each segment bears on both lateral surfaces two lateral papillae. The ventral surface lacking spinule fields bears instead large, pointed protuberances which are not arranged into distinct rows.

The dorsal surface of the terminal segment (fig. 79,5) is to a large extent (excluding the posterior margin) covered with oval platelets. The ventral surface also bears platelets of various shapes, some of which show clearly pronounced points. All six anal papillae are simple and lack bristles. The anal slit is fringed with longitudinal rows of spinules.

The stigmata are well-developed. The terminal pair is shifted to the dorsal side of the segment.

*Biology.* The larvae develop in galls of the parenchyma.

**Loewiola centaureae** (F. Loew)

*Morphology.* The dorsal surface of the prothorax bears oval or irregular platelets. On the other segments are predominantly oval verrucae (fig. 79,3) found. The verrucae on the anterior half of the segments bear clearly pronounced teeth, whereas the verrucae on the posterior half of the segments lack teeth. Similar verrucae cover the lateral surfaces of the segments (fig. 79,4). The ventral surface of the prothorax bears large, polygonal platelets (fig. 79,2). On the anterior half of the meso- and metathoracic, and the first seven abdominal segments pointed protuberances occur; on the posterior half of the same segments the platelets are in most cases transversely stretched and of irregular shape. The anterior ventral papillae are located between the last three or four rows of protuberances (fig. 79,7).

*Biology.* The larvae cause the development of parenchyma galls on the leaves of species of *Centaurea.*

*Material.* Vicinity of Khunzakh, Dagestanskaya A.S.S.R., 5 September 1955, three larvae.

38. Genus **Itonida** Meigen (fig. 5,1; 18,2; 80)

*Morphology.* When full-grown, the large, orange larvae measure up to five mm long. The dorsal papillae are markedly lengthened. The small, almost triangular head capsule (fig. 80,1) is provided with long and sturdy exocephalic rods. The rods are more than 2.5 times as long as the width of the head capsule. The proximal end of the rods is furcate. The antennae are small. The triangular second antennal segment is apically pointed. It is no more than 1.5 times as long as its width.

The sturdy spatula (fig. 80,2) lacks teeth. Its short, broad head is more or less spindle-shaped. The base of the broad manubrium is widened.

A considerable portion of the body is bare. The papillae are characteristic. On the

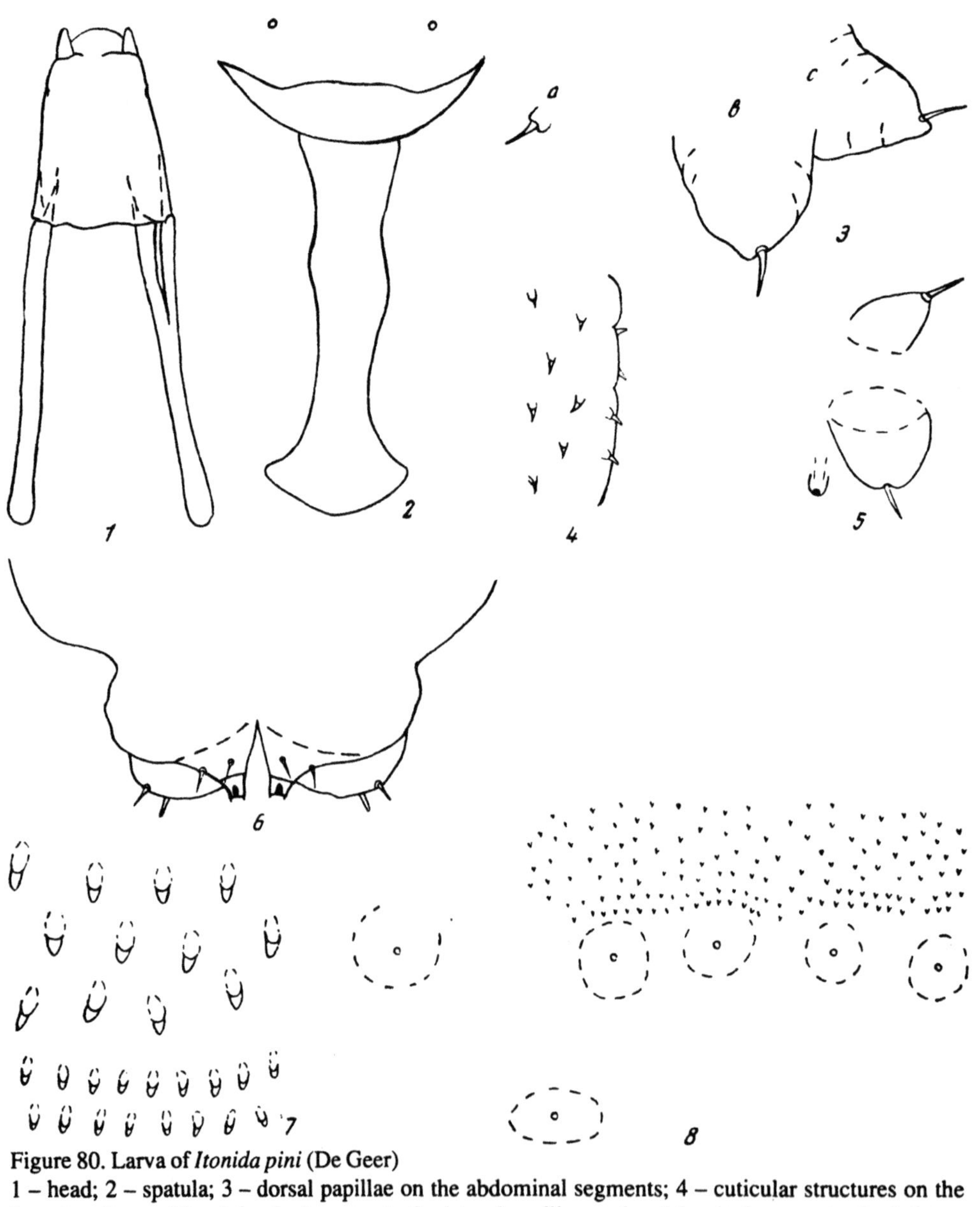


Figure 80. Larva of *Itonida pini* (De Geer)
1 – head; 2 – spatula; 3 – dorsal papillae on the abdominal segments; 4 – cuticular structures on the lateral surfaces of the abdominal segments; 5 – lateral papillae on the abdominal segments; 6 – last two segments, dorsal view; 7-8 – rows of spinules and ventral papillae on the abdominal segments (orig.)

thoracic segments the dorsal papillae are setiform and form a single transverse row. The papillae 'a' (fig. 1) that occur on the first seven abdominal segments have the same appearance as the ones on the thoracic segments. The papillae 'b' and 'c', however, though also setiform, are located on long outgrowths with fused bases (fig. 80,3). The dorsal papillae on the eighth abdominal segment, which are also setiform, stand pairwise in front of the stigmata and are found directly on the body surface. The terminal segment (fig. 80,6) bears four terminal papillae, two of which are setiform and two of which are rod-shaped.

One finds on the ventral surface of the thoracic segments: two simple sternal papillae, four groups of intercalary papillae (each group of which consists of one simple and two setiform papillae), and two simple sublateral papillae. The first seven abdominal segments bear four simple anterior ventral papillae (fig. 80,8), which are arranged into a single row behind the spinule fields. Two of the four simple posterior ventral papillae are found almost on the same level as the anterior ventral papillae. Just four simple papillae, arranged into a single transverse row, are found on the eighth abdominal segment. On the abdominal segments all ventral papillae are located on small swellings. Each of the lateral surfaces of the thoracic and the first seven abdominal segments bears two setiform lateral papillae (fig. 80,5). The lateral papillae on the thoracic segments stand directly on the surface, while those on the abdominal segments are situated on verruca-like swellings.

On the ventral surface of the meso- and metathoracic, and all abdominal segments small fields (fig. 80,8) occur, which are made up of randomly scattered, elongated spinules. The spinules tend to be smaller in the posterior part of the field than in the anterior part (fig. 80,7).

The eighth abdominal segment (fig. 80,6) is strikingly elongated so that its dorsal side covers the small terminal segment. A deep incision divides the tip of the eighth abdominal segment into two rounded lobes. The anal slit is surrounded by longitudinal rows of setiform structures. There are only two, simple, anal papillae.

The terminal pair of stigmata (fig. 18,2) occurs apically on the vicinal, large outgrowths of the eighth abdominal segment. The large, almost cylindrical stigmata are provided with elongated protuberances around the breathing opening. The stigmata on the other segments look like oval verrucae with a small protuberance at the tip.

*Biology*. The larvae live on coniferous trees.

**Itonida pini** (De Geer)

*Morphology*. Most of the dorsal surface is bare. The small, pointed spinules which are found on the anterior part and along the lateral surfaces of the segments (fig. 80,4), are smaller than the ones on the ventral surface.

On the first seven abdominal segments the dorsal papillae 'b' and 'c' (fig. 1) as well as the lateral papillae consist of large outgrowths with a bristle at the tip. The rather broad eighth abdominal segment is considerably shorter than its width. The terminal pair of stigmata is located on top of outgrowths which are no longer than their width (fig. 80,6). The stigmata (fig. 18,2) are provided with elongated protuberances around the breathing opening.

*Biology*. The larvae develop in the resin that collects at the bases of the needles of *Pinus*. Mass reproduction of the larvae may cause serious damage.

*Material.* Vicinity of Prague, Czechoslovakia, 7 November 1963, in cocoons on *Pinus* needles, two larvae (V. Skuhravý and B. Mamaev).

## 39. Genus **Stelechodiplosis** Möhn (fig. 18,3; 81)

*Morphology.* When full-grown, the very large larvae measure up to ten mm in length.

The small head capsule is almost triangular. The long, massive exocephalic rods are more than 2.5 times as long as the width of the head capsule at its base. The proximal end of the rods is furcate. The antennae are very small. The triangular second antennal segment is apically pointed.

The sturdy, toothless spatula bears a short and broad head. The thick manubrium has a widened base.

The body, which is to a considerable extent bare, bears simple as well as setiform papillae. The small dorsal papillae on the thoracic segments are well-separated from each

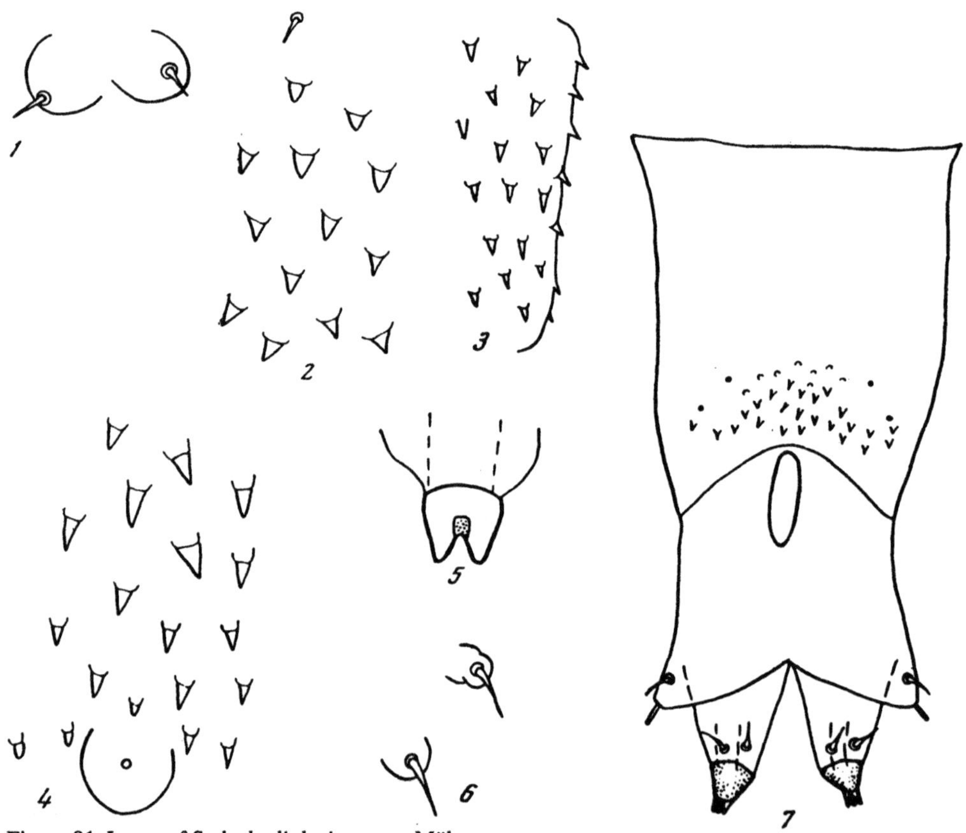


Figure 81. Larva of *Stelechodiplosis magna* Möhn
1 – dorsal papillae on the abdominal segments; 2-3 – cuticular structures on the abdominal segments, dorsal and lateral view; 4 – spinules and anterior ventral papilla on the abdominal segments; 5 – stigma on the seventh abdominal segment; 6 – lateral papillae on the abdominal segments; 7 – last two segments, ventral view (orig.)

other, and form a single transverse row. On the first seven abdominal segments the dorsal papillae 'a' (fig. 1) stand directly on the surface of the segments, whereas the other dorsal papillae are found on small, verruca-like swellings with contiguous bases (fig. 81,1). On the eighth abdominal segment two setiform papillae are situated on each outgrowth in front of the stigmata. The terminal segment is provided with four terminal papillae, two of which are setiform and two are rod-shaped.

On the ventral surface of the thoracic segments one finds: two simple sternal papillae, four groups of intercalary papillae (each group of which consists of one simple and two well-pronounced setiform papillae), and two simple sublateral papillae. A single transverse row of four anterior ventral papillae (fig. 81,4) is found behind the spinule field on the first seven abdominal segments. Two of the four posterior ventral papillae occur at nearly the same level as the anterior ventral papillae. The eighth abdominal segment bears just four ventral papillae the two central ones of which are markedly shifted to the anterior margin of the segment. All ventral papillae on the abdominal segments are located on small swellings. Two setiform lateral papillae are found on the thoracic and the first seven abdominal segments. On the thoracic segments these papillae stand directly on the surface, but on the abdominal segments they are situated on small verrucae (fig. 81,6).

The spinule fields form a ring on the anterior part of the meso- and metathoracic, and the first eight abdominal segments. A spinule field is also found on the ventral surface of the terminal segment in front of the anal slit.

The eighth abdominal segment (fig. 81,7) is strikingly elongated and covers the dorsal side of the small terminal segment. The tip of the eighth abdominal segment has a deep incision which forms two conical lobes. The anal slit is shifted to the anterior margin of the terminal segment. It is not surrounded by longitudinal rows of setiform structures.

The terminal pair of stigmata (fig. 18,3) is found on top of the large, vicinal outgrowths of the eighth abdominal segment. The large, almost conical terminal stigmata are apically provided with elongated protuberances. The other stigmata look like oval verrucae with short apical protuberances.

*Biology.* The larvae inhabit the resin canals of coniferous trees.

**Stelechodiplosis magna** Möhn

*Morphology.* The surface of the body is mostly bare. On the anterior margin of the dorsal surface of the meso- and metathoracic, and the first seven abdominal segments extensive spinule fields occur (fig. 81,2). The fields on the meso- and metathorax are clearly separated by a zone of bare cuticle. Similar broad fields are also found on the lateral and ventral surfaces of the segments (fig. 81,3). The fields form an almost closed ring around the anterior margin. A small number of spinules, the number of which increases towards the end of the body, is located on the posterior margin of the segments.

The dorsal and lateral papillae on the thoracic segments are small and setiform. On the first seven abdominal segments the papillae 'a' (fig. 1) are extremely small and located directly on the body surface, while the papillae 'b' and 'c' (which are also setiform) are found on small swellings. The markedly lengthened eighth abdominal segment (fig. 81,7) is at least twice as long as the width of the segment. The tip of the terminal segment is divided into two pointed, conical outgrowths.

The terminal pair of stigmata (fig. 18,3) is conical. The stigmata bear four finger-shaped protuberances at their tip. The other stigmata are slightly smaller and bear two shorter protuberances (fig. 81,5).

*Biology.* The larvae have been found in the resin flowing from a wounded *Picea* trunk.

*Material.* Pavlova Sloboda, Moskovskaya oblast, 14 September 1963, three larvae.

## 40. Genus **Echinella** Mamaev (fig. 82)

*Morphology.* The slightly dorso-ventrally flattened larvae bear large, setiform papillae.

The large, weakly sclerotized head capsule (fig. 82,1) is no longer than the width of the

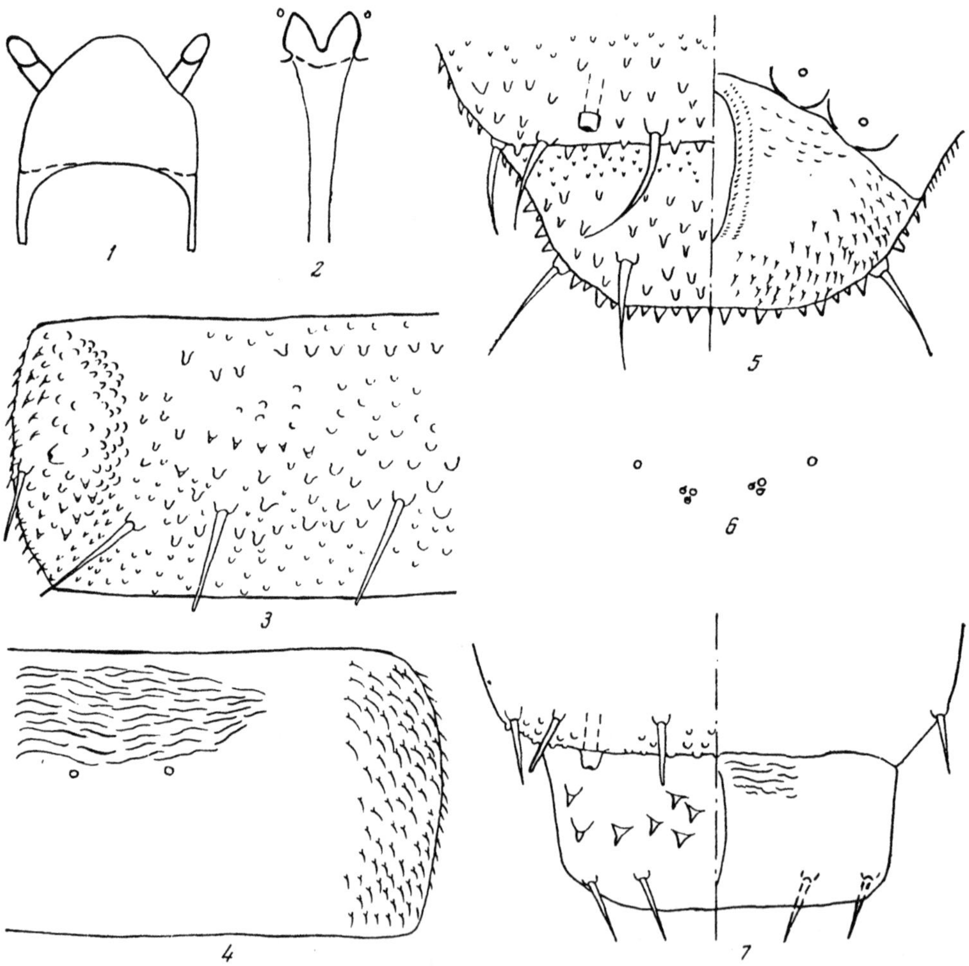


Figure 82. Larvae of the genus *Echinella* Mamaev
*E. albiventris* Mamaev: 1 – head; 2 – spatula; 3-4 – abdominal segment, dorsal and ventral view; 5 – last two segments, dorsal (left) and ventral (right) view; 6 – papillae on the prothorax, ventral view; *E. polypori* Mamaev: 7 – last two segments, dorsal (left) and ventral (right) view (orig.)

capsule at its base. The exocephalic rods are well-developed but comparatively short (no longer than half the width of the capsule), and not widened at the tip. The antennae have a peculiar shape: the length of the apical part of the second segment, which bears the sensorial unit, is no longer than the base.

The two-toothed spatula (fig. 82,2) is well-pronounced. The conical teeth of the spatula are separated by a deep emargination. The base of the weakly sclerotized manubrium is not widened.

The body is either bare, or the dorsal surface is covered with numerous rounded, pointed protuberances of various dimensions (fig. 82,3) while the lateral surfaces are covered with hair-like structures. A part of the papillae is simple, whereas the remaining ones are setiform. The dorsal surface of the thoracic and the first seven abdominal segments bears six large, setiform papillae (fig. 82,3) which are a quarter to half as long as the segment itself. On the eighth abdominal segment just two dorsal papillae occur.

The ventral surface of the thoracic segments bears: two simple sternal papillae, four groups of intercalary papillae (each group of which comprises two setiform and one simple papillae), and two simple sublateral papillae (fig. 82,6). Four simple anterior ventral papillae (fig. 82,4) (which are located behind the spinule fields) as well as two simple posterior ventral papillae are found on the first seven abdominal segments. All four ventral papillae on the eighth abdominal segment are simple. The terminal segment (fig. 82,5) bears four large, setiform terminal papillae, which are somewhat shifted to the dorsal side. Each lateral surface of the thoracic and the first seven abdominal segments bears a single large, setiform papilla. The lateral surfaces of the eighth abdominal segment bear two large lateral papillae.

The spinule fields on the ventral surface of the meso- and metathoracic, and all abdominal segments are well-pronounced and compact.

The terminal segment (fig. 82,5) is apically rounded. The dorsal surface bears distinct, pointed protuberances of various sizes, while the ventral surface may either be bare or possess long hair-like structures along its margin. The anal slit may be bare or surrounded by longitudinal rows of small hair-like structures.

*Biology*. The larvae are mycetophages and occur on the surface of vegetable remains that are covered with mycelium, or in the fruiting bodies of fungi.

*Comment*. The diagnosis of the genus is given for the first time.

Key to the species of *Echinella* Mamaev

1(2) The entire dorsal surface is covered with conical protuberances of various sizes, which are large, particularly in the vicinity of the dorsal papillae. The dorsal papillae consist of a long, sturdy, dark bristle (fig. 82,3), which is situated on an equally dark verruca. The bristles are at least 0.4 times as long as the corresponding segment. . . . . . . . . . . . . . . . . . . . . . . . . . . . . . *E. albiventris* Mamaev

2(1) The entire dorsal surface is bare. Rounded and pointed protuberances are only found on the dorsal surface of the posterior margin of the seventh and eighth abdominal segments, and on the terminal segment. The dorsal papillae consist of a long, light coloured bristle, which is situated on a still lighter coloured verruca. The length of the bristle is at most one third the length of the segment. *E. polypori* Mamaev

**Echinella albiventris** Mamaev

*Morphology.* The entire dorsal surface is uniformly covered with various cuticular structures. Large, conical protuberances, which are not arranged into distinct rows, are found in the vicinity of the dorsal papillae. The dimensions of the protuberances decrease strikingly towards the anterior and posterior margins of the segment. Protuberances of medium size form a single transverse row at the suture between the segments. The lateral surfaces of the segments bear tiny, markedly pointed protuberances, which pass into thin hair-like structures on the ventral surface (fig. 82,4).

The dorsal papillae on the thoracic and the first seven abdominal segments (fig. 82,3) are approximately equally long. Their length is no less than two fifths the length of a segment. Except on the eighth segment, where the dorsal and lateral papillae are equally long, the latter are slightly shorter than the former. On the midmost segments the spinule fields consist of thirteen or fourteen rows.

*Biology.* The larvae occur on the surface of small twigs that are covered with white mould.

*Material.* Volkovo, Yaroslavskaya oblast, 5 June 1961, three larvae.

*Comment.* The description of the larvae is given for the first time.

**Echinella polypori** Mamaev

*Morphology.* The dorsal surface of the body is predominantly bare. Rounded, distinct protuberances of various sizes occur only on the dorsal surface of the posterior margin of the seventh and eighth abdominal segments. On the central part of the dorsal surface of the terminal segment (fig. 82,7) transverse rows of large, pointed protuberances are found. The dorsal papillae on the thoracic and abdominal segments are approximately of equal length (no longer than one third the length of the segments). On the middle segments the spinule fields consist of thirteen or fourteen rows.

*Biology.* The larvae occur in the fruiting bodies of *Fomes marginatus* on *Betula* trunks.

*Material.* Volkovo, Yaroslavskaya oblast, 20 June 1961, five larvae.

*Comment.* The larvae are described for the first time.

41. Genus **Thomasiniana** Strand (fig. 83)

*Morphology.* The small, weakly sclerotized head capsule (fig. 83,1) is slightly longer than its width at the base. The dorsal side of the posterior edge of the head capsule has a shallow, oval emargination, while the ventral side is straight. The exocephalic rods are very short and hardly discernable. The well-developed, cylindrical antennae are apically rounded. The second antennal segment is almost three times as long as the width of its base.

The teeth of the two-toothed spatula (fig. 83,2) are blunt. The weakly sclerotized manubrium is not widened at the base.

The dorsal and lateral surfaces of the body are densely covered with oval and pointed protuberances and verrucae (fig. 83,4,6). Simple as well as setiform papillae occur. The dorsal surface of the thoracic and the first seven abdominal segments bears six setiform dorsal papillae (fig. 83,3). The bristles of the papillae 'c' (fig. 1) are longest, while those

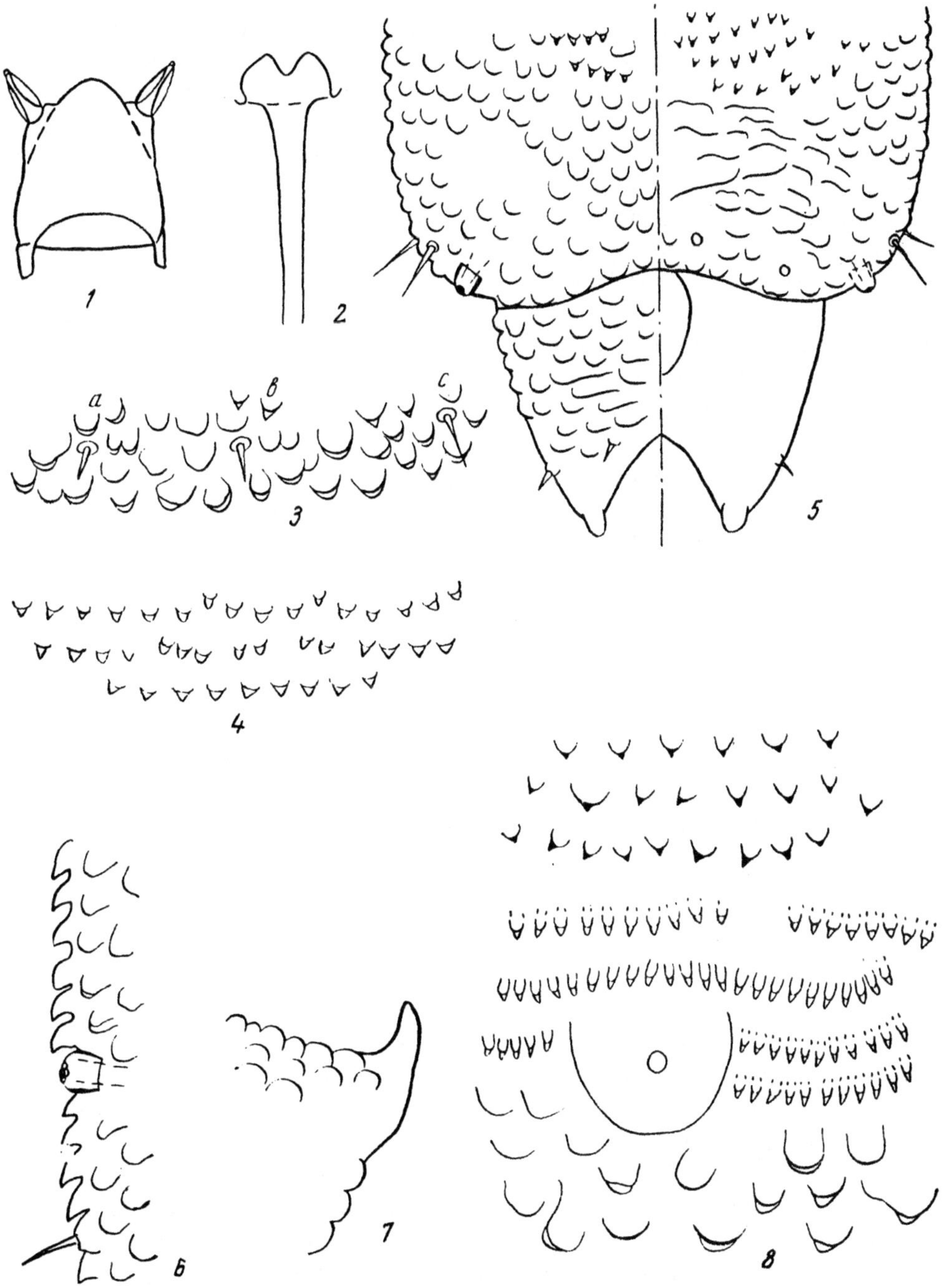


Figure 83. Larvae of *Thomasiniana oculiperda* (Rübsaamen)
1 – head; 2 – spatula; 3 – dorsal papillae on the abdominal segments; 4 – cuticular structures on the anterior half of the abdominal segments, dorsal view; 5 – last two segments, dorsal (left) and ventral (right) view; 6 – cuticular structures and lateral papilla on the abdominal segments; 7 – terminal segment, side view; 8 – rows of spinules and anterior ventral papilla on the abdominal segments (orig.)

of the papillae 'a' are shortest (no less than five times as short as the interstice between the papillae 'a'-'a'). The ventral surface of the thoracic segments is provided with: two simple sternal papillae, four groups of intercalary papillae (each group of which consists of one simple and two setiform papillae), and two simple sublateral papillae. Four simple anterior ventral papillae (fig. 83,8) (located behind the spinule fields) as well as two simple posterior ventral papillae occur on the ventral surface of the first seven abdominal segments. On the eighth abdominal segment four simple ventral papillae are found (fig. 83,5). The terminal segment (fig. 83,5) bears eight terminal papillae, two of which are large, hook-shaped, and apically sclerotized (fig. 83,7), whereas the remaining ones are simple. Each lateral surface of the segments is provided with two well-developed, setiform lateral papillae. Distinct spinule fields are found on the ventral surface of the meso- and metathoracic, and all abdominal segments. On all segments except the terminal one, the fields contain six distinct, oval bare spots, the four central ones of which are paired.

Although the small terminal segment (fig. 83,5) is noticeably narrower than the penultimate one, it is dorsally as well as ventrally clearly separated from the latter one. The terminal pair of stigmata is located in front of the posterior margin of the anal slit, and is not shifted to the dorsal side. The dorsal surface of the terminal segment is covered with comparatively large, oval platelets.

*Biology*. The larvae inhabit the cambial zone of twigs and stems.

**Thomasiniana oculiperda** (Rübsaamen)

*Morphology*. The dorsal surface of the body is densely covered with various cuticular structures (fig. 83,4). The anterior half of the segments bears numerous small, pointed verrucae, while on the posterior half larger, oval verruca-like structures are found (in the vicinity of the dorsal papillae). The verrucae on the last two segments are rounded, and lack teeth. The lateral surfaces of the segments (fig. 83,6) are densely covered with oval or slightly pointed verrucae which are most pronounced on the first seven abdominal segments. On the ventral surface the verrucae extend as far as the spinule fields. Behind the spinule fields the surface of the cuticle bears weakly outlined platelets of various shapes (fig. 83,8).

The spinule fields are large and consist of ten to fourteen rows on the thoracic segments, while sixteen to nineteen rows are found on the first seven abdominal segments. On the last two segments the fields comprise only five to eight rows. The spinules that make up the fields are not uniform: in the anterior rows the spinules are rounded and thick, whereas the ones in the posterior rows are considerably smaller and narrower. On the penultimate segment only five rows of scattered, blunt spinules are found. The anal slit is surrounded by three or four longitudinal rows of spinules.

*Biology*. Since the larvae cause the tissue in grafts of fruit trees to die off, and thus induce the loss of the scions, they bring about serious damage.

*Material*. Four larvae which had been put at the disposal of the authors by M. Skuhravá and V. Skuhravý (Entomological Institute of the Czechoslovakian Academy of Sciences, Prague, Czechoslovakia).

42. Genus **Profeltiella** Kieffer (fig. 14,5; 16,2; 84)

*Morphology*. The larvae have a small, short head, and two pointed outgrowths at the end of the body.

The small head capsule is only slightly longer than its width. The posterior edge of the head capsule has at its dorsal side an oval emargination, whereas the ventral side is straight. The short and thick exocephalic rods are no longer than half the width of the head capsule. The antennae are short. The apically rounded second antennal segment is only twice as long as its width at the base.

The teeth of the two-toothed spatula are short and wide. The weakly sclerotized manubrium has a widened base.

The dorsal and lateral surfaces bear oval, polygonal, or acicular protuberances (fig. 84,2). The papillae are setiform or simple. Six small, setiform papillae of approximately equal size are found on the dorsal surface of the prothorax and the first seven abdominal segments. On the meso- and metathorax the central pair of papillae bears no, or extremely short, bristles. The length of the papillae is equal to, or markedly less than, the width of the stigmata. The eighth abdominal segment is provided with two setiform dorsal papillae which are shorter than the terminal stigmata.

The ventral surface of the thoracic segments bears: two simple sternal papillae, four groups of intercalary papillae (each group of which consists of one simple and two setiform papillae), and two simple sublateral papillae. On the first seven abdominal segments four simple anterior ventral papillae (which are located behind the spinule fields) and two simple posterior ventral papillae occur. The eighth abdominal segment bears four simple ventral papillae. The terminal segment is provided with eight terminal papillae, two of which look like large, conical protuberances with a hook-like bent tip (fig. 16,2). The remaining six (setiform) papillae are situated near the bases of the hook-like ones. Two setiform lateral papillae are found on each lateral surface of the segments.

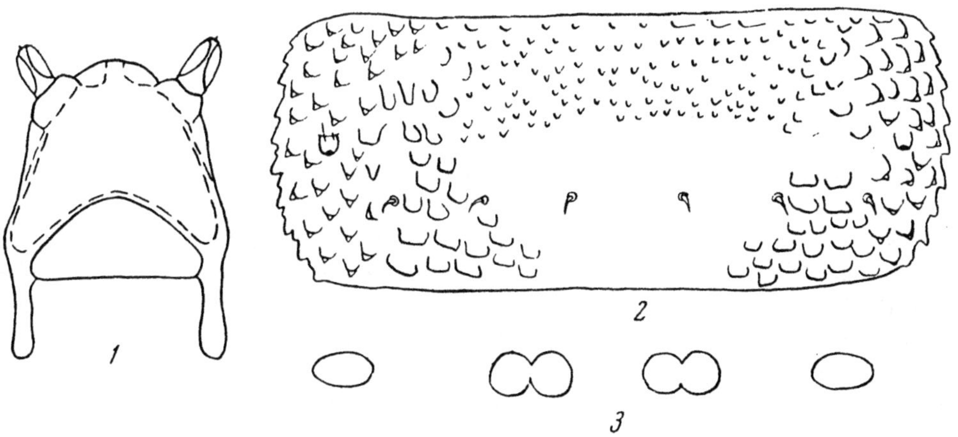


Figure 84. Larva of *Profeltiella meridionalis* Mamaev
1 – head; 2 – abdominal segments, dorsal view; 3 – outlines of the oval bare spots within the spinule fields on the ventral surface (orig.)

Spinule fields occur on the anterior third part of the ventral surface of the meso- and metathoracic, and all abdominal segments. The fields contain six oval bare spots, the four central ones of which are pairwise fused (fig. 84,3).

The last two segments (fig. 16,2) are not clearly marked off from each other, especially not on the dorsal side. The penultimate segment partly encloses the terminal one, so that the terminal pair of stigmata is located at the level of the anal slit. The dorsal and lateral surfaces of the penultimate segment are densely covered with polygonal and oval platelets. The ventral surface of the segment is bare. The dorsal surface of the small terminal segment is densely covered with oval platelets, except on the posterior margin. Six ventral papillae are found on the bare ventral surface. The anal slit is surrounded by a few rows of hair-like cuticular structures.

The stigmata look like well-developed cylindrical verrucae the length of which surpasses the width of the base. The terminal pair is found at the level of, or behind, the posterior edge of the anal slit.

*Biology*. The larvae generally inhabit the cambial zone of stems and twigs.

**Profeltiella meridionalis** Mamaev

*Morphology*. The anterior third part of the dorsal surface of the meso- and metathoracic, and the first seven abdominal segments is covered with relatively small, pointed protuberances. A few rows of similar protuberances are found on the posterior margin of the sixth and seventh abdominal segments. The posterior half of the segments is bare. The last two segments are entirely covered with protuberances. On the lateral surfaces of the segments (fig. 16,2) oval as well as pointed protuberances occur (the latter are mostly found on the anterior half of the segments).

The spinule fields (fig. 14,5) comprise: eight rows on the thoracic and the first abdominal segment, eleven to fourteen rows on the next five segments, and nine rows on the seventh abdominal segment. The spinule fields on the last two segments are small and consist of weakly sclerotized spinules.

*Biology*. The larvae develop within cracks in stubs of recently felled trees.

*Material*. Kudepsta, 21 June 1959, in a crack in a fresh stub of *Alnus*, five larvae.

*Comment*. The larvae are described for the first time.

## 43. Genus **Mycodiplosis** Rübsaamen (fig. 7,4; 11,5; 85)

*Morphology*. The flattened, orange larvae are densely covered with oval platelets.

The head capsule (fig. 7,4) is well-developed and clearly outlined. It is approximately as long as its width at the base. The capsule is considerably longer at the dorsal side than at the ventral side. The well-developed exocephalic rods are over their entire length rather broad. The posterior end is widened. The rods are noticeably longer than the width of the head capsule. Two short rods, the ends of which are fused with the exocephalic rods, originate on the dorso-posterior edge of the head capsule. The antennae are well-developed. The pointed second segment is approximately 3.5 times as long as its width at the base.

The spatula (fig. 11,5) is two-toothed. The manubrium is markedly more weakly sclerotized than the teeth and is not widened at the base.

The body is densely covered with oval platelets and setiform papillae. The dorsal surface of the thoracic and the first seven abdominal segments bears six setiform papillae (fig. 85,2). On the eighth abdominal segment just two setiform papillae occur. The ventral surface of the thoracic segments is provided with: two simple sternal papillae, two sublateral papillae, and four groups of intercalary papillae (each group of which consists of one simple and two setiform papillae). Four simple anterior ventral papillae (fig. 85,1) (located behind the spinule fields) as well as two simple posterior ventral papillae are found on the first seven abdominal segments. On the eighth abdominal segment four simple ventral papillae occur. The terminal segment (fig. 85,3) bears eight terminal papillae four of which are setiform, while the remaining ones are provided with a short, conical appendage. Two setiform lateral papillae are found on each of the lateral surfaces of the segment.

The spinules in the fields (which are located on the meso- and metathoracic, and all abdominal segments) are not uniform (fig. 85,1). The anterior three or four rows consist of large, broad, and pointed spinules which resemble the structures that are found on the lateral surfaces. The other rows are made up of elongated, narrowed spinules.

The terminal segment (fig. 85,3) ends in two wide processes, each of which bears two setiform and two rod-like papillae. The anal slit is surrounded by some longitudinal rows of spinules.

The stigmata look like distinct sclerotized verrucae which are provided with three or, on the terminal pair, four breathing openings. The terminal as well as the first pair of stigmata is shifted to the dorsal side of the segments.

Figure 85. Larva of *Mycodiplosis pucciniae* Rübsaamen
1 – rows of spinules and anterior ventral papillae on the abdominal segments; 2 – cuticular structures and dorsal papilla on the abdominal segments; 3 – last two segments, dorsal view (orig.)

*Biology.* The larvae are mycetophages. In the literature some species are recorded as predators in mite colonies. These records should, however, be verified.

**Mycodiplosis pucciniae** Rübsaamen

*Morphology.* The dorsal as well as the lateral surfaces of the body are densely covered with oval or slightly pointed verrucae. The verrucae are largest and oval on the prothorax. On the other segments the verrucae are smaller and variously shaped (fig. 85,3). Rows of small verrucae alternate with rows of larger, mostly oval (and nearer to the lateral surfaces, pointed) verrucae. The size of the protuberances increases towards the end of the body. On the ventral surface the protuberances extend as far as the spinule fields; behind the fields the surface is bare.

The spinule fields are compact (fig. 85,1). The tiny spinules form five or six rows on the mesothorax, nine rows on the metathorax, nine or ten rows on the first six abdominal segments, six or seven rows on the seventh abdominal segment, and five or six rows on the eighth abdominal segment. The terminal segment bears 15-17 rows of spinules.

*Biology.* The larvae develop in the spots of rust that are caused by *Puccinia* on the vegetative organs of various plants.

*Material.* Danki, Moskovskaya oblast, 1 June 1962, from rust spots on the leaves of *Rhamnus*, five larvae; Zavety Il'icha, Moskovskaya oblast, 10 June 1962, from rust spots on *Urtica*, five larvae; idem, 13 June 1962, from rust spots on the needles of *Pinus*, three larvae.

44. Genus **Sitodiplosis** Kieffer (fig. 86)

*Morphology.* The second antennal segment is twice as long as it is wide. The comparatively small teeth of the spatula (fig. 86,1) are separated by a wide emargination.

The dorsal surface bears large, distinctly outlined, rounded verrucae. The prothorax is provided with polygonal platelets. From the metathorax onwards the spinules have small points. All six dorsal papillae are setiform. The two dorsal papillae on the eighth abdominal segment are also setiform. The ventral surface of the thoracic segments (fig. 86,1) bears two simple sternal and sublateral papillae as well as four groups of ordinary intercalary papillae. Four simple anterior ventral papillae, which are located behind the spinule fields, and two simple posterior ventral papillae occur on the first seven abdominal segments. Four simple ventral papillae are found on the eighth abdominal segment. The terminal segment is provided with eight terminal papillae, six of which are conical (two bear a short apical bristle), whereas the two remaining ones are setiform. Both lateral surfaces of the segments bear two lateral papillae. Spinule fields are found on the ventral surface of the body.

The dorsal surface of the terminal segment is provided with some rows of verrucae (fig. 86,2) which bear small teeth. The ventral surface is bare. On the lateral surfaces of the segment rounded verrucae occur. The anal slit is surrounded by longitudinal rows of spinules. A few transverse rows are found in front of the slit. All six anal papillae are simple.

*Biology.* The larvae develop inside the ears of cereals where they cause serious damage.

Figure 86. Larva of *Sitodiplosis mosellana* (Géhin)
1 – spatula and papillae; 2 – last two segments, dorsal view (after Möhn 1955b)

**Sitodiplosis mosellana** (Géhin)
*Morphology*. The dorsal surface is provided with rounded verrucae. The prothorax bears large, polygonal platelets. From the metathorax onwards, the first three or four rows of verrucae bear teeth. The spinule fields consist of 16-18 rows of spinules. The anal slit is surrounded by six to eight longitudinal rows of spinules; in front of the slit three or four transverse rows are found.

*Biology*. Since they prevent the grains in the ears to complete their development, the larvae are a serious pest of rye, wheat, and barley.

*Comment*. The authors made use of the description by Möhn (1955b).

45. Genus **Clinodiplosis** Kieffer (fig. 87-90)

*Morphology*. The body of the larvae is densely covered with numerous, oval platelets. On both sides the terminal segment bears three processes of unequal size.

The small, weakly sclerotized head capsule (fig. 89,1; 90,1) is only slightly longer than it is wide. The posterior edge of the head capsule has a shallow, oval emargination. The exocephalic rods are apically widened and somewhat longer than the width of the capsule at its base. The antennae are short. The second antennal segment has a rounded tip and is more than three times as long as its width at the base.

The two-toothed spatula (fig. 89,2; 90,3) has a well-pronounced manubrium.

The dorsal as well as the lateral surfaces of the body are densely covered with oval platelets and protuberances (fig. 87,2). Setiform as well as simple papillae are found. The dorsal surface of the thoracic and the first seven abdominal segments bears six setiform dorsal papillae the papillae ‘b’ (fig. 1) of which are provided with extremely short bristles. On the eighth abdominal segment two setiform dorsal papillae occur. The ventral surface of the thoracic segments (fig. 90,5) bears: two simple sternal papillae, four groups of intercalary papillae (each group of which consists of one simple and two setiform papillae), and two simple sublateral papillae. On the ventral surface of the first seven abdominal segments a single row of four simple anterior ventral papillae (fig. 87,5) is found behind the spinule fields. The two simple posterior ventral papillae (fig. 90,6) are also well-pronounced. The eighth abdominal segment is provided with four simple ventral papillae. The terminal segment (fig. 88,4) bears eight terminal papillae, two of which bear long bristles, while the six remaining ones are conical and apically sclerotized. These six papillae are arranged into two rows. Four papillae are found on the tip of the segment, the two remaining ones are more proximally situated. Both lateral surfaces of the segment bear two setiform lateral papillae.

Spinule fields occur on the meso- and metathoracic, and all abdominal segments. The spinules (fig. 88,3) in the anterior part of the fields are large and conical. In the posterior part of the fields the spinules are strikingly elongated and almost hair-like. The fields contain four indistinct, small, oval bare spots.

The dorsal surface of the terminal segment (fig. 88,4) is bare, not counting one or two transverse rows of pointed protuberances. On the lateral surfaces of the segments the protuberances are more densely distributed. The ventral surface of the segment bears six simple anal papillae and is otherwise bare. The anal slit is found on the ventral side of the segment, where it is surrounded by longitudinal rows of hair-like structures.

The stigmata look like small cylindrical verrucae. The terminal pair is shifted to the dorsal side.

*Biology*. In most cases the larvae live as inquilines in the galls of gall midges or in flowers; more rarely they make galls of their own.

Key to the species of *Clinodiplosis* Kieffer

1(4) The dorsal papillae 'a' and 'c' (fig. 1) are comparatively short. The length of the papillae 'a' is no more than one fifth the distance between them.

2(3) The dorsal surface is over a considerable extension bare. The pointed protuberances are arranged into transverse rows only on the anterior halves of the segments (fig. 87,2). A part of the protuberances bear teeth, the remaining ones are rounded. The dorsal surface of the terminal segment is bare and lacks protuberances. The spinule fields on the ventral surface of the first seven abdominal segments consist of ten to fourteen rows. . . . . . . . . . . . . . . . . . . . . . . . . . *C. papaveri* (Kjellander)

3(2) The dorsal and lateral surfaces of the segments are entirely covered with clearly separated platelets and verrucae of various shapes and sizes (fig. 88,1). The body segments bear on the anterior half of the dorsal surface six or seven rows of pointed protuberances. The pointed protuberances on the dorsal surface of the terminal segment (fig. 88,4) are arranged into two rows which are divided into two halves by the midline. The spinule fields (fig. 88,3) on the ventral surface of the first seven segments consist of eight to twelve rows. . . . . . . . . . . . . *C. cilicrus* (Kieffer)

4(1) The dorsal papillae 'a' and 'c' (fig. 1) are comparatively long. The length of the papillae 'a' is at least one third the distance separating them.

5(6) Small, pointed protuberances which are not contiguous (fig. 89,4) occur on the anterior half of the dorsal surface of the first seven abdominal segments. The spinule fields on the ventral surface of the first seven abdominal segments consist of eleven to fourteen rows. A single row of tiny, pointed protuberances which is divided into two halves by the midline, is found on the dorsal surface of the terminal segment. *C. fagopyri* Kh. Mamaeva

6(5) Oval verrucae (fig. 90,2), a part of which bear small, pointed teeth occur on the anterior half of the dorsal surface of the first seven abdominal segments. The spinule fields on the ventral surface of the first seven abdominal segments consist of eighteen to twenty rows. The terminal segment bears on its dorsal surface a single, uninterrupted transverse row of pointed protuberances. . . . . . . *Clinodiplosis* sp.

**Clinodiplosis papaveris** (Kjellander)

*Morphology*. The papillae 'c' (fig. 1) are longest among the six dorsal papillae, the papillae 'a' are somewhat shorter, while the papillae 'b' bear extremely short bristles. The length of the central papillae is at most one fifth the interstice separating them. The body is over a considerable extension bare; only the rows of pointed protuberances on the anterior half of the dorsal surface, and the spinule fields on the ventral surface of the segments are well-pronounced (fig. 87,2-3). The pointed protuberances on the dorsal surface are arranged into four or five rows. The posterior half of the segments is smooth (fig. 87,5) and bears no distinct platelets and verrucae. The spinule fields on the ventral surface of the segments contain four oval bare spots. The fields on the thoracic and the first seven abdominal segments consist of eight to fourteen rows of spinules. The spinules

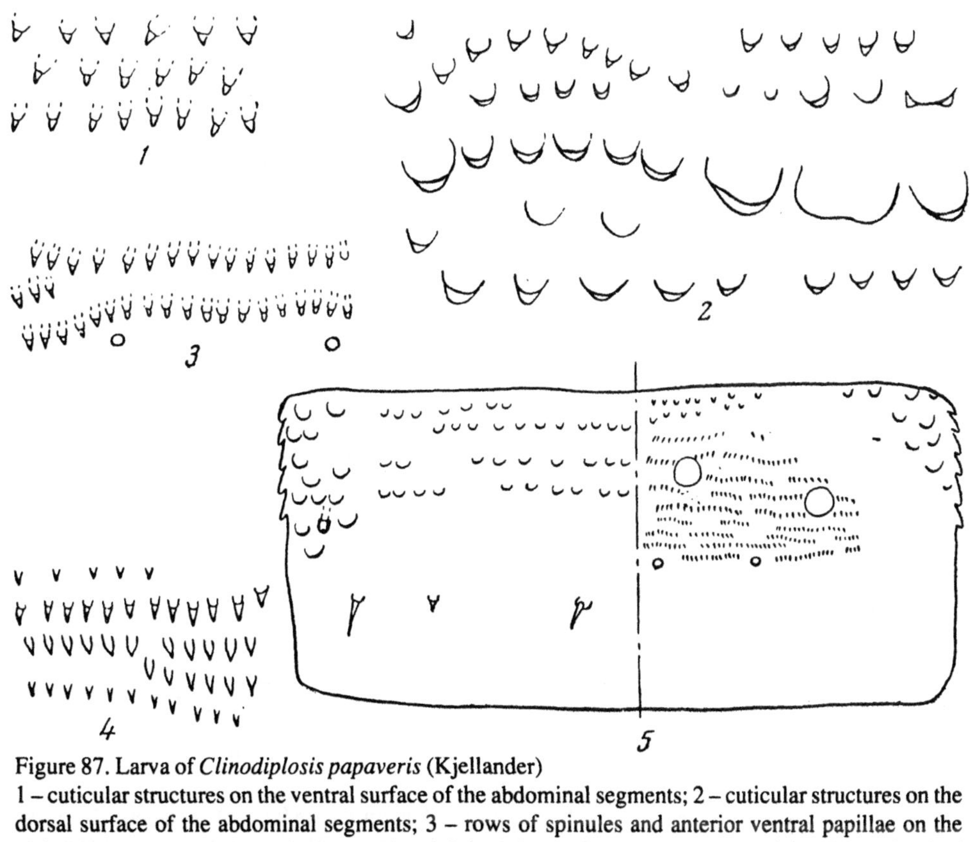


Figure 87. Larva of *Clinodiplosis papaveris* (Kjellander)
1 – cuticular structures on the ventral surface of the abdominal segments; 2 – cuticular structures on the dorsal surface of the abdominal segments; 3 – rows of spinules and anterior ventral papillae on the abdominal segments; 4 – spinules on the eighth abdominal segment, ventral view; 5 – abdominal segment, dorsal (left) and ventral (right) view (orig.)

in the first five or six rows are large and thick (fig. 87,1), while the spinules in the other ones are long and thinner (fig. 87,3).

On the eighth abdominal segment occur just four or five rows of weakly sclerotized, pointed spinules (fig. 87,4). The spinules on the terminal segment are similar to the ones on the first seven abdominal segments. The lateral surfaces of the segments are covered with large, mostly oval, platelets. Only on the anterior half of the first six abdominal segments the platelets are provided with small teeth. The dorsal surface of the terminal segment is bare.

*Biology.* The larvae damage, and cause the rotting of, the bolls of *Papaver*.

*Material.* A series of larvae that had been put to the disposal of the authors by M. Skuhravá and V. Skuhravý (Czechoslovakia, Prague, Entomological Institute of the Czechoslovakian Academy of Sciences).

*Comment.* The species was initially described as a species of *Carpodiplosis* Kjellander. Afterwards, however, the genus *Carpodiplosis* was shown to be synonymous with the genus *Clinodiplosis* Kieffer (Skuhravá and Nijveldt 1963).

**Clinodiplosis cilicrus** (Kieffer)

*Morphology.* Among the six dorsal papillae, the papillae 'c' (fig. 1) are longest, the papillae 'a' are somewhat shorter, while the papillae 'b' are reduced and bear extremely short bristles (fig. 88,2). The interstice separating the papillae 'a' is no less than five times

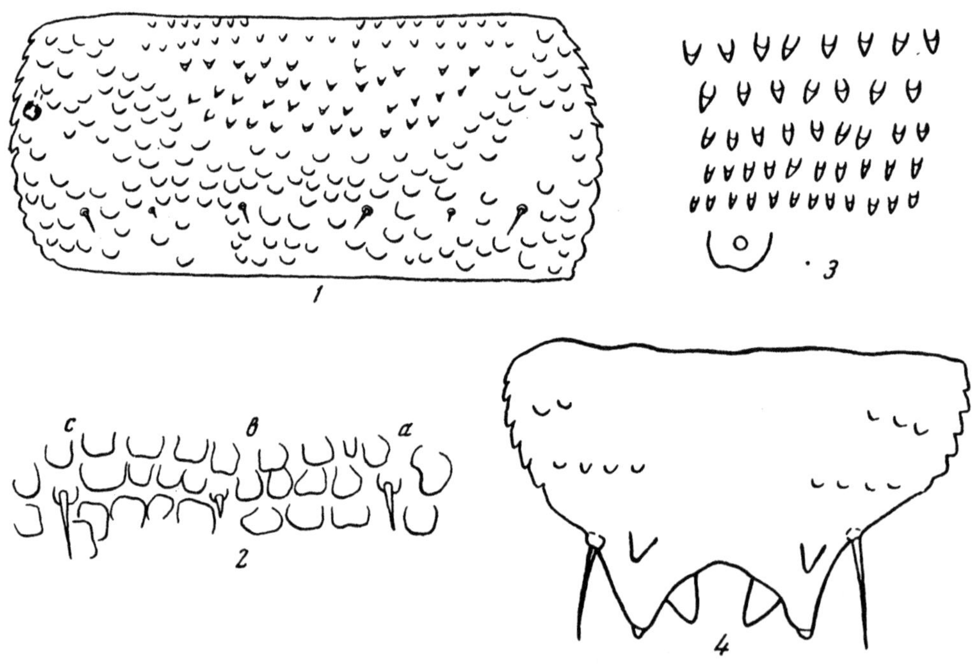


Figure 88. Larva of *Clinodiplosis cilicrus* (Kieffer)
1 – abdominal segment, dorsal view; 2 – dorsal papillae on the abdominal segments; 3 – rows of spinules and anterior ventral papilla on the abdominal segments; 4 – terminal segment, dorsal view (orig.)

as large as the length of the papillae themselves. The dorsal as well as the lateral surfaces of the body are densely covered with pointed protuberances, oval verrucae and platelets (fig. 88,1). The anterior half of the dorsal surface of the segments is provided with six or seven rows of small, pointed protuberances. The posterior half of the segments bears flat platelets of irregular shape.

The spinule fields on the ventral surface of the meso- and metathoracic, and the first seven abdominal segments consist of eight to twelve rows of spinules of various shapes (fig. 88,3). The fields contain four oval bare spots. On the eighth abdominal segment occur just four or five rows, the spinules of which are similar to those found on the first seven abdominal segments. In the anterior half of the fields the spinules are wide and blunt (fig. 88,3); in the posterior half they are long and pointed. The lateral surfaces of the segments are densely covered with oval, pointed platelets of various sizes. The platelets on the anterior half of the segments are mostly pointed, whereas the ones on the posterior half are rounded. The midline divides the two rows of pointed protuberances which are found on the dorsal surface of the terminal segment (fig. 88,4) into two halves.

*Biology*. The larvae develop in the inflorescences of Compositae and Leguminosae.

*Material*. Pavlova Sloboda, Moskovskaya oblast, 8 July 1962, from the flower heads of *Crepis pilosa*, three larvae.

**Clinodiplosis fagopyri** Kh. Mamaeva

*Morphology*. Among the six dorsal papillae the papillae 'c' (fig. 1) are longest (as long as

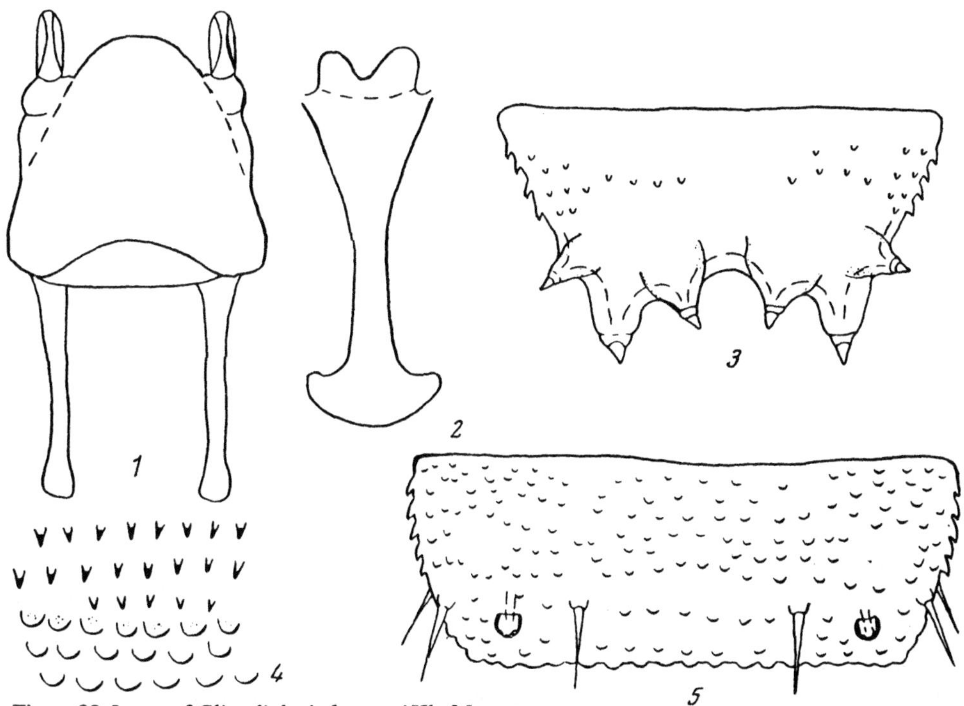


Figure 89. Larva of *Clinodiplosis fagopyri* Kh. Mamaeva
1 – head; 2 – spatula; 3 – terminal segment, dorsal view; 4 – cuticular structures on the abdominal segments, dorsal view; 5 – eighth abdominal segment, dorsal view (orig.)

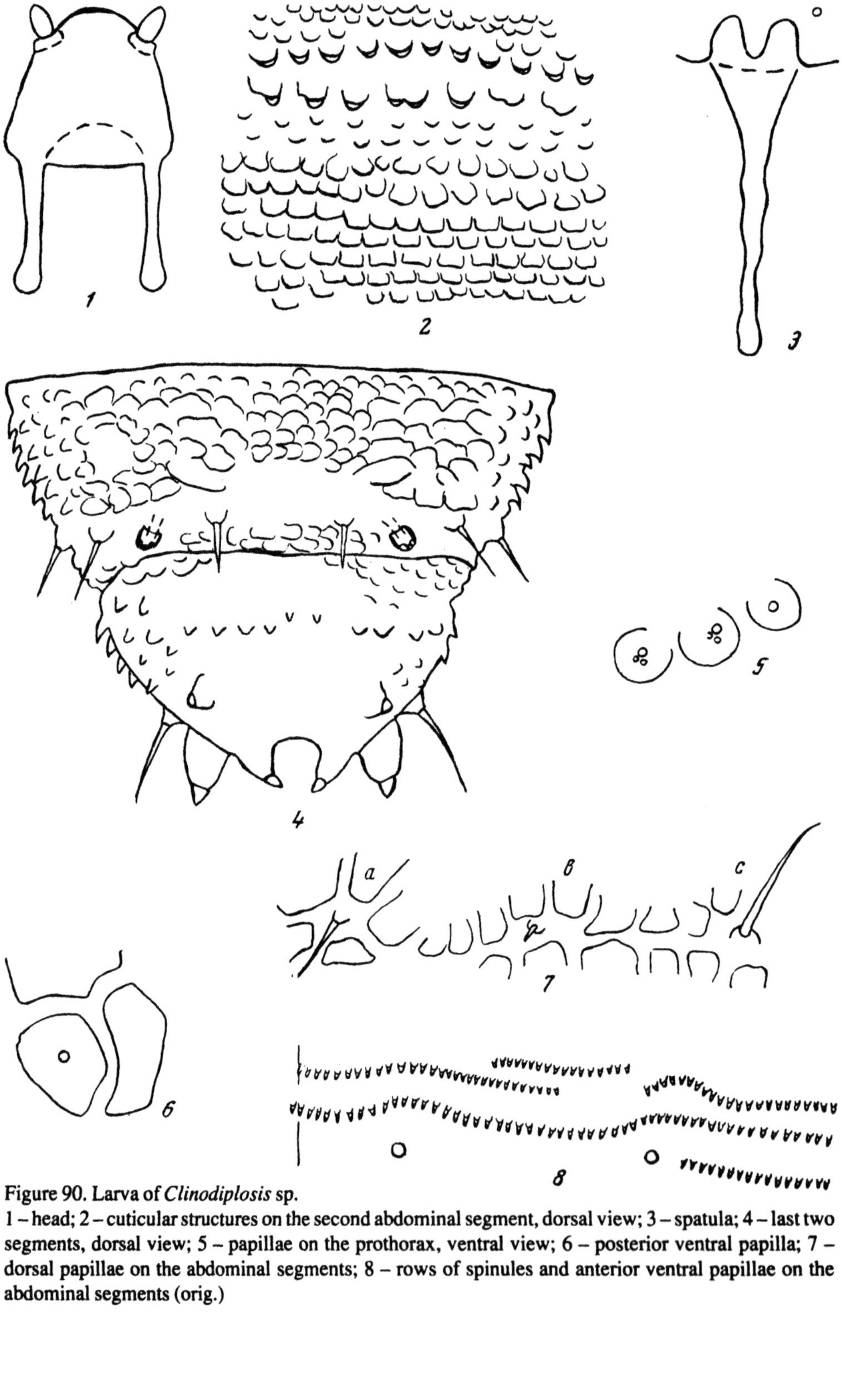


Figure 90. Larva of *Clinodiplosis* sp.
1 – head; 2 – cuticular structures on the second abdominal segment, dorsal view; 3 – spatula; 4 – last two segments, dorsal view; 5 – papillae on the prothorax, ventral view; 6 – posterior ventral papilla; 7 – dorsal papillae on the abdominal segments; 8 – rows of spinules and anterior ventral papillae on the abdominal segments (orig.)

or slightly shorter than the interstice 'b'-'c'). The papillae 'a' are somewhat shorter, while the papillae 'b' are reduced and bear extremely short bristles. The papillae 'a' are no longer than one third the distance separating them. The dorsal and lateral surfaces of the body are entirely covered with comparatively small protuberances (fig. 89,5) which may be rounded as well as pointed. On the anterior half of the dorsal surface of the segments the protuberances resemble spinules, whereas they resemble oval platelets (fig. 89,4) on the posterior half of the segments.

On the meso- and metathoracic, and the first seven abdominal segments, the spinule fields comprise eleven to fourteen rows, while only five rows are found on the eighth abdominal segment. On the terminal segment the number of rows is generally somewhat raised. The fields, which contain four oval bare spots, consist of elongated, pointed spinules. The spinules on the anterior and the posterior parts of the fields are only slightly different. The terminal segment bears just a single row of pointed protuberances. This row is divided into two halves by the midline.

*Biology*. The larvae develop in the flowers of *Polygonum*.

*Material*. Pavlova Sloboda, Moskovskaya oblast, 17 August 1962, five larvae.

*Comment*. The diagnosis of the larvae is given for the first time.

**Clinodiplosis** sp.

*Morphology*. The papillae 'c' (fig. 1) are longest, the papillae 'a' are somewhat shorter. The papillae 'b' are strongly reduced and bear an extremely short bristle (fig. 90,7). On the thoracic and the first eight abdominal segments the papillae 'a' are 2-2.5 times as long as the interstice between them. The papillae 'c' are as long as, or longer than, the distance 'b'-'c'. The dorsal and lateral surfaces of the body are densely covered with oval and polygonal platelets as well as with oval verrucae (fig. 90,2). On the anterior half of the dorsal surface of the segments oval verrucae occur, a part of which is provided with small, pointed teeth. On the lateral surfaces of the segments pointed and oval verrucae are found standing close together.

On the first seven abdominal segments, the very large spinule fields (fig. 90,8) contain four oval bare spots. The spots are distinct and well-separated. The composition of the spinule fields is as follows: fourteen or fifteen rows on the meso- and metathorax, eighteen to twenty rows on the first seven abdominal segments, only seven rows on the eighth, and eight or nine rows on the terminal (ninth) abdominal segment. The dorsal surface of the terminal segment (fig. 90,4) bears a distinct transverse row of pointed protuberances.

*Biology*. The larvae are found in swollen, shrivelled and decaying rose-hips.

*Material*. Tellermanov forestry, Voronezhskaya oblast, 26 September 1960, three larvae.

46. Genus **Ametrodiplosis** (Rübsaamen) (fig. 91)

*Morphology*. The body of the larvae is densely covered with numerous platelets of various shapes. The terminal segment is provided with six processes. Four of these processes have equal sizes and are found at the tip, whereas the two remaining ones are somewhat shifted to the anterior margin of the segment.

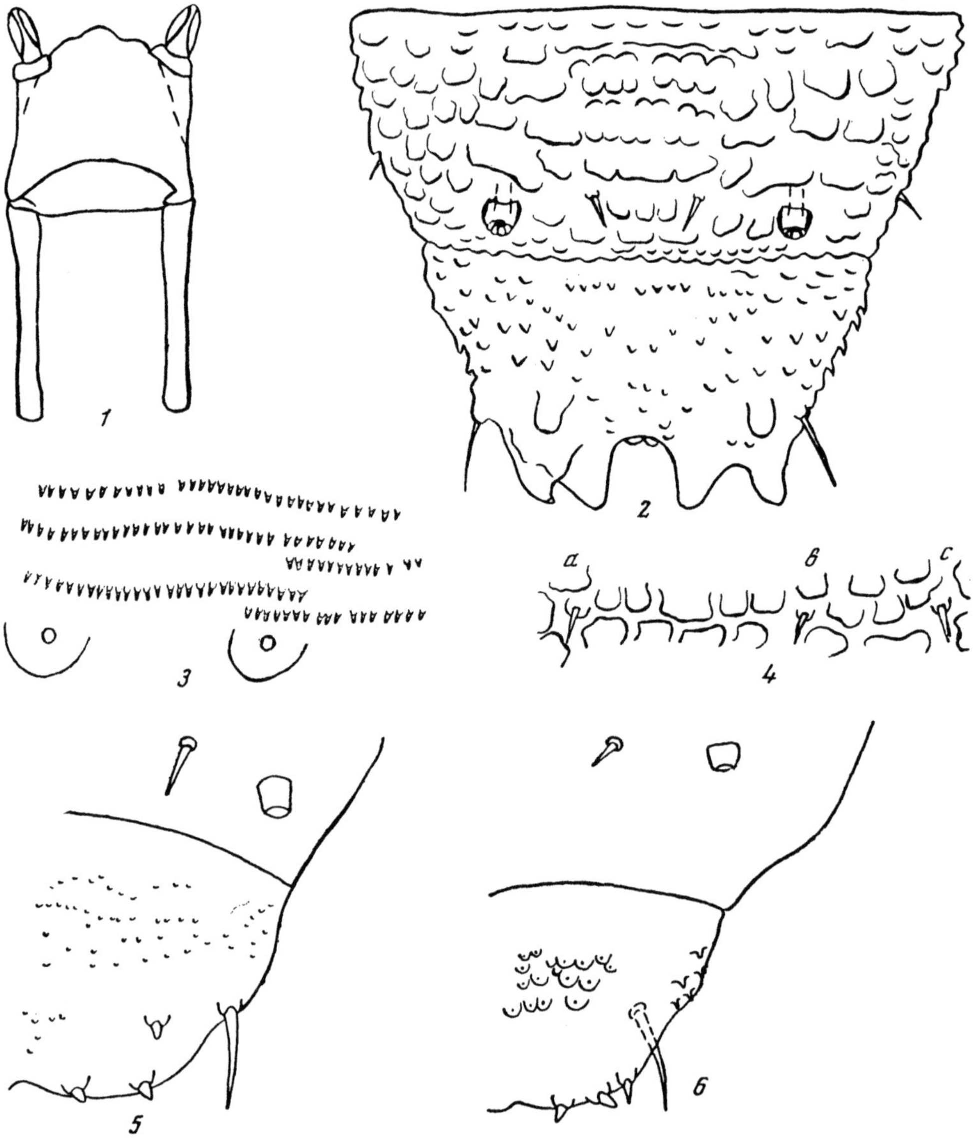


Figure 91. Larvae of the genus *Ametrodiplosis* Rübsaamen
*A. medialis* Mamaev: 1 – head; 2 – last two segments, dorsal view; 3 – rows of spinules and anterior ventral papillae on the abdominal segments; 4 – dorsal papillae on the abdominal segments; 5 – terminal segment, dorsal view; *A. auripes* (F. Loew): 6 – terminal segment, dorsal view (1-4 – orig.; 5-6 – after Möhn 1955b)

The weakly sclerotized head capsule (fig. 91,1) is relatively short. It is no longer than the width of its base. The posterior edge of the capsule shows a shallow, oval emargination at the dorsal side, whereas its ventral side is almost straight. The well-developed, thick exocephalic rods have a slightly widened tip. They are 1.5 times as long as the width of their base. The thick, short antennae are apically rounded. The length of the second antennal segment surpasses its width at the base.

The head of the two-toothed spatula is noticeably widened as compared with the manubrium. The base of the manubrium is weakly sclerotized.

The dorsal surface of the body, including the prothorax, is densely covered with oval and polygonal platelets (fig. 91,2). On the lateral surfaces a part of the platelets passes into weakly sclerotized protuberances. Simple as well as setiform papillae occur. Six setiform dorsal papillae are found on the dorsal surface of the thoracic and the first seven abdominal segments. The papillae 'c' (fig. 1) are somewhat longer than the remaining ones. On the eighth abdominal segment just two dorsal papillae occur. The ventral surface of the thoracic segments bears: two simple sternal papillae, four groups of intercalary papillae (each group of which consists of one simple papilla and two papillae with a short bristle), and two simple sublateral papillae. On the ventral surface of the first seven abdominal segments four simple anterior ventral papillae (located behind the spinule fields) (fig. 91,3) as well as two simple posterior ventral papillae are found. Four simple ventral papillae occur on the eighth abdominal segment. The terminal segment bears eight terminal papillae, the two outer ones of which are provided with long bristles, whereas the remaining ones look like conical verrucae. The eight terminal papillae are all of approximately the same size. The lateral surfaces each bear one simple and two setiform papillae.

The spinule fields are distributed on the ventral surface of the meso- and metathoracic, and all abdominal segments. The fields contain four bare spots. The spinules in the fields are not uniform. The first four or five rows consist of large, thick spinules, whereas the remaining rows are formed by markedly smaller, pointed spinules.

The well-developed terminal segment (fig. 91,2) is somewhat shorter than the other segments. Its dorsal surface is covered with small verrucae which are arranged into distinct rows. The ventral surface of the terminal segment is over a considerable extension bare; only on its anterior half rows of small spinules are found. The elongated anal slit is surrounded by longitudinal rows of small spinules. All six anal papillae are simple and lack bristles.

The well-developed stigmata are comparatively large. The terminal pair is shifted to the dorsal side of the segment.

Key to the species of *Ametrodiplosis* Rübsaamen

1(2) The dorsal papillae on the abdominal segments are short; the papillae 'a' and 'b' (fig. 1) are no longer than the second antennal segment. The dorsal and lateral surfaces of the terminal segment bear sparse verrucae. On the anterior half of the segment the verrucae are somewhat pointed, on the posterior half they are blunter (fig. 91,2). . . . . . . . . . . . . . . . . . . . . . . . . . . . . . . *A. medialis* Mamaev

2(1) The dorsal papillae on the abdominal segments are comparatively long. The papillae 'a' and 'b' are twice as long as the second antennal segment. Though the dorsal surface of the terminal segment is over a considerable extension bare, it bears clearly pronounced protuberances or verrucae on its central part.

3(4) The dorsal papillae 'c' are on the abdominal segments somewhat longer than the other papillae. The dorsal surface of the terminal segment bears pointed protuberances on its central part. . . . . . . . . . . . . . . . . . *A. thalictricola* (Rübsaamen)

4(3) All dorsal papillae on the abdominal segment have the same size. The dorsal surface of the terminal segment bears rounded protuberances on its central part.
*A. auripes* (F. Loew)

**Ametrodiplosis medialis** Mamaev
*Morphology*. The dorsal surface of the body, except the terminal segment, is densely covered with oval and polygonal platelets, which are not pointed (fig. 91,4). On the lateral surfaces a part of the platelets is replaced by weakly pointed protuberances. The platelets on the prothorax are very large and polygonal. The small dorsal papillae (fig. 91,4) bear short bristles. The papillae 'c' (fig. 1) are slightly longer than the other ones. The papillae 'a' and 'b' are of the same size. They are approximately as long as the second antennal segment and as one fifth the interstice 'a'-'a'. The bristles of the dorsal papillae on the eighth abdominal segment are only slightly longer than the stigmata of the terminal pair.

The spinule fields (fig. 91,3), which are distributed on the meso- and metathoracic, and all abdominal segments, are not uniform. The first two or three rows consist of broad, conical spinules, whereas the spinules in the remaining rows are thicker. The number of rows in the fields is distributed as follows: mesothorax – 7, metathorax – 12, first abdominal segment – 13, middle abdominal segments – 15 or 16, seventh abdominal segment – 13, penultimate segment – 8, terminal segment – 6 rows. On the dorsal surface of the terminal segment sparsely distributed rows of verrucae are found. The verrucae are pointed on the anterior and lateral parts of the segment and relatively blunt on the posterior part.

*Biology*. The larvae live as inquilines in the galls of *Dasyneura crataegi* (Winnertz).

*Material*. Kudepsta, Krasnodarskiy kray, 25 June 1961, three larvae.

*Comment*. The diagnosis of this species is given for the first time.

**Ametrodiplosis thalictricola** (Rübsaamen)
*Morphology*. The dorsal papillae 'c' are somewhat longer than the other ones. The papillae 'a' and 'b' are approximately equally long (almost twice as long as the second antennal segment). The dorsal papillae on the eighth abdominal segment are noticeably longer than the stigmata of the terminal pair (fig. 91,5). The spinule fields consist of 20-22 rows. The dorsal surface of the terminal segment is over a considerable extension bare; only on its central part six to nine rows of spinules are found.

*Biology*. The larvae deform the fruits of *Thalictrum flavum*.

*Comment*. The authors made use of the description by Möhn (1955b).

**Ametrodiplosis auripes** (F. Loew)
*Morphology*. The dorsal papillae are all of the same size. Their length surpasses that of the second antennal segment by one third. The dorsal papillae on the eighth abdominal segment are long (noticeably longer than the length of the terminal stigmata. (fig. 91,6)). The spinule fields consist of twelve to fourteen rows. The dorsal surface of the terminal segment is over a considerable extension bare (fig. 91,6). Its central part is provided with closely adjoining rounded platelets which have a weakly pronounced acumen.

*Biology*. The larvae develop in subterreanean bud galls on *Galium*.

*Comment*. The authors made use of the description by Möhn (1955b).

47. Genus **Karschomyia** Felt (fig. 92)

*Morphology.* The larvae have a very small head and a slightly dorso-ventrally flattened body. The terminal segment is provided with eight wide, blunt outgrowths.

The head capsule (fig. 92,1) is quite small and weakly sclerotized. Its length slightly surpasses the width of its base. Though the dorsal side of the posterior edge of the head capsule shows a shallow emargination, its ventral side is straight. The long exocephalic rods are approximately 1.5 times as long as the width of the head capsule at the base. The second antennal segment is apically rounded and twice as long as the width of its base.

The teeth of the two-toothed spatula (fig. 92,2) are rounded. The comparatively long and narrow manubrium is weakly sclerotized. Its base is indistinct.

The dorsal and lateral body surfaces are partially covered with blunt as well as pointed spinules, and are partially bare (fig. 92,5-6). The papillae may be simple or may bear a short bristle. Six short setiform dorsal papillae (fig. 92,3) are found on the dorsal surface of the thoracic and the first seven abdominal segments. On the eighth abdominal segment just two setiform dorsal papillae occur. The ventral surface of the thoracic segments bears: two simple sternal papillae, four groups of intercalary papillae (each group of which consists of one simple and two setiform papillae), and two simple sublateral papillae. The ventral surface of the first seven abdominal segments is provided with four simple anterior ventral papillae (the two central ones of which are found behind the spinule fields (fig. 92,4)), and two simple posterior ventral papillae. All four ventral papillae on the eighth abdominal segment are simple. The lateral surfaces of the body segments each bear two setiform papillae.

The ventral body surface is bare, not counting the spinule fields (fig. 92,4), which are found on the anterior part of the meso- and metathoracic, and all abdominal segments. The spinule fields on the thoracic and the last two segments are compact. The fields on the first seven abdominal segments contain six bare spots the marginal ones of which are rounded and clearly outlined. The central bare spots are paired and may be very small.

The dorsal surface of the terminal segment (fig. 92,7) bears distinctly pronounced oval platelets. On the lateral surfaces the platelets pass into pointed protuberances. The ventral surface of the segment is bare. The posterior margin of the segment is provided with two wide, blunt outgrowths, each of which bears four terminal papillae. These papillae look like oval verrucae with a rod-shaped appendage at the tip (fig. 92,8). On the ventral surface of the segment six simple anal papillae are found. The anal slit is located on the ventral side of the segment, and is surrounded by numerous transverse rows of hair-like structures.

The stigmata are well-developed. The terminal pair is shifted to the dorsal side of the segment.

*Biology.* The larvae develop in decaying vegetable remains.

*Comment.* The diagnosis of the genus is given for the first time.

**Karschomyia aceris** Mamaev

*Morphology.* The short antennae are apically rounded. The second antennal segment is twice as long as its width (fig. 92,1). The dorsal surface of the pro- and mesothorax is bare. The anterior part of the metathorax and the first six abdominal segments bears rows of spinules which occupy three areas (fig. 92,5): a central one and two lateral ones. The

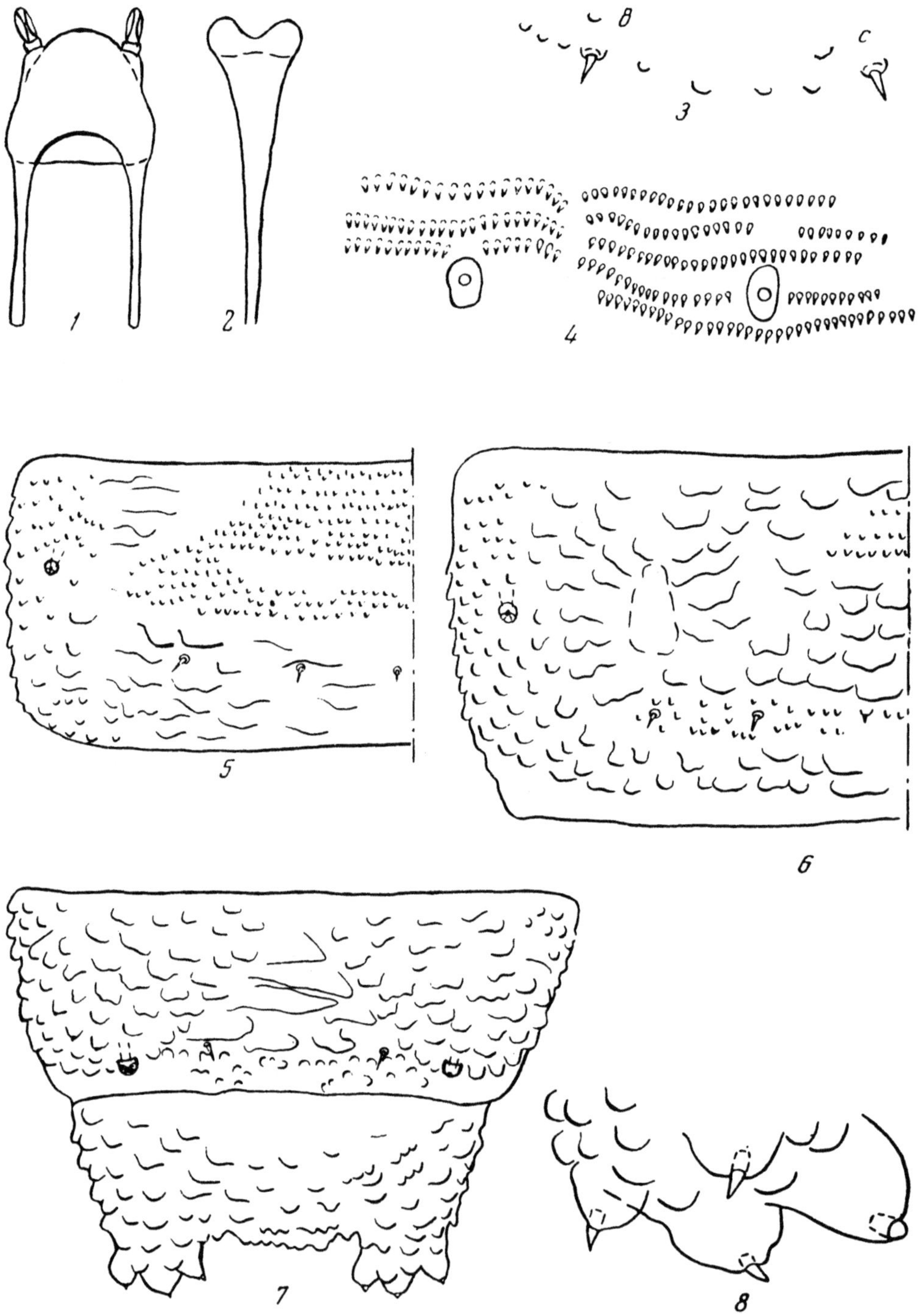

Figure 92. Larva of *Karschomyia aceris* Mamaev
1 – head; 2 – spatula; 3 – dorsal papillae on the abdominal segments; 4 – rows of spinules and anterior ventral papillae on the abdominal segments; 5 – fifth abdominal segment, dorsal view; 6 – seventh abdominal segment, dorsal view; 7 – last two segments, dorsal view; 8 – terminal papillae (orig.)

lateral spinule fields on the sixth abdominal segment are very small. The seventh abdominal segment bears on the anterior part of its dorsal surface a small number of spinules which form just two or three rows (fig. 92,6). The lateral surfaces of the thoracic segments are covered with flat, polygonal platelets of irregular shape. The lateral surfaces of the abdominal segments bear pointed protuberances which pass into flat, polygonal platelets on the dorsal and ventral surfaces.

The spinule fields are distributed on the ventral surface of the body and are comparatively large, in particular on the thoracic segments, where up to 30 rows of spinules can be counted. The composition of the spinule fields is as follows: first two abdominal segments – 18 or 19 rows, next five segments – 23-28 rows, last two segments – only 11 and 6 rows respectively.

*Biology.* The larvae are found on films of mycelium below strongly decaying *Acer* bark that is coming off.

*Material.* Tellermanov forestry, Voronezhskaya oblast, 3 October 1959, five larvae.

*Comment.* The diagnosis of the species is given for the first time.

### 48. Genus **Putoniella** Kieffer (fig. 10; 93)

*Morphology.* The larvae are provided with characteristic, wide and blunt processes along the posterior margin of the terminal segment.

The relatively small head capsule (fig. 93,1) is slightly shorter than its width at the base. The well-developed, long exocephalic rods are considerably longer than the width of the head capsule at its base. The antennae are well-developed. The second antennal segment is no less than 2.5 times as long as its width at the base.

The spatula (fig. 93,2) has two wide, spreading teeth which are separated by a comparatively narrow, rounded emargination. The tips of the teeth are pointed. The manubrium is wide. Its width is no less than the width of the teeth at the base. The base of the manubrium is weakly sclerotized and not widened.

Though the otherwise smooth body is partially covered with spinules it is over a considerable extension bare. Simple as well as setiform papillae can be found. The thoracic and the first seven abdominal segments bear six small, setiform dorsal papillae (fig. 93,3). The eighth abdominal segment is provided with just two dorsal papillae, the bristles of which are somewhat shorter than the terminal stigmata. The ventral surface of the thoracic segments bears: two simple sternal papillae, four groups of intercalary papillae (each group of which consists of one simple and two setiform papillae), and two simple sublateral papillae. On the first seven abdominal segments four simple anterior ventral papillae, which are located behind the spinule fields (fig. 93,4), as well as two simple posterior ventral papillae are found. On the ventral surface of the eighth abdominal segment four simple ventral papillae occur. The terminal segment (fig. 93,6) is provided with eight terminal papillae six of which are conical, whereas the two remaining ones bear well-developed bristles. Two setiform lateral papillae occur on each of the lateral surfaces of the segments.

The compact spinule fields are distributed on the ventral surface of the meso- and metathoracic, and all abdominal segments.

The dorsal and lateral surfaces of the terminal segment (fig. 93,6) bear pointed spinules which stand in a checkerboard pattern. The ventral surface of the segment is over a considerable extension bare. The anal slit is surrounded by longitudinal rows of spinules. The six anal papillae are simple.

The large stigmata have the appearance of cylindrical verrucae. The terminal pair is shifted to the dorsal side of the eighth abdominal segment.

*Biology*. The larvae develop within galls.

**Putoniella marsupialis** (F. Loew)

*Morphology*. On the dorsal surface of the body, pointed protuberances are only found on the anterior part of the segments. The remaining part of the body is bare. The protuberances are comparatively small, most of them are arranged into a checkerboard pattern (fig. 93,5), in some cases two or three protuberances are connivent. The lateral surfaces of the thoracic segments are mainly bare, whereas the lateral surfaces of the abdominal

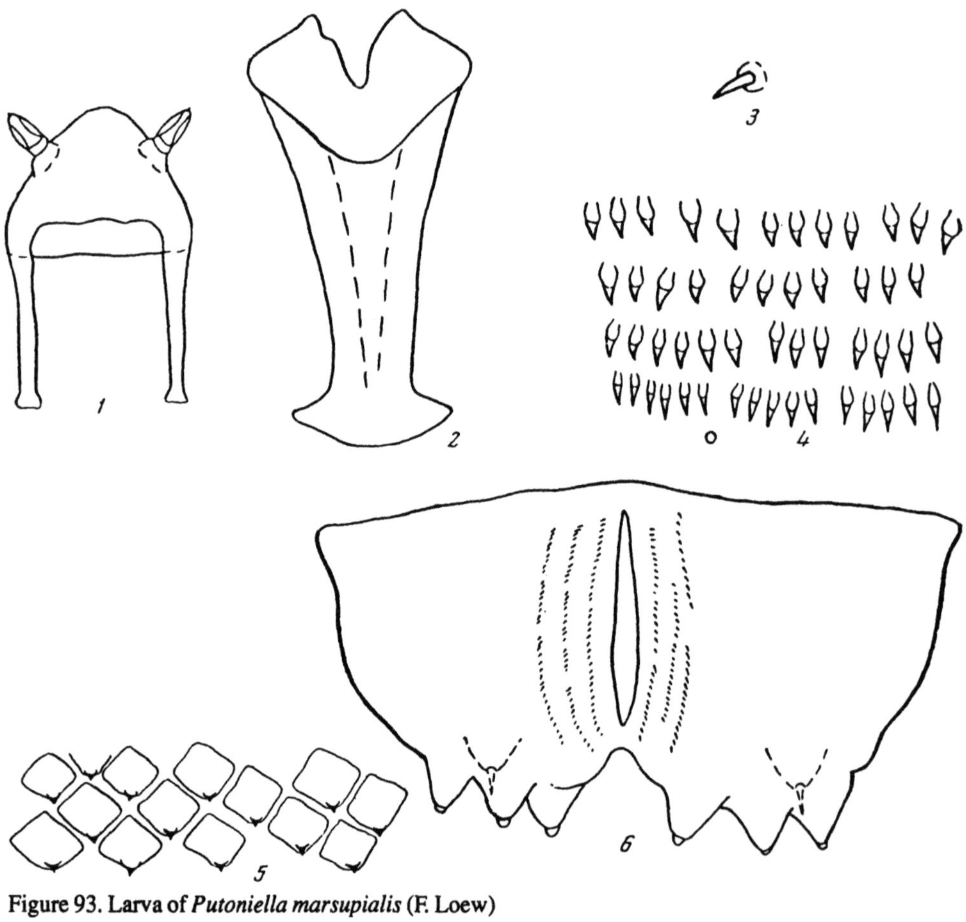


Figure 93. Larva of *Putoniella marsupialis* (F. Loew)
1 – head; 2 – spatula; 3 – dorsal papilla; 4 – rows of spinules and anterior ventral papilla on the abdominal segments; 5 – cuticular structures on the anterior half of the abdominal segments, dorsal view; 6 – terminal segment, ventral view (orig.)

segments are over a noticeable extension (up to two thirds of the surface) covered with spinules. The dorsal surface of the terminal segment is completely covered with protuberances.

The large spinule fields (fig. 93,4) on the ventral surface are concentrated on the anterior half of the segments and consist of 20-24 rows on most segments. Up to 13 rows of spinules occur on the last two segments. The spinules are not uniform. The wide spinules in the anterior part of the fields resemble the protuberances on the dorsal surface of the segments, while the spinules in the posterior part are longer and more pointed. The dorsal papillae on the thoracic segments have very short bristles. The papillae on the abdominal segments are larger and bear bristles which are no less than twice as long as the diameter of the verrucae at the base of the papillae. The anal slit is surrounded by twelve to fourteen longitudinal rows of spinules (fig. 93,6).

*Biology.* The larvae damage plum trees and blackthorn shrubs causing thickened folds on the leaves.

*Material.* Gorodishche, Luganskaya oblast, 15 June 1952, on blackthorn, five larvae.

## 49. Genus **Macrodiplosis** Kieffer (fig. 94)

*Morphology.* The heavily sclerotized head capsule (fig. 94,1) is approximately as long as it is wide. The dorsal side of the posterior margin of the head capsule is slightly emarginate, whereas the ventral side is straight. The well-pronounced exocephalic rods are widened at the tip and somewhat longer than the width of the head capsule at its base. The antennae are comparatively short. The second antennal segment is apically rounded and no more than 2.5 times as long as the width of its base.

The teeth of the two-toothed spatula (fig. 94,2) are wide and pointed. The width of the sturdy manubrium is about one third the width of the head of the spatula. The manubrium has a widened base.

The body is over a considerable extension bare. Simple as well as setiform papillae can be found. Six setiform papillae (fig. 94,3), the sizes of which increase from the central pair towards the marginal pair, occur on the thoracic and the first seven abdominal segments. The length of the bristles on the papillae does not surpasss the diameter of the stigmata. On the eighth abdominal segment just two well-developed setiform papillae are found. The ventral surface of the thoracic segments (fig. 94,7) bears: two simple sternal papillae, four groups of intercalary papillae (each group of which consists of one simple and two short setiform papillae), and two simple sublateral papillae. The ventral surface of the first seven abdominal segments is provided with four simple anterior ventral papillae (located within the spinule fields) (fig. 94,4), and two simple posterior ventral papillae. On the eighth abdominal segment four simple ventral papillae (fig. 94,5) occur. The terminal segment bears eight terminal papillae (fig. 94,8), four of which are setiform and four are conical. The lateral surfaces of the segments each bear two setiform lateral papillae.

The spinule fields are distributed on the ventral surface of the meso- and metathoracic, and the first eight abdominal segments. The fields on the thoracic segments are compact, while the fields on the abdominal segments contain four oval bare spots. The spinules in the fields are not uniform: in the anterior part of the fields they are shorter and thicker than in the posterior part.

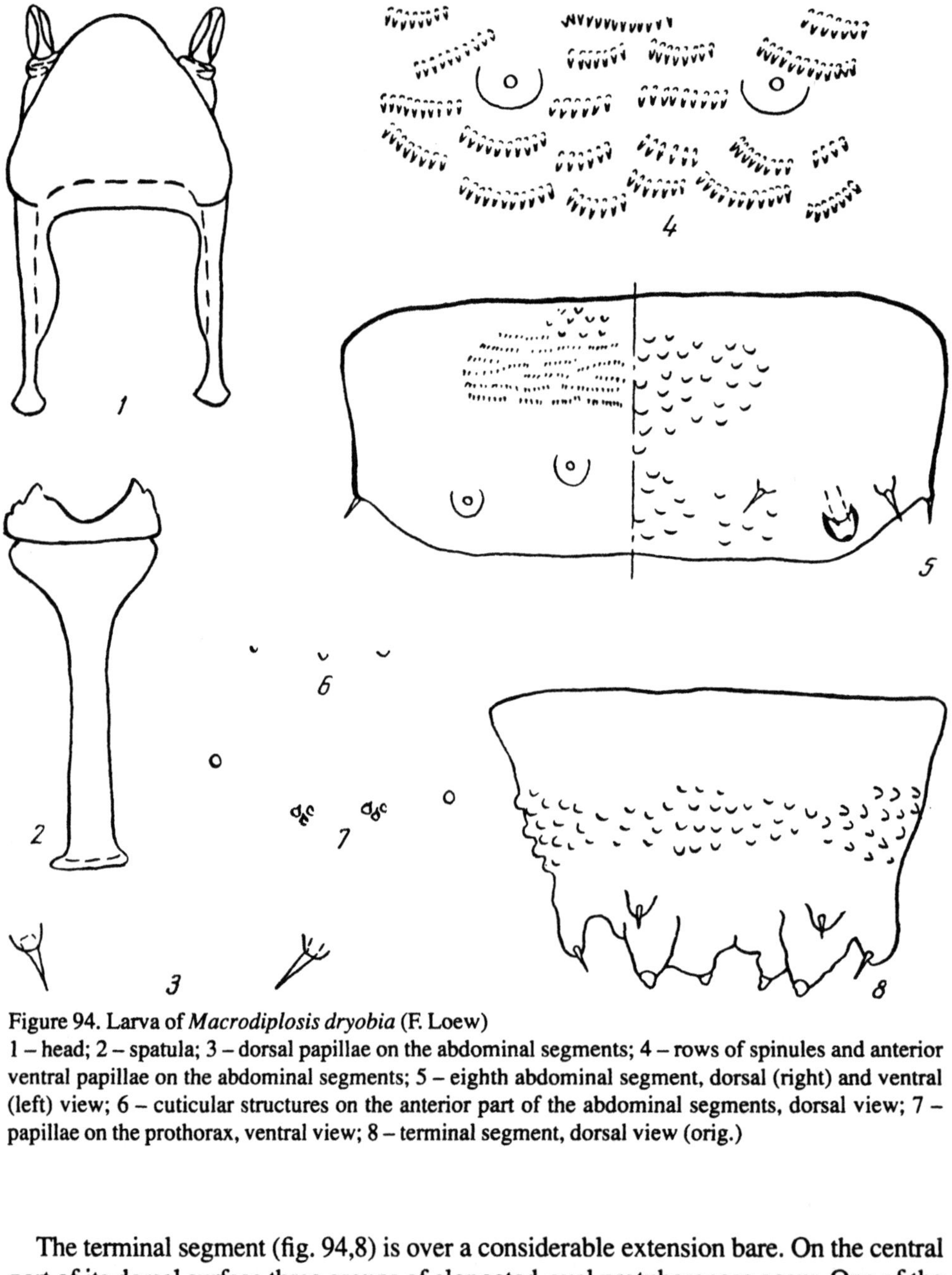


Figure 94. Larva of *Macrodiplosis dryobia* (F. Loew)
1 – head; 2 – spatula; 3 – dorsal papillae on the abdominal segments; 4 – rows of spinules and anterior ventral papillae on the abdominal segments; 5 – eighth abdominal segment, dorsal (right) and ventral (left) view; 6 – cuticular structures on the anterior part of the abdominal segments, dorsal view; 7 – papillae on the prothorax, ventral view; 8 – terminal segment, dorsal view (orig.)

The terminal segment (fig. 94,8) is over a considerable extension bare. On the central part of its dorsal surface three groups of elongated-oval protuberances occur. One of the groups holds a central position, whereas the two remaining groups are shifted to the lateral surfaces of the segment. The ventral surface of the segment is bare. All six anal papillae are simple. The anal slit is surrounded by longitudinal rows of weakly sclerotized spinules.

The well-developed stigmata have the appearance of short, cylindrical verrucae. The terminal pair of stigmata is shifted to the dorsal surface of the eighth abdominal segment.

*Biology*. The species belonging to this genus develop within galls on the leaves of *Quercus*.

**Macrodiplosis dryobia** (F. Loew)

*Morphology*. Two symmetrical rows of extremely small spinules (fig. 94,6) are found along the anterior margin of the dorsal side of the first seven abdominal segments. On the central part of the dorsal surface of the last two segments more distinctly outlined small protuberances (fig. 94,8) are found. Similar protuberances may be found in the vicinity of the dorsal papillae on the seventh abdominal segment.

The composition of the spinule fields is as follows: on the first two abdominal segments – 14 and 18 rows respectively, on the next four segments – 20 rows, on the seventh and eighth segment – 17 and 11 rows respectively. The terminal segment lacks spinule fields. The anterior ventral papillae are located among the last four or five rows of spinules. Around the anal slit six or seven longitudinal rows of spinules occur.

*Biology*. The larvae cause a recurvation of the lobes of oak leaves and develop in the resulting folds.

*Material*. Pavlova Sloboda, Moskovskaya oblast, 26 June 1962, five larvae.

APPENDIX

# Descriptions of the new species of gall midges mentioned in this monography

In this publication descriptions are given of the larvae of thirteen new species of gall midges, five of which represent the type species of new genera. In Chapter 1: Outline of the biology, a reference is also made to new species of which the descriptions have not been published before. In order to complete the characteristics of the material, the imaginal phases of the new species have been briefly described in this Appendix.

## SUBFAMILY LESTREMIINAE

1. **Peromyia monilis** Mamaev, sp. n. This species is closely related to *P. trimera* (Edwards). The species is clearly distinct by the fact that the costal vein runs beyond the junction with $R_5$ (over a distance equal to five times the thickness of $R_5$), and by the sharp widening of the gonofurca at its base.

Holotype – male; allotype – female; preparation No. 685; Khust, Zakarpatskaya oblast, 5 July 1963.

2. **Aprionus dentifer** Mamaev, sp. n. According to its character-set this species is closely related to *A. flaviventris* Winnertz), but is easily distinguished from the latter species by the arrangement of the spines on the central part of the hypopygium. The spines spread fan-like and are situated on a heavily sclerotized basal part.

Holotype – male; allotype – female; preparation No. 686; Pavlova Sloboda, Moskovskaya oblast, 21 May 1963 (Kh. P. Mamaeva).

3. **Mycopriona** Mamaev, gen. n.; type species *Mycopriona caucasica* (Mamaev) (Mamaev 1963). The genus is distinguished by the following combination of characters: the antennae of the male have more than 2+11 antennal segments, the empodium is reduced, the larvae are provided with apical hooks.

## SUBFAMILY CECIDOMYIINAE

4. **Solntsevia** Mamaev, gen. n. The representatives of this genus are distinguished by the elongated wings with a simple cubital vein, the two-segmented palps, and the toothed claws on all tarsi. In both sexes the antennae consist of 2+11 segments with long, cylindrical, basal thickenings.

The type species, *Solntsevia nigripes* Mamaev, sp. n. is distinguished by the strikingly bent gonostyles and the peculiar gonofurca which bears two apical, spreading spines.

Holotype – male; allotype – female; preparation No. 219; Pavlova Sloboda, Moskovskaya oblast, 27 May 1963. The genus is named in honour of the zoologist E. L. Solntseva.

5. **Gynapteromyia** Mamaev, gen. n. This genus holds an isolated position within the tribe Porricondylini. It is readily distinguished from the other genera by the reduced eye-bridge, the absence of wings in the females, the simple claws, and the reduced empodium.

The type species, *Gynapteromyia carpatica* Mamaev, sp. n. is distinguished by the one-segmented palps, the strikingly hairy hypopygium, and the gonostyles which are provided with a black apical claw. The cerci of the ovipositor are strikingly widened.

Holotype – male; preparation No. 221; Kvasy, Zakarpatskaya oblast, 30 June 1963.

6. **Dicroneurus montanus** Mamaev, sp. n. This species is readily distinguished from the known European species by the strikingly widened gonostyle. In this character the species approaches the Nearctic species *D. pini* Felt. The latter species is, however, noticeably smaller (length of *D. pini* – 1.5 mm; length of *D. montanus* – 3 mm), while its antennae have 2+14 segments.

Holotype – male; allotype – female; preparation No. 222; Kvasy, Zakarpatskaya oblast, 30 June 1963.

7. **Porricondyla lata** Mamaev, sp. n. This species is only distinguished from the related species by the structural peculiarities of the hypopygium: the gonocoxites have an apical lobe, the gonostyles have a hyaline spine at the tip, the ends of the parameres are not crossed, but contiguous.

Holotype – male; allotype – female; preparation No. 22; Pavlova Sloboda, Moskovskaya oblast, 12 August 1963.

8. **Porricondyla neglecta** Mamaev, sp. n. This species is closely related to the preceding one. It is distinguished by the crossed parameres, the black apical claw of the gonostyles, and the presence of a straight hyaline appendage on the gonocoxites.

Holotype – male; preparation No. 223; Zavety Il'icha, Moskovskaya oblast, 13 June 1962.

9. **Echinella** Mamaev, gen. n. This genus is characterized by the following combination of characters: the antennae of the males are provided with three whorls of loop-shaped sensillae, as is the rule in the tribe Cecidomyiini. $R_5$ meets the edge of the wing behind the tip, and the sickle-shaped, emarginate ninth and tenth tergites (of the hypopygium) are short and simple.

The type species is *Echinella albiventris* Mamaev, sp. n. (holotype – male; allotype – female; preparation No. 573; Volkova, Yaroslavskaya oblast, 5 July 1961). A second species is *Echinella polypori* Mamaev, sp. n. (holotype – male; allotype – female; preparation No. 574; Yaroslavskaya oblast, 20 June 1961). These two species are only clearly distinct in the larval stage.

10. **Profeltiella meridionalis** Mamaev, sp. n. This species is closely related to *P. dizygomyzae* Barnes. The species is distinguished by the absence of a distinct emargination on the tenth tergite (which is a part of the hypopygium).

Holotype – male; allotype – female; preparation No. 575; Kudepsta, Krasnodarskiy kray, 26 June 1959.

11. **Profeltiella quercivora** Mamaev, sp. n. This species is easily distinguished by its large (up to three mm.) size, the sharply outlined black rings on the legs and the antennal segments, and the shortened ninth tergite (which is a part of the hypopygium).

Holotype – male; preparation No. 576; collection of A.L. Shcherbin-Parfenenko from the Krasnodarskiy kray (Maykop).

12. **Dichaetia** Mamaev, gen. n. A peculiarity of the wing venation is the weakly curved $R_5$ which, however, meets the edge behind the tip of the wing. The antennal segments of the males have long stalks and normally developed loop-shaped sensillae. The claws of the anterior, and sometimes of the mid-tarsus as well, have a tooth near the base; the hind tarsi are simple.

The type species is *Dichaetia pusilla* Mamaev, sp. n. Its gonocoxites have a weakly pronounced lobe at the base, the gonostyles are very long and thin, the lobes of the ninth tergite are triangular, and the tenth tergite is emarginate.

Holotype – male; preparation No. 577; Vicinity of Tula, 1 June 1958.

13. **Therodiplosis beglarovi** Mamaev, sp. n. This species is distinguished from *Th. persica* Kieffer by the very short stalks of the antennal segments of the males, and by the thicker gonofurca.

Holotype – male; from the collection of G.A. Beglyarov (Leningradskaya oblast). The species is named in honour of G.A. Beglyarov.

# References

Recent key references on nomenclature are added in square brackets by the editor.


Anderson, J.A.T. 1935a. The morphology and anatomy of the immature and adult stages of *Oligotrophus oleariae* Mask. (Cecidomyidae, Diptera). *Proc. Zool. Soc. Lond.* 2: 405-420.

Anderson, J.A.T. 1935b. The description, bionomica, morphology and anatomy of a new *Dryomyia* (Cecidomyidae, Diptera). *Proc. Zool. Soc. Lond.* 2: 421-430.

Barnes, H.F. 1927. British gall midges. I. *Ent. Mon.Mag.* 63: 164-221.

Barnes, H.F. 1928. British gall midges. II. *Lestodiplosis* Kieffer. *Ent. Mon. Mag.* 64: 68-148.

Barnes, H.F. 1929. Gall midges (Diptera, Cecidomyidae) as enemies of aphids. *Bull. Ent. Res.* 20: 433-442.

Barnes, H.F. 1930. Gall midges (Cecidomyidae) as enemies of the Tingidae, Psyllidae, Aleyrodidae and Coccidae. *Bull. Ent. Res.* 21: 319-329.

Barnes, H.F. 1933. Gall midges (Cecidomyidae) as enemies of mites. *Bull. Ent. Res.* 24: 215-228.

Barnes, H.F. 1935. Studies of fluctuations in insect populations. V. The leaf-curling pear midge, *Dasyneura pyri* (Cecidomyidae). *J. Anim. Ecol.* 4: 224-253.

Barnes, H.F. 1937. Insects and other pests injurious to the production of seed in herbage and forage crops. *Bull. Bur. Pl. Genet., Aberistw.* 20.

Barnes, H.F. 1943. Studies of fluctuations in insect populations. X. Prolonged larval life and delayed subsequent emergence of the adult gall midge. *J. Anim. Ecol.* 12: 137-138.

Barnes, H.F. 1946-1956. *Gall midges of economic importance*, vol. 1-7. London.

Barnes, H.F. 1951. A new gall midge predaceous on the flour mite *Tyroglyphus farinae* (Deg.). *Bull. Ent. Res.* 42: 265-266.

Barnes, H.F. 1952. Studies of fluctuations in insect populations. XII. Further evidence of prolonged larval life in the wheat blossom midges. *Ann. Appl. Biol.* 39: 370-373.

Barnes, H.F. 1954. Gall midge larvae as endoparasites, including the description of a species parasitizing aphids in Trinidad, B.W.I. *Bull. Ent. Res.* 45: 769-775.

Barnes, H.F. and Arnold, M.K. 1960. Delayed emergence of the pea midge. *Plant pathol.* 9: 52-54.

Barnes, H.F. and Nijveldt, W. 1954. A new gall midge, *Coccomyza leefmansi* sp. n., predaceous on the eggs of *Pulvinaria polygonata* in Indonesia. *Ent. Ber. Amst.* 15: 91-93.

Batra, L.R. and Lichtward, R.W. 1963. Association of fungi with some insect galls. *J. Kansas Ent. Soc.* 36: 262-278.

Borg, A. 1960. *Tricholaba trifolii* Rübs. (Dipt., Cecidomyidae) a gall maker on red clover. *Ent. Tidskr.* 81: 30-34.

[Buhr, H. 1964-1965. *Bestimmungstabellen der Gallen (Zoo- und Phyto cecídien) an Pflanzen Mittel- und Nordeuropas* 1-2. Jena.]

Buxton, P.A. and Barnes, H.F. 1953. British Diptera associated with fungi. I. Gall midges (Cecidomyidae) reared from the larger fungi. *Proc. R. Ent. Soc. Lond.* 22: 195-200.

Camenzind, R. 1962. Untersuchungen über die bisexuelle Fortpflanzung einer paedogenetischen Gallmücke. *Rev. suisse Zool.* 69: 377-384.

Camenzind, R. 1966. Die Zytologie der bisexuellen und parthenogenetischen Fortpflanzung von *Heteropeza pygmaea* Winnertz, einer Gallmücke mit pädogenetischer Vermehrung. *Chromosoma (Berl.)* 18: 123-152.

Colightly, W.H. 1952. Soil sampling for wheat blossom midges. *Ann. Appl. Biol.* 39: 379-384.

Dombrovskaya, E.V. 1934. *Instruktsiya po sboru i issledovaniyu gallits (Cecidomyidae) i vyzyvaemykh imi povrezhdeniy.* [Instructions for the collection and investigation of gall midges and the damage caused by them]. Leningrad. (Russ.)

Dombrovskaya, E.V. 1940. Cecidomyidae Kuybyshevskoy oblasti. [The Cecidomyidae of the Kuybyshevskaya oblast]. *Trudy Zool. Inst. Akad. Nauk SSSR* 6: 184-221. (Russ.)

Edwards, F.W. 1938. On the British Lestremiinae, with notes on exotic species (Diptera, Cecidomyidae). *Proc. R. Ent. Soc. Lond.* B7: 18-24, 25-32, 102-108, 173-182, 199-210, 229-243, 253-265.

Elago, L.F. 1940. Ob ispol'zovanii gallitsy *Silvestrina tyrophagi* Dombr. v bor'be s udlinennym kleshchom. [On the use of the gall midge *S. tyrophagi* in the struggle with the mite *Tyrophagus castellanii*]. *Vestn. zashch. rast.* 3: 85-86. (Russ.)

Felt, E.P. 1925. Key to the gall midges. *Bull. N.Y. St. Mus.*: 257.

Frediani, D. 1955. Note sulla genesi della galla, con cenni di morfologia e biologia, della *Dryomyia lichtensteinii* Fr. Lw. (Diptera, Cecidomyidae). *Redia* 40: 141-180. (Ital.)

Fröhlich, G. 1960. *Gallmücken-Schädlinge unserer Kulturpflanzen.* Die neue Brehm-Büch, Wittenberg.

Gabritschewsky, E. 1928. Sénescence embryonnaire, rajeunissement et déterminisme des formes larvaires de *Miastor metraloas* (Cecidomyidae, Diptera). *Bull. biol. France et Belg.* 62: 478-524.

Gabritschewsky, E. 1930. Der umkehrbare Entwicklungszyclus bei *Miastor metraloas. Arch. Entwickl.-Mech. Org.* 121: 450-465.

Gilyarov, M.S. 1945. Vliyanie kharaktera rasseleniya na khod ontogeneza nasekomykh. [The influence of the mode of dispersal on the course of the ontogenesis of insects]. *Zhurn. Obshch. Biol.* 6: 26-36. (Russ.)

Gilyarov, M.S. 1949. *Osobennosti pochvy kak sredy obitaniya i eye znachenie v evolyutsii nasekomykh.* [The properties of the soil as a habitat and their meaning for the evolution of the insects]. Moskva-Leningrad. (Russ.)

Gilyarov, M.S. (ed.) 1964. *Opredelitel' obitayushchikh v pochve lichinok nasekomykh.* [A determination key to the insect larvae living in soil]. Nauka, Moskva. (Russ.)

Giraud, J. 1863. Mémoire sur les insectes qui vivent sur le Roseau commun, *Phragmites communis* Trin. (*Arundo phragmites* L.) et plus spécialement sur ceux de l'ordre des Hyménoptères. *Verh. Zool.-bot. Ges. Wien* 13: 1251-1288.

Harris, R.G. 1923. Sur la culture des larves de Cécidomyies pédogénétiques (*Miastor*) en milieu artificiel. *C.R. Soc. Biol.* 88: 256-258.

Haseman, L. 1931. The anatomy of the Hessian fly larva. *J. Econ. Ent.* 24: 417-420.

Hauschteck, E. 1959. Über die Zytologie der Parthenogenese und der Geschlechtsbestimmung bei der Gallmücke *Oligarces paradoxus* Mein. *Experientia (Basel)* 15: 260-265.

Hauschteck, E. 1962. Die Cytologie der Pädogenese und der Geschlechtsbestimmung einer heterogenen Gallmücke. *Chromosoma* 13: 163-182.

Hennig, W. 1948. *Die Larvenformen der Dipteren.* I Teil. Berlin.

Hennig, W. 1954. Flügelgeäder und System der Dipteren. *Beitr. Ent.* 4: 245-388.

Ivanova-Kazas, O.M. 1963. Embrional'nye prisposobleniya k gemotsel'nomu zhivorozhdeniyu u pedogeneticheskikh dvukrylikh. [The embryonic adaptations for vivipary in paedogenic Diptera]. *Chetvertoe soveshchanie embriologov, tezisy dokl.*: 67-68. [Fourth Conference of Embryologists, theses and communications]. LGU Press. (Russ.)

Jensen, P.B. 1948. Formations of galls by *Mikiola fagi. Physiol. Plantarum* I: 95-108.

Jensen, P.B. 1952. Untersuchungen über die Bildung der Galle von *Mikiola fagi. Dan. Biol. Medd.* 18: 1-19.

Kieffer, J.J. 1894. Sur le groupe *Epidosis*. *Ann. Soc. Ent. France* 63: 311-350.

Kieffer, J.J. 1895. Essai sur le groupe *Campylomyza*. *Misc. Ent.* 3: 46-47, 57-63, 73-79, 91-97, 109-113, 129-132.

Kieffer, J.J. 1896a. Neuer Beitrag zur Kenntnis der *Epidosis*-Gruppe.

Kieffer, J.J. 1896b. Observations sur les *Diplosis* et diagnose de cinq espèces nouvelles. *Bull. Soc. Ent. France* 65: 382-384.

Kieffer, J.J. 1898. Synopse des Cécidomyies d'Europe et d'Algérie descrites jusqu'à ce jour. *Bull. Soc. Hist. Nat. Metz (sér. 2)* 8: 1-64.

Kieffer, J.J. 1900 (1901). Monographie des Cécidomyies d'Europe et d'Algérie. *Ann. Soc. Ent. France* 69: 181-472.

Kieffer, J.J. 1904. Nouvelles Cécidomyies xylophiles. *Ann. Soc. Sci. Brux.* 28: 367-410.

Kieffer, J.J. 1907. Eine neue endoparasite Cecidomyide. *Z. Hymen. Dipt.* 7: 129-130.

Kieffer, J.J. 1913. Diptera. Fam. Cecidomyidae. *Genera Insectorum,* fasc. 152. Bruxelles.

Kirkpatrick, T.W. 1954. Notes on *Pseudoendaphis maculans* Barnes, a cecidomyid endoparasite of aphids in Trinidad. *B.W.I. Bull. Ent. Res.* 45: 777-781.

Krivosheina, N.P. 1964a. Stroenie golovy lichinok i estestvennaya sistema otryada dvukrylykh. [The larval head structure in relation to the natural system of the order Diptera]. *Zool. Zhurn.* 43: 193-205. (Russ.)

Krivosheina, N.P. 1964b. Otryad Diptera – dvukrylye. [Order Diptera – two-winged insects]. In: *Opredelitel' obitayuschikh v pochve lichinok nasekomykh:* 605-608. [Key to the soil-inhabiting insect larvae]. Moskva. (Russ.)

Köster, E. 1911. *Die Gallen der Pflanzen.* Leipzig.

Lal, K.B. 1934. Insect parasites of Psyllidae. *Parasitology* 26: 325-334.

Lawson, J.W. 1951. The anatomy and morphology of the early stages of *Culicoides nubeculosus* Meigen (Diptera: Ceratopogonidae-Heleidae). *Trans. R. Ent. Soc. Lond.* 102: 511-570.

MacDugall, R.S. 1913. The red clover gall gnat, *Amblyspatha ormerodi* nov. sp. Kieffer. *J. Bd. Agric.* 2: 225-230.

Mamaev, B.M. 1955. Sootnoshenie polov i vozmozhnost' partenogeneza u gallits (Itonididae, Diptera). [Sex-ratio and the possibility of parthenogenesis in the gall midges]. *Zool. Zhurn.* 34: 564-569. (Russ.)

Mamaev, B.M. 1960a. Opisanie dvukh novykh rodov i vida gallits (Itonididae, Diptera), razvivayushchikhsya v gniloy drevesine. [Description of two new genera and species of gall midges developing in decaying wood]. *Zool. Zhurn.* 39: 1521-1524. (Russ.)

Mamaev, B.M. 1960b. Gallitsa *Micropteromyia ghilarovi* gen. et spec. n. (Itonididae, Diptera) s redutsirovannymi kryl'yami. [A gall midge, *M. ghilarovi*, with reduced wings]. *Ent. Obozr.* 39: 951-955. (Russ.)

Mamaev, B.M. 1961a. Novye predstaviteli rodov gallits, izvestnykh iz Nearktiki, v faune Evropeyskoy chasti SSSR (Itonididae, Diptera). [New representatives of gall midge genera, known from the Nearctic, in the fauna of the European part of the USSR]. *Dokl. AN SSSR* 139: 227-229. (Russ.)

Mamaev, B.M. 1961b. Opisanie gallitsy *Aprionus smirnovi* Mamaev, sp. n. (Itonididae, Diptera). [Description of the gall midge *A. smirnovi*]. *Zool. Zhurn.* 40: 614-615. (Russ.)

Mamaev, B.M. 1961c. Novye vidy rodov *Triopsis* Kieffer i *Triommatomyia*, gen. n. (Itonididae, Diptera). [New gall midges of the genera *Triopsis* and *Triommatomyia*]. *Ent. Obozr.* 40: 405-412. (Russ.)

Mamaev, B.M. 1961d. Novye gallitsy roda *Ametrodiplosis* Rübs. (*Löwodiplosis* Kieff.) (Diptera, Itonididae). [New gall midges of the genus *Ametrodiplosis*]. *Acta Soc. Ent. Cechosloveniae* 58: 384-388. (Russ.)

Mamaev, B.M. 1961e. Gallitsy SSSR, Novye vidy roda *Camptomyia* Kieffer (Itonididae, Diptera). [The gall midges of the USSR, New species of the genus *Camptomyia*]. *Zool. Zhurn.* 40: 1677-1690. (Russ.)

Mamaev, B.M. 1961f. Neue *Hastatus*-Arten (Diptera: Itonididae). *Beitr. Ent.* 11: 446-450.

Mamaev, B.M. 1962. *Gallitsy, ikh biologiya i khozyaystvennoe znachenie.* [Gall midges, their biology and agricultural importance]. Moskva. (Russ.)

Mamaev, B.M. 1963a. Gallitsy SSSR. 2. Triba Micromyiini (Itonididae, Diptera). [The gall midges of the USSR. 2. Tribe Micromyiini]. *Ent. Obozr.* 42: 195-213. (Russ.)

Mamaev, B.M. 1963b. Gallitsy SSSR. 3. Novye vidy roda *Winnertzia* Rondani, razvivayushchiesya v pochve, v plenkach mitseliya gribov i pod gniyushchey koroy khvoynykh derev'ev (Itonididae, Diptera). [The gall midges of the USSR. 3. New species of the genus *Winnertzia* Rondani, developing in soil, in films of mycelium and under rotting bark of conifers]. *Zool. Zhurn.* 42: 562-573. (Russ.)

Mamaev, B.M. 1963c. Gallitsy SSSR. 5. Novye sredneaziatskie vidy iz trib Lestremiini, Mycromyiini, Porricondylini (Itonididae, Diptera). [The gall midges of the USSR. 5. New Central Asian species of the tribes Lestremiini, Mycromyiini, and Porricondylini]. *Uzbekskiy biol. Zhurn.* 2: 70-77. (Russ.)

Mamaev, B.M. 1963d. Perekhod gallits k obitaniyu v pustyne i ikh dal'neyshaya evolutsiya. [The transition of gall midges to desert life and their subsequent evolution]. *Pyatoe soveshchanie VEO, Tezisy Dokl. Tashkent*: 29-30. [Fifth Communication of the All Union Ent. Soc.]. (Russ.)

Mamaev, B.M. 1964a. Gallitsy SSSR. 4. Novye vidy triby Lestremiini (Itonididae, Diptera). [The gall midges of the USSR. 4. New species of the tribe Lestremiini]. *Zool. Zhurn.* 43: 776-778. (Russ.)

Mamaev, B.M. 1964b. O sistematicheskom polozhenii triby Leptosyni Enderlein v semeystve gallits (Diptera, Itonididae). [On the systematic position of the tribe Leptosyni within the gall midge family]. *Acta Soc. Ent. Cechosloveniae* 61: 89-99. (Russ.)

Mamaev, B.M. 1964c. Ekologicheskie aspekty gallits (Diptera, Itonididae – Cecidomyiidae). [Ecological aspects of the gall midges]. *Zool. Zhurn.* 43: 1169-1180. (Russ.)

Mamaev B.M. 1964d. Gallitsy SSSR. 6. Novye vidy triby Porricondylini (Cecidomyiidae, Diptera). [The gall midges of the USSR. 6. New species of the tribe Porricondylini]. *Ent. Obozr.* 43: 894-913. (Russ.)

[Mamaev, B.M. 1968. *Evolution of gall forming insects – Gall midges*. Boston Spa, Wetherby. Translated by A. Crozy (1975).]

Mamaeva, Kh.P. 1964. Gallitsy (Itonididae, Diptera), razvivayushchiesya v koloniakh tley. [Gall midges developing within aphid colonies]. *Ent. Obozr.* 43: 447-457. (Russ.)

Mamaeva, Kh.P. 1965. Vidovoy sostav i ekologia gallits Moskovskoy oblasti. [The species composition and ecology of the gall midges of the Moskovskaya oblast]. Avtoref. kand. diss. Moskva. (Russ.)

Marchal, P. 1897. Les Cécidomyies des céréales et leur parasites. *Ann. soc. Ent. France* 66: 1-105.

Marikovskiy, P.I. 1953a. Novye rody i vidy gallits (Diptera, Cecidomyidae) yugo-vostochnogo Kazakhstana. [New gall midge genera and species from South-East Kazakhstan]. *Izv. AN Kazakhskoy SSR, seria biol.* 8: 128-139. (Russ.)

Marikovskiy, P.I. 1953b. Novye vidy gallits (Diptera, Itonididae), vredyashchie saksaulam. [New gall midge species damaging *Haloxylon*]. *Ent. Obozr.* 33: 331-341. (Russ.)

Marikovskiy, P.I. 1955a. Novye gallitsy roda *Asiodiplosis* Marik. (Diptera, Itonididae) iz pustyni Kazakhstana. [New gall midges of the genus *Asiodiplosis* from the desert of Kazakhstan]. *Zool. Zhurn.* 34: 336-346. (Russ.)

Marikovskyi, P.I. 1955b. Materialy po faune i biologii gallits (Diptera, Itonididae) zony pustyni. [Contributions to the gall midge fauna and biology of the desert zone]. *Trudy In-ta Zool. Parazitol. AN Kirgizskoy SSR* 3: 103-109. (Russ.)

Marikovskiy, P.I. 1955c. Novye gallitsy (Diptera, Itonididae) s saksaulov 2. [New gall midges from *Haloxylon* 2]. *Ent. Obozr.* 34: 298-312. (Russ.)

Marikovskiy, P.I. 1955d. Novye vidy gallits (Diptera, Itonididae) fauny Sredney Azii. [New gall midge species to the fauna of Central Asia]. *Trudy In-ta Zool. Parazitol. AN Kirgizskoy SSR* 4: 21-26. (Russ.)

Marikovskiy, P.I. 1957. Novye vidy gallits (Diptera, Itonididae) iz kereuka *Salsola rigida* Pall. v severo-vostochnykh Karakumakh. [New gall midge species from *S. rigida* in North-East Karakum]. *Ent. Obozr.* 36: 935-943. (Russ.)

Marikovskiy, P.I. 1960. Novye vidy gallits (Diptera, Itonididae) fauny Sredney Azii. [New gall midge species to the fauna of Central Asia]. *Trudy Tomskogo Gos. Un-ta* 148: 183-189. (Russ.)

Marikovskiy, P.I. 1961. Novye vidy gallits (Diptera, Itonididae) fauny yugo-vostochnogo Kazakhstana. [New gall midge species to the fauna of South-East Kazakhstan]. *Ent. Obozr.* 40: 37-50. (Russ.)

Meijere, J.C.H. de 1907. Eerste supplement op de nieuwe naamlijst van Nederlandsche Diptera. *Tijdschr. Ent.* 50: 151-195. (Dutch)

Metcalfe, M.E. 1933. The morphology and anatomy of the larva of *Dasyneura leguminicola* Lint. (Diptera). *Proc. Zool. Soc. Lond.* 1: 119-130.

Milne, D.L. 1960. The gall midges (Diptera: Cecidomyidae) of clover flower-heads. *Trans R. Ent. Soc.* 112: 73-108.

Milne, D.L. 1961. The function of the sternal spatula in gall midges. *Proc. R. Ent. Soc. Lond.* A36: 126-131.

Mityaev, I.D. 1961. Materialy po biologii gallits (Diptera, Itonididae) – vrediteley tamariksov Yugo-Vostoka Kazakhstana. [Contributions to the biology of the gall midges damaging *Tamarix* in South-East Kazakhstan]. *Ent. Obozr.* 40: 51-62. (Russ.)

Möhn, E. 1954. Eine neue zoophage Gallmücken-Art an Tannenläusen. *Z. angew. Ent.* 36: 462-468.

Möhn, E. 1955a. Neue freilebende Gallmücken-Gattungen. *Dtsch. ent. Z.* 2 (n.F.): 127-151.

Möhn, E. 1955b. Beiträge zur Systematik der Larven der Itonididae Mitteleuropas. *Zoologica* 38, 105, Lf. 2: 129-247.

Möhn, E. 1955c. Eine neue Gallmücken-Gattung aus Harzwunden an Tanne. *Z. angew. Ent.* 38: 97-102.

Möhn, E. 1955d. Neue zoophage Gallmücken-Gattungen (Diptera: Itonididae). *Beitr. Ent.* 5: 415-425.

Möhn, E. 1960. Studien an paedogenetischen Gallmückenarten (Diptera, Itonididae). *Stuttgarter Beitr. Naturk.* 31: 1-11.

Nijveldt, W. 1952. Over de levenswijze van *Phaenobremia urticariae* Kffr. (Diptera, Itonididae) n. sp. *Ent. Ber.* 14: 8-13, 23-29. (Dutch)

Nijveldt, W. 1954. The range of prey of *Phaenobremia urticariae* Kffr. and *Ph. aphidivora* Rübs. (Diptera, Itonididae). *Ent. Ber.* 15: 207-211.

Nijveldt, W. 1955. Additional notes on my paper on the range of prey of *Phaenobremia aphidivora* Rübs. and *Ph. urticariae* (Diptera, Itonididae). *Ent. Ber.* 15: 436-437.

Nijveldt, W. 1957. Aphid-eating gall midges (Cecidomyidae) with special reference to those in the Barnes collection. *Ent. Ber.* 17: 233-239.

Nijveldt, W. 1963. Observations on gall midges of white and red clover. *Neth. J. Plant Pathol.* 69: 161-187.

Nikolei, E. 1958. Untersuchungen über den Generationswechsel pädogenetischer Gallmücken. *Rev. suisse Zool.* 65: 390-396.

Nikolei, E. 1961. Vergleichende Untersuchungen zur Fortpflanzung heterogoner Gallmücken unter experimentellen Bedingungen. *Z. Morph. Ökol. Tiere* 50: 281-329.

Nolte, H.W. 1957. Die Wirt-Parasit Beziehungen bei Gallmücken Befall. IV. Intern. *Pflanzenschutz-Kongr., Hamburg.*

Otter, G.W. 1934. On the early stages of *Lestodiplosis alvei* Barnes, especially in relation to the larval head capsule. *Parasitology* 26: 582-593.

Otter, G.W. 1938. On the morphology of the larvae of three species of Cecidomyiidae from knapweed (*Centaurea*) flowers. *Trans. R. Ent. Soc. Lond.* 87: 39-68.

[Panelius, S. 1965. A revision of the European gall midges of the subfamily Porricondylinae (Diptera: Itonidae). *Act. Zool. Fenn.* 113: 3-157.]

Pavlov, I.F. 1958. Vyzhivanie lichinok i kolichestvo pokoleniy gessenskoy mukhi. [Larval survival and generation size in hessian fly]. *Zool. Zhurn.* 37: 1831-1841. (Russ.)

Principi, M.M. 1958. Ricerche di morfologia e di etologia su di un dittero Cecidomyiidae galligeno, la *Putoniella marsupialis* F. Loew, vivente su piante del gen. '*Prunus*'. Primi resultati. *Boll. Inst. Ent. Univ. Bologna* 23: 35-68. (Ital.)

Pritchard, A.E. 1947. The North American gall midges of the tribe Micromyini, Itonididae (Cecidomyiidae), Diptera. *Ent. Amer.* 27: 1-87.

Pritchard, A.E. 1948. *Clinodiplosis pucciniae*, a new gall midge feeding on rust (Diptera, Itonididae). *Pan-Pac. Ent.* 24: 29-30.

Pritchard, A.E. 1951. The North American gall midges of the tribe Lestremiini, Itonididae (Cecidomiidae), Diptera. *Univ. California Publ.* 8: 239-275.

Pritchard, A.E. 1952. A new gall midge predator of mealybugs, *Coccomyza donaldi* (Diptera, Cecidomyiidae). *Bull. Ent. Res.* 43: 477-478.

Pritchard, A.E. 1960. A new classification of the paedogenic gall midges formerly assigned to the subfamily Heteropezinae (Diptera: Cecidomyiidae). *Ann. Ent. Soc. Amer.* 53: 305-316.

Roberti, D. 1946. La *Phaenobremia aphidimyza* (Rond.) (Diptera, Cecidomyidae) predatrice di *Aphis (Doralis) frangulae* Koch. *Boll. Inst. Ent. Univ. Bologna* 15: 233-256. (Ital.)

Roberti, D. 1953. Contributo all conoscenza delle specie italiane di *Mayetiola* Kieffer (Diptera, Cecidomyidae). *Boll. Lab. Ent. Agr. Portici 'F. Silvestri'* 12: 98-153. (Ital.)

Roberti, D. 1954. I simbionti degli acari fitofagi. *Therodiplosis persicae* Kieff. *Boll. Lab. Ent. Agr. Portici 'F. Silvestri'* 13: 285-302. (Ital.)

Rohdendorf, B.B. 1964. Istorecheskoe razvitie dvukrylykh nasekomykh. [The historical development of the Diptera]. Moskva. (Russ.)

Romeys, B. 1953. *Mikroskopicheskaya tekhnika.* [Microscopic methods]. Moskva, Foreign Publ. (Russ. transl.)

Ross, H. 1932. *Praktikum der Gallenkunde (Cecidologie).* Berlin.

Ross, H. and Hedicke, H. 1927. *Die Pflanzengallen Mittel- und Nordeuropas, ihre Erreger und Biologie und Bestimmungstabellen.* Jena.

Rübsaamen, E.H. 1899. Über die Lebensweise der Cecidomyiden. *Biol. Centralbl.* 19: 529-549, 561-570, 593-607.

Rübsaamen, E.H. 1911. Über Deutsche Gallmücken and Gallen. *Z. wiss. Ins.-Biol.* 7: 278-282.

Rübsaamen, E.H. and Hedicke, H. 1926-1938. Die Cecidomyiden (Gallmücken) und ihre Cecidien. *Zoologica* 29: 1-350.

[Skuhravá, M. 1986. Cecidomyiidae. In: Á. Soós and L. Papp (eds), *Catalogue of Palaearctic Diptera* 4: 72-297. Elsevier, Amsterdam.]

Skuhravá, M. and Skuhravý, V. 1960. *Beljomorky.* [The gall midges]. Praha. (Czech.)

Slepyan, E.I. 1960. O razlichiyakh v stroenii pishchevaritel'nogo trakta u lichinok gallits (Diptera, Itonididae). [On the structural differences of the digestive tract in gall midge larvae]. *Zool. Zhurn.* 39: 1362-1370. (Russ.)

Slepyan, E.I. 1961a. O strukturnykh skhodstvakh gallov i kallyusov. [On the structural resemblances of galls and calluses]. *Dokl. AN SSSR* 136: 241-244. (Russ.)

Slepyan, E.I. 1961b. Priobretenie sposobnosti vyzyvat' obrazovanie gallov i terat kak etap evolyutsii pishchevoy spetsializatsii na primere gallits Itonididae s.l. (Diptera, Nematocera). [The aquisition of the capability to induce the formation of galls and terata as an evolutionary stage of food specialization, illustrated by the gall midges]. *Zool. Zhurn.* 40: 1495-1509. (Russ).

Slepyan, E.I. 1961c. Sopostavlenie gallov i terat, vyzyvaemykh nasekomymi, s plodami i semenami. [A comparison of insect galls and terata from fruits and seeds]. *Botan. Zhurn.* 46: 1702-1717. (Russ.)

Slepyan, E.I. 1962. Nomenklatura i klassifikatsiya vyzyvaemykh chlenistonogimi gallov i pochkovykh teratozov v svyazi s ikh mestom sredi patologicheskikh yavleniy. [Nomenclature and classification of galls and bud teratoses caused by arthropods in connexion with their position among the pathological formations]. *Botan. Zhurn.* 47: 721-753. (Russ.)

Sorokin, S. 1938. Novyy khishchnik ambarnykh kleshchey. [A new predator of flour mites]. *Zashchita Rast.* 17: 122.(Russ.)

Springer, F. 1915. Über den Polymorphismus bei den Larven von *Miastor metraloas*. *Zool. Jahrb. Syst.* 40: 57-118.

Stebaev, I.B. 1963. Izmenenie zhivotnogo naseleniya pochv v khode ikh razvitiya na skalakh i na rykhlykh produktakh vyvetrivaniya v leso-lugovykh landshaftakh Yuzhnogo Urala. [The change of the animal population during the development of soil on rocks and crumbly products of wheathering in forest-meadow landscapes in the Southern Urals]. *Pedologia* 2: 265-309. (Russ.)

Stelter, H. 1960. Untersuchungen über Gallmücken. 5. *Buhriella* n. g., *rubicola* n. sp. *Ent. Mitt. Zool. Staatsinst. u. Zool. Museum Hamburg* 25: 1-6.

Stelter, H. 1963. Untersuchungen über Gallmücken. VIII. Bemerkungen zu einigen Arten der Gattung *Tricholaba* Rübs., 1917. *Dtsch. Ent. Z.* 10 (n.F.): 104-115.

Ulrich, H. 1934. Experimentelle Untersuchungen über den Generationswechsel einer pädogenetischen Gallmücke. *Rev. suisse Zool.* 41: 423-428.

Ulrich, H. 1936. Experimentelle Untersuchungen über den Generationswechsel der heterogonen Cecidomyide *Oligarces paradoxus*. *Z. indukt. Abstamm.- und Vererb.-Lehre* 71: 1-60.

Ulrich, H. 1938. Untersuchungen über Morphologie und Physiologie des Generationswechsels von *Oligarces paradoxus*. *Verh. Int. Kongr. Ent. Berlin*: 955-974.

Ulrich, H. 1940. Über den Generationswechsel und seine Bedingungen. *Naturwissenschaften* 28: 569-576, 587-591.

Ulrich, H. 1943. Über den Einfluss verschiedener, den Ernährungsgrad bestimmender Kulturbedingungen auf Entwicklungsgeschwindigkeit, Wachstum und Nachkommenzahl der lebendgebärenden Larven von *Oligarces paradoxus* (Cecidom., Dipt.). *Biol. Zbl.* 63: 109-142.

Vimmer, A. 1925. Larvy a kukly dvojkřidlêho hmuzustředoevropského se zvláštním zřetelem na škûdce rostlin kulturnìch. [The larvae and pupae of Central European Diptera with special reference to the pests of cultivated plants]. Praha. (Czeh.)

Wagner, N. 1862. *Samoproizvol'noe razmnozhenie gussenits u nasekomykh*. [Spontaneous reproduction in insect larvae]. Kazan'. (Russ.)

Wagner, N. 1863. Beitrag zur Lehre von der Fortpflanzung der Insektenlarven, *Z. wiss. Zool.* 13: 513-527.

Wagner, N. 1865. Über die viviparen Gallmückenlarven. *Z. wiss. Zool.* 15: 106-117.

White, M.J.D. 1946. The cytology of the Cecidomyidae (Diptera). I. Polyploidy and polyteny in salivary gland cells *Lestodiplosis* spp. *J. Morphol.* 78.

White, M.J.D. 1948. The cytology of the Cecidomyidae (Diptera). IV. The salivary gland chromosomes of several species. *J. Morphol.* 82: 53-80.

Williams, F.X. 1910. The anatomy of the larvae of *Cecidomyia resinicoloides* Will. *Ann. Ent. Soc. Amer.* 3: 45-60.

Wissmann H. 1926. Über ein stärkeres Auftreten von freilebenden Gallmilben (*Phyllocoptes*) an Obstbäumen und über neue natürliche Feinde der Gallmücken aus der Familie der Cecidomyiden. *Z. Pflanzenkrankh. u. Pflanzenschutz* 36: 98-106.

Wyatt, I.J. 1961. Pupal paedogenesis in the Cecidomyiidae (Diptera). I. *Proc. R. Ent. Soc. Lond.* A36: 133-143.

Wyatt, I.J. 1963a. Pupal paedogenesis in the Cecidomyiidae (Diptera). II. *Proc. R. Ent. Soc. Lond.* A38: 136-144.

Wyatt, I.J. 1963b. A root-infesting gall midge (Diptera: Cecidomyiidae) from New Zealand. *Proc. R. Ent. Soc. Lond.* B32: 103-107.

# Index of scientific names

References to descriptions of taxa are in **bold**; references to pages with figures are in *italics*

For Product Safety Concerns and Information please contact our EU representative GPSR@taylorandfrancis.com Taylor & Francis Verlag GmbH, Kaufingerstraße 24, 80331 München, Germany

**Batch number: 10731490**

Printed by Printforce, the Netherlands